3

Topics in Organometallic Chemistry

Topics in Organometallic Chemistry

Forthcoming volumes:

The Metal-Carbon Bond: Theoretical and Fundamental Studies
Volume Editors: J.M. Brown, P. Hofmann

Springer-Verlag
Berlin Heidelberg GmbH

Activation of Unreactive Bonds and Organic Synthesis

Volume Editor: S. Murai

With contributions by
H. Alper, R.A. Gossage, V.V. Grushin, M. Hidai, Y. Ito, W.D. Jones, F. Kakiuchi, G. van Koten, Y.-S. Lin, Y. Mizobe, S. Murai, M. Murakami, T.G. Richmond, A. Sen, M. Suginome, A. Yamamoto

Springer

The series *Topics in Organometallic Chemistry* presents critical overviews of research results in organometallic chemistry, where new developments are having a significant influence on such diverse areas as organic synthesis, pharmaceutical research, biology, polymer research and materials science. Thus the scope of coverage includes a broad range of topics of pure and applied organometallic chemistry. Coverage is designed for a broad academic and industrial scientific readership starting at the graduate level, who want to be informed about new developments of progress and trends in this increasingly interdisciplinary field. Where appropriate, theoretical and mechanistic aspects are included in order to help the reader understand the underlying principles involved.
The individual volumes are thematic and the contributions are invited by the volumes editors.
In references Topics in Organometallic Chemistry is abbreviated
Top. Organomet. Chem. and is cited as a journal

Springer WWW home page: http://www.springer.de

ISSN 1436-6002

Cataloging-in-Publication Data applied for
Die Deutsche Bibliothek - CIP-Einheitsaufnahme
Activation of unreactive bonds and organic synthesis / vol. ed.: S. Murai.
With contributions by H. Alper ...
(Topics in organometallic chemistry)
DOI 10.1007/978-3-540-68525-8

Originally published by Springer-Verlag Berlin Heidelberg New York in 1999
MyCopy version of the original edition 1999

Cover: Friedhelm Steinen-Broo, Pau/Spain; MEDIO, Berlin
Typesetting: Data conversion by MEDIO, Berlin

SPIN: 10678944 66/3020 - 5 4 3 2 1 0 - Printed on acid-free paper.
www.springer.com/mycopy

Volume Editor

Prof. Shinji Murai
Department of Applied Chemistry
Faculty of Engineering
Osaka University
Yamadaoka 2-1, Suita shi
Osaka 565-0817
e-mail: murai@chem.eng.osaka-u.ac.jp

Editorial Board

Preface

This is the first book of its kind which deals with the chemistry of transition metal-mediated activation of unreactive bonds not only from the inorganic point of view but also from that of synthetic organic chemistry.

Progress in this area has been remarkably rapid; so much so that a mere 10 years ago, a compilation such as this would not have been possible. The authors of this volume have, by enlisting the aid of transition-metal complexes, been able to activate many otherwise unreactive bonds such as C–H, C–C, C–O, C–F, C–Cl, C–Si, Si–Si, and even N–N triple bonds. In this volume, overviews of these subjects from the perspectives of organic and inorganic chemistry are provided. This comprehensive collection of catalytic and stoichiometric transformations should prove invaluable to a wide range of chemists. For practicing synthetic chemists, a marvelous new world awaits discovery in which previously unheard of transformations make their synthetic schemes shorter and more efficient. Of great importance is this volume's detailed treatment of fundamental principles underlying the chemistry will facilitate extrapolation to a wide variety of systems. Synthetic inorganic and organometallic chemists will similarly learn the key targets and applications of interest to the organic chemistry community.

The term "activation" has been used among organometallic chemists to describe the activation, or energization, or perturbation of a stable bond irrespective of whether complete or only partial cleavage of the bond in question is achieved. This term has gained widespread approval among organometallic and inorganic chemists. However, because of this imprecision and other problems, the editor has been somewhat hesitant to use this term for organic transformations. However, the term is convenient and will be employed in this volume.

Finally the authors wish to dedicate this book to the students who will become, by reading this volume, the first generation of chemists to consider non-acidic C–H bonds, although there are only a few so far, as reactive and useful for organic transformations.

Osaka, January 1999 — Shinji Murai

Contents

A General Survey and Recent Advances in the Activation of Unreactive Chemical Bonds by Metal Complexes
R.A. Gossage, G. van Koten 1

Activation of C–H Bonds. Stoichiometric Reactions
W.D. Jones 9

Activation of C–H Bonds: Catalytic Reactions
F. Kakiuchi, S. Murai 47

Catalytic Activation of Methane and Ethane by Metal Compounds
A. Sen 81

Cleavage of Carbon–Carbon Single Bonds by Transition Metals
M. Murakami, Y. Ito 97

Activation of Si–Si Bonds by Transition-Metal Complexes
M. Suginome, Y. Ito 131

Activation of C–O Bonds: Stoichiometric and Catalytic Reactions
Y.-S. Lin, A. Yamamoto 161

Activation of Otherwise Unreactive C–Cl Bonds
V.V. Grushin, H. Alper 193

Activation of the N–N Triple Bond in Molecular Nitrogen: Toward its Chemical Transformation into Organo-Nitrogen Compounds
M. Hidai, Y. Mizobe 227

Metal Reagents for Activation and Functionalization of Carbon–Fluorine Bonds
T.G. Richmond. 243

Author Index 271

A General Survey and Recent Advances in the Activation of Unreactive Chemical Bonds by Metal Complexes

Robert A. Gossage[a], Gerard van Koten*

Debye Institute, Department of Metal-Mediated Synthesis, Utrecht University, Padualaan 8, 3584 CH, Utrecht, The Netherlands
E-mail: g.vankoten@chem.uu.nl
[a]AnorMED Inc., #100 20353–64th Ave., Langley, British Columbia, Canada, V2Y 1N5

1	**General Introduction**	1
2	**The Activation of C–H Bonds**	2
3	**The Activation of C–Cl and C–F Bonds**	3
4	**The Activation of C–C and Si–Si Bonds**	4
5	**The Activation of C–O Bonds**	4
6	**The Activation of Molecular Nitrogen**	5
7	**Concluding Remark**	5
References		6

1 General Introduction

The activation of chemical bonds is perhaps the most important area of modern chemistry. In our never ending search for cheap raw materials, the use of abundant but traditionally unreactive molecules as synthetic precursors is becoming increasingly more attractive. A major goal of this endeavor is to devise not only better methods to produce bulk (commodity) chemicals but also to synthesize new materials for industry, medicine and research. Unreactive chemicals include compounds such as hydrocarbons and other media which, under normal circumstances, do not react with other substrates (or themselves). Two primary examples are molecular nitrogen and saturated alkanes. These substances represent very inexpensive potential sources of nitrogen and carbon, respectively. Hydrocarbons (i.e., oil and petroleum products) are the largest fraction of world primary energy production and are thus readily available starting materials. The same is true for dinitrogen as it is a major component of the earth's atmosphere. In addition, the activation of general classes of inert bonds, such as the C–Cl, C–F

Topics in Organometallic Chemistry, Vol. 3
Volume Editor: S. Murai

or C–O bonds, has importance in the destruction of certain man-made environmental toxins (PCB's, CFC's, etc.) and in the potential application of much cheaper chlorinated compounds as reagents (e.g., in Grignard reactions). The activation of specific C–C bonds has great potential in speciality chemical synthesis as does, to a lesser extent, the Si–Si bond in materials science.

This volume of Topics in Organometallic Chemistry is devoted to recent advances in the activation, by metal complexes, of what are termed unreactive chemical bonds. In this introduction, a brief overview of each topic will be presented. This is certainly not intended as a comprehensive review of each subject but merely as a stepping stone to the more detailed chapters that follow within.

2
The Activation of C–H Bonds

The synthetic utility of activating C–H bonds has long been recognized in chemistry. For example, the combustion of hydrocarbons as an energy source with the concomitant formation of water and carbon dioxide is a fundamental reaction in most machines. The *selective* activation of specific types of C–H bonds is however, by no means a simple problem. The strength of the C–H bond in for example, methane (440 kJ/mol; 105 kcal/mol) or benzene (461 kJ/mol; 110 kcal/mol), is a factor which alone does not dominate the chemistry, but it does make activation of this type of bond relative to weaker bonds present in a molecule more difficult [1]. Catalytic activation by metal containing compounds has therefore been an area of intense research.

The study of the reactivity of aryl C–H bonds began in the last century. However, direct bond activation was only observed in the last 35 years, despite the quite early successes of scientists like Charles Friedel and James Crafts who discovered (1877) that aluminum chloride catalysts can promote the alkylation of benzene. Specifically, *orthometallation* of aryl groups attached to coordinated donor atoms was among the first examples of direct C–H bond activation in a transition metal complex. This work demonstrated that an M–C(aryl) fragment and a metal hydride could be formed directly by oxidative addition of an *ortho* H–C(aryl) group [2–8]. An early example of this is the orthometallation of a coordinated triphenylphosphine ligand bound to an Ir(I) metal center [2–6]. Intermediates in this reaction have been isolated and all show close interatomic contact distances between the metal and a hydrogen nucleus. Examples include the very early crystallographic work of Laplaca and Ibers [9] of the Ru(II) complex $RuCl_2(PPh_3)_3$. Related chemistry includes examples of hydrogen atoms of alkyl groups in close proximity to a metal centre, such as the early disclosure by Maitlis and coworkers concerning a Pt phosphine compound with a close intramolecular Pt···H(alkyl) distance [10]. Crabtree and others have coined the phrase agostic to describe this type of interaction and have used these complexes as models of the intimate first stages of C–H bond activation [11, 12]. However, the direct use of aromatic compounds in synthesis is generally restricted to the activation of groups other then the C–H fragment (e.g., C–X bonds: X=Br, Cl).

Since aryl halides are fairly cheap reagents, there has been less recent emphasis (see Section 3 below) on the development of aryl relative to that of alkyl C–H bond activation [13–17]. However, the manufacture of aryl halides is not an environmentally friendly process and thus the future of bulk aromatic synthesis may lie in the direct activation of C–H bonds. For example, the formation of benzaldehyde from the insertion of CO into a C–H bond of benzene is a recent development in this area [17].

Bergman has referred to the selective activation of C–H bonds of saturated alkanes as one of last remaining Holy Grails of synthetic chemistry [18–20]. The initial (perceived) breakthrough in alkyl C–H bond activation came from the simultaneous [21, 22] but independent work of Graham and coworkers (University of Alberta, Canada) and by Bergman et al. (University of California at Berkeley, U.S.A.). Both of these groups discovered that cyclopentadienyl complexes of Ir and Rh can, under photochemical conditions, oxidatively add alkanes to yield hydridometal alkyls. Although none of these systems have yet been shown to operate catalytically, detailed study of the fundamental aspects of this chemistry have continued in earnest [23–29]. Kinetic analysis has revealed that initial agostic interactions of the alkane with the metal center is a key (reversible) intermediate step and that later oxidative addition likely occurs via a simple three-center transition state [30]. Further work in this important area of research will be disclosed in the later chapters of this text by Prof. W. D. Jones (Stoichiometric Activation of C–H Bonds), Prof. F. Kakiuchi and Prof. S. Murai (Catalytic Activation of C–H Bonds) and the chapter by Prof. A. Sen (Catalytic Activation of Methane and Ethane by Metal Compounds).

3
The Activation of C–Cl and C–F Bonds

The activation of chlorinated hydrocarbons is an area of study that has direct environmental consequences in relation to the facile destruction of polychlorinated biphenyls (PCB's C–Cl bond strength 402 kJ/mol; 96 kcal/mol for C_6H_5–Cl) and other pollutants. The high reactivity of simple polyhalogenated alkanes such as tetrachloromethane [31–33], is generally facilitated by the ease of formation of the trichloromethane radical. However, other chlorocarbons are not so easily activated. There are a few examples of metal complexes which contain coordinated alkyl halides [34–37] and these may be considered as agostic interactions (i.e., M···X–R) between the metal and an alkyl halide atom. Direct oxidative addition of unreactive haloalkanes such as dichloromethane has been reported [38, 39]. The activation of C–Cl bonds will be described in the chapter by Dr. V. V. Grushin and Prof. H. Alper (Activation of Otherwise Unreactive C–Cl Bonds) which details the use of chlorocarbons for a variety of applications in synthesis. The environmental (e.g., ozone depletion) and potential pharmaceutical relevance of fluorocarbons has made C–F bond activation an area of quite active research. Despite the high energy of C–F bonds (e.g., 644 kJ/mol; 154 kcal/mol for the C–F bond in C_6F_6), the activation of this fragment by a number of complexes has been reported [40, 41]. Many metal complexes can activate C–F bonds in a stoichiometric fashion and the list includes reagents containing Ti [42], Fe [43], Ni [44], Ru [45, 46], Rh [47–50], Ir [47–50], W [51], U [52] and Yb [53]. Recently, this area has expanded to the catalytic activation of C–F bonds as reported by Aizenberg and Milstein [54, 55]. The use of electron-rich Rh phosphine compounds enables the activation of fluorobenzenes, although turnover numbers are low (<1 turnover per hour at 94°C) [54]. Further details on this chemistry can be found in the relevant chapter by Prof. T. G. Richmond (Metal Reagents for Activation and Functionalization of C–F Bonds).

4 The Activation of C–C and Si–Si Bonds

The activation of C–C bonds (specifically cleavage reactions) by metal compounds has been extensively studied and there are many reagents that can be used to perform the selective manipulation of alkynes and alkenes to produce a plethora of new compounds [56–59]. Transition metal and lanthanide complexes are playing an ever increasing role in the selective formation and cleavage of C–C bonds [60]. Considerable effort has been applied to the application of metal compounds in such diverse areas as ring forming reactions [61–64], regio- and enantio-selective addition reactions [65–71], direct C–C bond formation [72–76] and other processes [56, 57, 77]. This represents a very large cross section of chemical reactivity and the review in this text by Dr. M. Murakami and Prof. Y. Ito (Cleavage of C–C Single Bonds by Transition Metals) will emphasize this selected area of organometallic chemistry and (homogeneous) catalysis.

A related area of this chemistry is the selective reactivity of Si–Si bonds [78]. The use of disilanes in organic synthesis for the creation of new (and reactive) Si–C bonds is an area of current interest and application [79–81]. This work has stemmed from earlier studies which have demonstrated that disilanes readily oxidatively add to a variety of metal centres [82] to produce transition metal silyl complexes [83]. The investigation of this area will be detailed by Prof. M. Sugimone and Prof. Y. Ito (Activation of Si–Si Bonds: Stoichiometric and Catalytic Reactions).

5 The Activation of C–O Bonds

The cleavage of C–O bonds has significant applications in a number of important areas of synthetic organic and organometallic chemistry [56, 57, 83, 84]. The activation of carbon-oxygen single, double and triple bonds has been an area of active research in relation to catalysis and to many fundamental chemical reactions [85]. The reactions of carbon monoxide or dioxide has long been recognized in organometallic chemistry and thus can not be considered unreactive substrates by the definition of this book. Other organic compounds which contain the C=O functionality such as aldehydes and ketones can also not be considered unreactive as per the definition of this manusript [84, 85]. However, asymmetric reactions at this functional group is an area of great current interest. Specifically, Prof. Y.-S. Lin and Prof. A. Yamamoto will describe recent advances in the activation of the important allylic C–O bond with special detail paid to allyl metal complexes. The allylic functionality has been studied extensively because of its importance in a number of specialty organic syntheses [86–92]. In addition, the chapter "Activation of C–O Bonds. Stoichiometric and Catalytic Reactions" will review recent applications of transition metal compounds to the cleavage of C–O bonds of esters, ethers [91, 92] and anhydrides. The rupture of the C–O multiple bonds is also detailed.

6 The Activation of Molecular Dinitrogen

Catalytic dinitrogen activation has been one of the most difficult challenges in the field of organometallic chemistry. Nitrogen gas is used as the N-containing feedstock in the (high temperature and pressure) Haber process for the production of ammonia. The drastic conditions that are necessary for efficient catalysis has lead to a vast amount of study directed towards the design of a mild ammonia synthesis from N_2. The very high strength of the N–N bond (945 kJ/mol; 226 kcal/mol) and the differences in energy distribution between the bonds of N_2 compared with acetylene for example, lead to few similarities in the reactivity of these compounds [93–96]. Dinitrogen is a very weak base and hence does not easily (or strongly in most cases) coordinate to metals. The first example of an N_2 complex was not isolated until 1965 by Allen and Senoff [97]. Since that time, dinitrogen complexes have been isolated for almost all of the transition metals and lanthanides. However, simple coordination of N_2 did not immediately lead to any activation of the molecule. It took several years before a simple N_2-compound could even be stoichiometrically converted to hydrazine [98]. The most promising results in this area have been reported by Laplaza and Cummins [99]. Their 1995 disclosure was the first example of a mild, catalytic conversion of N_2 to ammonia using a high valent Mo complex as catalyst [100]. The chapter presented by Prof. M. Hidai and Prof. Y. Mizobe entitled "Activation of the N–N Triple Bond in Molecular Nitrogen: Toward its Chemical Transformation into Organo-nitrogen Compounds" will discuss recent work concerning direct organonitrogen synthesis using transition metal catalysts.

7 Concluding Remark

The use of metal compounds in the activation of chemical bonds is an area of useful and fascinating research [101]. The following chapters will detail many of the recent advances in this area of chemistry with the emphasis on the use of traditionally unreactive substrates.

References

1. CRC Handbook of Chemistry and Physics. 72nd edn. (1991). CRC, Boston, Sect. D
2. Duff JM, Shaw BL (1972) J Chem Soc Dalton Trans 2219
3. Valentine JS (1973) J Chem Soc Chem Commun 857
4. Ryabov AD (1990) Chem Rev 90:403 and references cited therein
5. Parshall GW (1970) Acc Chem Res 3:139 and references cited therein
6. Omae I (1986) Organometallic intramolecular-coordination compounds. Elsevier, Amsterdam (and references cited therein)
7. Steenwinkel P, Gossage RA, van Koten G (1998) Chem Eur J 4:159
8. Bruce MI (1977) Angew Chem 89:75, Angew Chem Int Ed Engl 16:73
9. La Placa SJ, Ibers JA (1965) Inorg Chem 4:778
10. Roe DM, Bailey PM, Moseley K, Maitlis PM (1972) Chem Commun 1273
11. Crabtree RH, Hamilton DG (1988) Adv Organomet Chem 28:299
12. Richardson TB, Koetzle TF, Crabtree RH (1996) Inorg Chim Acta 250:69
13. Arndtsen BA, Bergman RG (1995) Science 270:1970
14. Gutiérrez E, Monge A, Nicasio MC, Poveda ML, Carmona E (1994) J Am Chem Soc 116:791 and references cited therein
15. Fukuyama T, Chatani N, Kakiuchi F, Murai S (1997) J Org Chem 62:5647
16. Murai S, Chatani N, Kakiuchi F (1997) Catalysis Surveys from Japan 1:35 and references cited therein
17. Sakakura T, Sodeyama T, Sasaki K, Wada K, Tanaka M (1990) J Am Chem Soc 112:7221 and references cited therein
18. Arndtsen BA, Bergman RG, Mobley TA, Peterson TH (1995) Acc Chem Res 28:154
19. Jones WD (1989) Alkane activation by cyclopentadienyl complexes of rhodium, iridium and related species. In: Hill CL (ed) Activation and functionalization of alkanes. John Wiley, NY, chap 4
20. Shilov AE (1994) Activation of saturated hydrocarbons by transition metal complexes. Reidel, Boston
21. Hoyano JK, Graham WG (1982) J Am Chem Soc 104:3723
22. Janowicz AH, Bergman RG (1982) J Am Chem Soc 104:352
23. Lohrenz JCW, Jacobsen H (1996) Angew Chem 108:1403, Angew Chem Int Ed Engl 35:1305
24. Pérez PJ, Poveda ML, Carmona E (1995) Angew Chem 107:242, Angew Chem Int Ed Engl 34:231
25. Schneider JJ (1996) Angew Chem 108:1132, Angew Chem Int Ed Engl 35:1068
26. Cundai TR (1994) J Am Chem Soc 116:340
27. Hoveya AH, Morken JP (1996) Angew Chem 108:1378, Angew Chem Int Ed Engl 35:1262
28. Jonas RT, Stack TDP (1997) J Am Chem Soc 119:8566
29. Bromberg SE, Yang H, Asplund AT, McNamara BK, Kotz KT, Yeston JS, Wilkens M, Frei H, Bergman RG, Harris CD (1997) Science 278:260
30. Bengali AA, Arndtsen BA, Burger PM, Schultz RH, Weiller BH, Kyle KR, Bradley S, Moore C, Bergman RG (1995) Pure Appl Chem 67:281
31. Knapen JWJ, van der Made AW, de Wilde JC, van Leeuwen PWNM, Wijkens P, Grove DM, van Koten G (1994) Nature 372:359
32. van de Kuil LA, Grove DM, Gossage RA, Zwikker JW, Jenneskens LW, Drenth W, van Koten G (1997) Organometallics 16:4985
33. Gossage RA, van de Kuil LA, van Koten G (1998) Acc Chem Res 31:423
34. Friedrich HB, Moss JR (1991) Adv Organomet Chem 33:235
35. Kulawiec RJ, Crabtree RH (1990) Coord Chem Rev 99:89
36. Peng T-S, Winter CH, Gladysz JA (1994) Inorg Chem 33:2534
37. Pathak DD, Adams H, White C (1994) J Chem Soc Chem Commun 733

38. Huser M, Youinou M-T, Osborn J (1989) Angew Chem 101:1427
39. Olivèn M, Caulton KG (1997) Chem Commun 1733
40. Kiplinger JL, Richmond TG, Osterberg CE (1994) Chem Rev 94:373
41. Hughes RP (1990) Adv Organomet Chem 31:183
42. Kiplinger JL, Richmond TG (1996) J Am Chem Soc 118:1805
43. Kiplinger JL, Richmond TG (1993) J Am Chem Soc 115:5303
44. Bach I, Pörchke K-R, Goddard R, Kopiske C, Krüger C, Rufinska A, Seevogel K (1996) Organometallics 15:4959
45. Whittlesey MK, Perutz RN, Greener B, Moore MH (1997) Chem Commun 187
46. Whittlesey MK, Perutz RN, Moore MH (1996) Chem Commun 787
47. Edelbach BL, Jones WD (1997) J Am Chem Soc 119:7734
48. Jones WD, Partridge MG, Perutz RN (1991) J Chem Soc Chem Commun 264
49. Belt ST, Helliwel M, Jones WD, Partridge MG, Perutz RN (1993) J Am Chem Soc 115:1429
50. Blum O, Frolow F, Milstein D (1991) J Chem Soc Chem Commun 258
51. Kiplinger JL, King MA, Fechtenkotter A, Arif AM, Richmond TG (1996) Organometallics 15:5292
52. Weydert M, Anderson RA, Bergman RG (1993) J Am Chem Soc 115:8837
53. Deacon GB, Forsyth CM, Sun J (1994) Tetrahedron Lett 35:1095
54. Aizenberg M, Milstein D (1995) Science 265:359
55. Aizenberg M, Milstein D (1995) J Am Chem Soc 117:359
56. Schlosser M (ed, 1994) Organometallics in synthesis: a manual. John Wiley, New York
57. Davies SG (1982) Organotransition metal chemistry: applications to organic synthesis. Pergamon, Frankfurt
58. Keim W (1994) New J Chem 18:93
59. Schwarz H (1989) Acc Chem Res 22:282
60. Trost BM (1983) Science 219:245
61. Trost BM, Parquette JR, Marquart AL (1995) J Am Chem Soc 117:3284
62. Trost BM, Higuchi RI (1996) J Am Chem Soc 118:10094
63. Negishi E-I, Takahashi T (1994) Acc Chem Res 27:124
64. Trost BM, Grese TA (1991) J Am Chem Soc 113:7363
65. van Koten G, Gossage RA, Grove DM, Jastrzebski JTBH (1998) Selective product formation with organometallic radicals of nickel and zinc. In: Matyjaszewski K (ed) Controlled radical polymerization. ACS Pub, Washington, chap 5
66. Wissing E, van der Linden S, Rijnberg E, Boersma J, Smeets WJJ, Spek AL, van Koten G (1994) Organometallics 13:2602
67. Trost BM, Indolese AF, M ller TJJ, Treptow B (1995) J Am Chem Soc 117:615
68. Trost BM, Imi K, Davies IW (1995) J Am Chem Soc 117:5371
69. Trost BM, Toste FD (1996) J Am Chem Soc 118:6305
70. van der Boom ME, Kraatz *H*-B, Ben-David Y, Milstein D (1996) Chem Commun 2167 and references cited therein
71. Donkervoort JG, Vicario JL, Jastrzebski JTBH, Gossage RA, Cahiez G, van Koten G (1998) J Organomet Chem 558:61
72. de Vries AHM, Meetsma A, Feringa BL (1996) Angew Chem 108:2526, Angew Chem Int Ed Engl 35:2374
73. RajanBabu TV, Casalnuovo AL (1994) Pure Appl Chem 66:1535
74. Rijnberg E, Hovestad NJ, Kleij AW, Jastrzebski JTBH, Boersma J, Janssen MD, Spek AL, van Koten G (1997) Organometallics 16:2847
75. Hoveya AH, Moreken JP (1996) Angew Chem 108:1378, Angew Chem Int Ed Engl 35:1263
76. van Koten G (1994) Pure Appl Chem 66: 1455
77. Tsuji J (1975) Organic synthesis by means of transition metal complexes. Springer, Berlin Heidelberg New York (and references cited therein)

78. Sharma HK, Pannell KH (1995) Chem Rev 95:1351
79. Tamao K, Okazaki S, Kumada M (1978) J Organomet Chem 146: 87
80. Hayashi T, Matsumoto Y, Ito Y (1988) J Am Chem Soc 110:5579
81. Ito H, Ishizuka T, Tateiwa J, Sonoda M, Hosomi A (1998) J Am Chem Soc 120:11196
82. Sugimone M, Ito Y (1998) J Chem Soc Dalton Trans 1925
83. See for example: Gossage RA, McLennan G, Stobart SR (1996) Inorg Chem 35:1729
84. van Leeuwen PWNM, van Koten G (1993) Rhodium Catalyzed Hydroformylation. In: Moulijn JA, van Leeuwen PWNM, van Santen RA (eds) Catalysis: an integrated approach to homogeneous, heterogeneous and industrial catalysis. Elsevier, Amsterdam, Sect. 6.2
85. Elschenbroich Ch, Salzer A (1992) Organometallics: a concise introduction. VCH, Basel, chap 17
86. Mandai T, Matsumoto T, Kawada M, Tsuji J (1994) Tetrahedron 50:475
87. Shimizu I, Ishii H (1994) Tetrahedron 50:487
88. Yamamoto A, Ozawa F, Osakada K, Huang L (1991) Pure Appl Chem 63:687
89. Trost B (1981) Pure Appl Chem 53:2357
90. Trost B (1980) Acc Chem Res 13:385
91. Djakovich L, Moulines F, Astruc D (1996) New J Chem 20:1071
92. Maercher A (1987) Angew Chem 99:1002, Angew Chem Int Ed Engl 26:972
93. Bazhenova TA, Shilov AE (1995) Coord Chem Rev 144:69
94. Leigh GJ (1992) Acc Chem Res 25:177
95. Dilworth JR (1996) Coord Chem Rev 154:163
96. Richards RL (1996) Coord Chem Rev 154:83
97. Allen AD, Senoff CV (1965) J Chem Soc Chem Commun 621
98. Hidai M, Mizobe Y (1995) Chem Rev 95:1115
99. Laplaza CE, Cummins CC (1995) Science 268:861
100. For a critical evaluation of the Laplaza/Cummins system [99] see: Leigh GJ (1995) Science 268:827.
101. Herrmann WA, Cornils B (1997) Angew Chem 109:1074, Angew Chem Int Ed Engl 36:1048

Activation of C–H Bonds: Stoichiometric Reactions

William D. Jones

Department of Chemistry, University of Rochester, NY 14627 USA
E-mail: jones@chem.rochester.edu

The activation of hydrocarbon C–H bonds by way of oxidative addition and other pathways has been found to be possible with a wide variety of homogeneous transition metal complexes. Studies have provided detailed information about the intermediates involved and the mechanism(s) of activation. Thermodynamic and kinetic selectivities have been established with several reactive metal fragments. An overview of these developments is given and new examples of complexes that activate C–H bonds are also described.

Keywords: C–H activation, Alkane complexes, Oxidative addition, Selectivity

1	**Introduction**	10
2	**Alkane Activation by Cp*ML and Tp'ML Systems; L=CO, PMe_3, Cp*=Pentamethylcyclopentadienyl, Tp'=Tris-(3,5-dimethylpyrazolyl)borate**	11
2.1	Early Examples of Hydrocarbon Activation	11
2.2	Trispyrazolylborate Complexes in Alkane Activation	12
2.3	Selectivity in C–H Activation	14
2.4	C–H Bond Activation in Matrices	18
2.5	Evidence for Alkane Sigma Complexes	19
2.5.1	Observation of Intermediates by Transient Absorption	19
2.5.2	Indirect Detection in Solution	21
2.6	Stereochemistry of Oxidative Addition of C–H Bonds	24
2.7	Theoretical Treatment of C–H Activation	25
3	**Alkane Activation by Ir^{III} and Pt^{II} Complexes**	25
4	**Electrophilic Alkane Activation**	30
5	**Addition of C–H Bonds Across M=X Bonds**	32
6	**Other Alkane Activations**	34

7 Arene Activation 37

8 Alkene Activation. 40

References 43

1 Introduction

Hydrocarbons serve as the ultimate resource for organic chemicals. Despite their abundance, however, hydrocarbons are used mainly as fuels. Heterogeneous catalysts are commonly employed for the redistribution of the hydrocarbon chains and the production of chemical intermediates. By comparison, there are relatively few examples of alkane and arene functionalizations catalyzed by homogeneous transition metal compounds. Examples include the production of terephthalic acid and esters, adipic acid synthesis, and acetic acid production from butane oxidation [1].

Homogeneous catalysis offers several advantages over heterogeneous processes. Typical improvements can be seen in the control of regiochemistry, stereoselectivity, and even enantioselectivity using chiral catalysts. Control of temperature and mixing is more facile, as is control of catalyst and ligand concentrations. Finally, the nature of the catalytic species can be better regulated to effect the desired reaction specifically. Opportunities arise for improvements in both the economics of current processes as well as in the environmental factors surrounding existing processes.

This chapter will examine fundamental aspects of stoichiometric hydrocarbon activation, focusing on alkanes. Examples of metal complexes that have been found to react by various mechanisms will be summarized accordingly. For some of the more thoroughly studied examples, trends in kinetic and thermodynamic selectivities will be discussed, and experimental evidence for the intermediates in the activation reaction(s) will be presented. New complexes that activate hydrocarbons will be given. Examples of arene and alkene activation will also be summarized. A few examples where alkane and arene functionalization have been accomplished stoichiometrically will be given.

In examining the reactions of homogeneous metal complexes with hydrocarbons, the various examples can be classified according to four general types of reaction. These types are shown in Fig. 1. Reaction 1a represents the most common type, in which a metal with a vacant coordination site undergoes oxidative addition (formation of two new bonds, using $2e^-$ from the metal+$2e^-$ from the C–H bond) to the metal center. Reaction 1b shows a homolytic or radical process that is quite rare, but nonetheless has been observed. Reaction 1c represents the use of an electrophilic metal center to break the C–H bond, and there is debate as to whether this mechanism is concerted or actually proceeds via an oxidative addition pathway. Finally, reaction 1d shows the reversible addition of a C–H bond to a M=X bond where X can be either a heteroatom containing ligand or

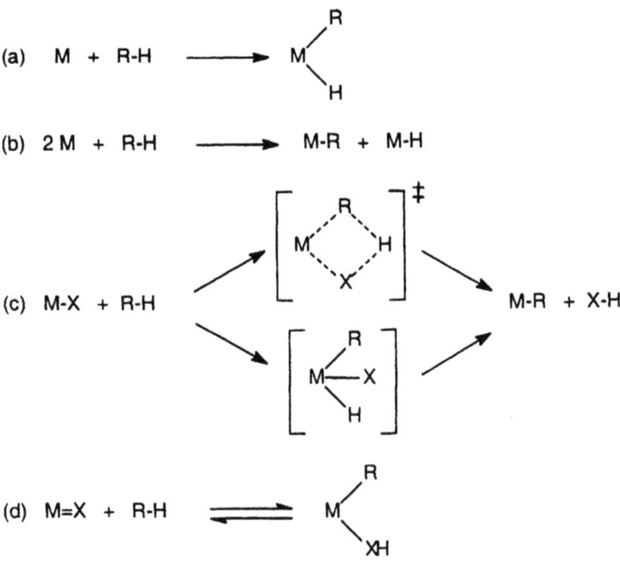

Fig. 1. Mechanisms for C–H activation by transition metal complexes

an alkylidene. In the sections that follow, all hydrocarbon activation examples can be assigned as belonging to one of these classes of reaction.

2
Alkane Activation by Cp*ML and Tp'ML Systems; L=CO, PMe_3, Cp*= pentamethylcyclopentadienyl, Tp'=tris-(3,5-dimethylpyrazolyl)borate

2.1
Early Examples of Hydrocarbon Activation

Prior to 1982, Crabtree's report of the reaction of cyclopentane with a solvated $IrH_2(PPh_3)_2^+$ species to give a cyclopentadienyl-iridium product stood as the only well characterized example of a reaction of an alkane with a homogeneous transition metal, in contrast to the widespread reactivity of arenes [2]. Based upon the instability of the platinum methyl hydride complex $Pt(PPh_3)_2(CH_3)H$, it was believed that alkane oxidative addition might not be a thermodynamically feasible process, and consequently few attempts were made to attempt such a reaction [3]. It was not until the discovery of the formation of stable alkane oxidative addition products in 1982 that it was realized that reactions of hydrocarbons were in fact feasible.

The first reports were based upon reactions of the [Cp*IrL] fragment where L=PMe_3 or CO and Cp*=η^5-C_5Me_5 by Bergman and Graham, respectively [4, 5]. A less stable alkane oxidative addition product to the fragment [Cp*Rh(PMe_3)] was also reported by Jones [6]. In all of these cases, irradiation was used to either labilize a coordinated ligand or induce reductive elimination of dihydrogen, thereby generating a reactive 16-electron Cp^*M^IL fragment that could then undergo a thermal reaction with the alkane (Eq. 1). All of these fragments are so reactive that the substrate must also serve as the solvent for the reaction.

$Cp^*Ir(CO)_2$ or $Cp^*M(PMe_3)H_2$ $\xrightarrow[RH]{h\nu}$ $Cp^*M(L)(R)H$ (1)

M = Rh, Ir

R = *c*-hexyl, *n*-propyl, neopentyl
L = CO, PMe_3

Later work with these systems extended the reactivity to include a variety of other hydrocarbons, including arenes, methane, *n*-alkanes, and cycloalkanes. The use of perfluorohexane as an inert solvent allowed for the examination of the reaction of $Cp^*Ir(CO)_2$ with methane [7]. Liquid xenon also proved to be an effective inert solvent in which to examine reactions of the fragment [Cp*Ir (PMe_3)] with hydrocarbons [8]. The iridium complexes $Cp^*Ir(PMe_3)$(alkyl)H were found to be stable up to ~110°C, at which temperature reversible reductive elimination/oxidative addition reactions occur. For example, heating the complex $Cp^*Ir(PMe_3)$(*c*-hexyl)H in C_6D_6 at 110°C leads to the formation of $Cp^*Ir(PMe_3)(C_6D_5)D$ plus *c*-C_6H_{12}. Kinetic studies showed that the rate of cyclohexane loss was independent of the benzene concentration, implying first order loss of cyclohexane in or prior to the rate determining step. A kinetic isotope effect for loss of *c*-C_6H_{12} vs. *c*-C_6D_{12} was found to be 0.7 [9]. The thermal lability of the carbonyl analog, Cp*Ir(CO)(alkyl)H, was not reported. In contrast, the rhodium analogs $Cp^*Rh(PMe_3)$(alkyl)H are quite unstable at ambient temperature, undergoing facile reductive elimination of alkane at –20°C [10]. The phenyl derivatives of these complexes are substantially more stable, with $Cp^*Rh(PMe_3)$ PhH losing benzene at 60°C and $Cp^*Ir(PMe_3)$PhH not losing benzene even at 200°C.

2.2 Trispyrazolylborate Complexes in Alkane Activation

In view of the bonding similarities between the Cp ligand and the trispyrazolylborate ligand as have been popularized by Trofimenko [11], several investigations of the tris-(3,5-dimethylpyrazolyl)borate (Tp') complexes have been reported. Graham reported that irradiation of the dicarbonyl complex $Tp'Rh(CO)_2$ leads to the efficient loss of CO and the activation of hydrocarbon solvents (Eq. 2). Product formation could be driven to completion if the CO was removed by a purge of N_2 or Ar. Upon treatment with benzene, the cyclohexyl hydride addition product Tp'Rh(CO)(*c*-hexyl)H is converted to the phenyl hydride adduct within 10 min at 25°C, indicating that the system is more stable than its $Cp^*Rh(PMe_3)$ counterpart but less stable that its Cp*Ir(CO) analog.

Alkane/alkane exchange can also be effected as demonstrated by purging a solution of the cyclohexyl hydride complex with methane to generate the methyl hydride product. The equilibrium constant for this exchange (1 atm CH_4, cyclohexane solvent, 25°C) is 190, indicating that activation of methane is preferred over cyclohexane [12]. Reaction of Tp'Rh$(CO)_2$ with N_2O at room temperature generates the 16-electron coordinatively unsaturated intermediate [Tp'Rh(CO)], which then reacts with benzene or cyclohexane to give the oxidative addition products [13]. Alternatively, the ethylene derivative Tp'Rh(CO)(C_2H_4) undergoes photochemical reaction in benzene to give a mixture of Tp'Rh(CO)PhH and Tp'Rh(CO)(Ph)Et, in which ethylene has inserted into the Rh-H bond [14].

hν, RH (2)

L, L' = CO
L, L' = CNR, PhN=C=NR

Jones investigated the analogous tris-(3,5-dimethylpyrazolyl)borate isocyanide rhodium fragment using a photochemically labile carbodiimide leaving group. Again, reaction was seen with alkanes and arenes to give adducts of similar stability [15–17]. In contrast to the Tp'Rh$(CO)_2$ complex, which contained an η^3-Tp' ligand, Tp'Rh$(CNR)_2$ complexes (R=neopentyl) have been found to contain η^2-Tp' ligands [18], so that differences in hapticity cannot be taken for granted.

Tolman has also reported an interesting chiral C–H activation by a menthol substituted trispyrazolylborate rhodium dicarbonyl complex. In this example, an 85:15 mixture of two intramolecular activation products resulting from oxidative addition of the menthyl methyl group is observed (Eq. 3) [19]. Irradiation at –78°C yielded a 40:60 ratio of the two products, which adjusted to an 85:15 ratio at room temperature, demonstrating the lability of the C–H activation adduct. Furthermore, NOESY spectroscopy showed exchange only between the unactivated isopropyl methyl group and the hydride ligand, indicating that equilibration between the major and minor photoproducts involves the reversible cyclometalation of the two methyls on the same isopropyl group.

$Tp^{menth}Rh(CO)_2$ → hν, -CO (3)

2.3
Selectivity in C–H Activation

Several of the systems that have been found to activate hydrocarbon C–H bonds by way of oxidative addition have also been investigated with regard to their selectivity for different types of C–H bonds. There are two types of selectivity that can be considered, the *kinetic selectivity*, which describes the rate at which a certain type of bond reacts, and the *thermodynamic selectivity*, which describes the energetic preference for cleavage of a particular type of bond. The main features of these distinct types of selectivity can be seen in Fig. 2, which is a diagram of the change in free energy (ΔG) as a function of reaction coordinate for the reaction of a metal fragment with two different hydrocarbons. The kinetic selectivity of a metal fragment [M] towards a mixture of two hydrocarbons RH and R'H is reflected by the difference in the barrier heights for the oxidative addition reaction, $\Delta\Delta G^{\ddagger}$, where $\Delta\Delta G^{\ddagger} = RT\ln\left(\frac{\text{rate of formation of M(R)(H)}}{\text{rate of formation of M(R')(H)}}\right)$ The thermodynamic selectivity of a metal towards two hydrocarbons refers to the ultimate stability of the C–H insertion products, $\Delta G°$, where $\Delta G° = -RT\ln\left(\frac{\text{M(R)(H)}}{\text{M(R')(H)}}\right)$.
If an experiment is conducted that gives a mixture of two products, there is oftentimes difficulty in determining if the observed product ratio reflects the kinetic or themodynamic selectivity of the system. If the products are known to be stable towards reductive elimination under the reaction conditions, then the product ratio reflects the kinetic selectivity. Alternatively, if the initially formed product ratio is observed to change over time to give a constant ratio of prod-

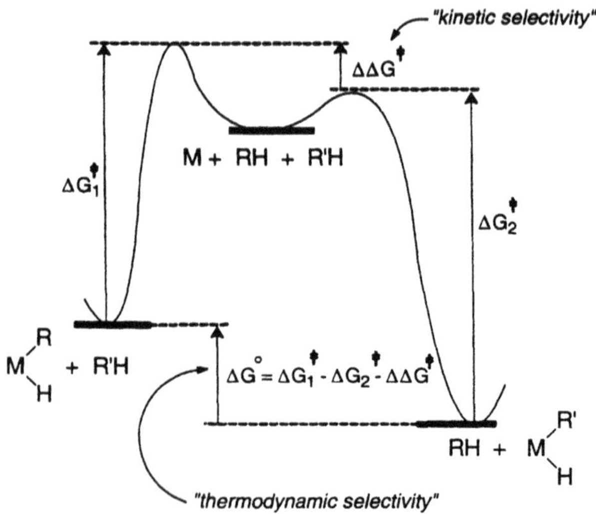

Fig. 2. Free energy diagram showing kinetic vs. thermodynamic selectivity

ucts, then the initial distribution reflects the kinetic selectivity and the final distribution reflects the thermodynamic selectivity.

Three of the above described systems have been investigated in some detail for kinetic and thermodynamic selectivities. $Cp^*Ir(PMe_3)H_2$ was reacted photochemically with mixtures of two alkanes in order to determine kinetic selectivities, since the products are not labile under the reaction conditions [20, 21]. Similar competitive experiments were conducted with $Cp^*Rh(PMe_3)H_2$ [10] and Tp'Rh(CNR)(PhN=C=Nneopentyl) (R=neopentyl) [17] under conditions where the kinetic products of the reaction are stable. The results of these studies are shown in Table 1. There are several points worth noting. First, the relative rates of reaction are listed on a *per hydrogen* basis; that is, a 2:1 mole ratio favoring benzene over cyclohexane corresponds to a 1:1 ratio on a per hydrogen basis. It can be seen that for all complexes, benzene is preferred over alkanes. Smaller cycloalkanes are preferred over larger cycloalkanes, and normal alkanes are preferred over cycloalkanes. Within a normal alkane, there is a preference for activation of the terminal methyl group C–H bonds. Finally the rhodium complexes display a much higher kinetic selectivity over the iridium complex, and the Tp'Rh complex displays a higher selectivity than the Cp*Rh complex. With both rhodium complexes, activation of the internal C–H bonds of a normal alkane was *not* observed, while both terminal and internal activation was seen with iridium.

Table 1. Relative kinetic selectivities by various metal fragments for different types of C–H bonds on a per-hydrogen basis. Product distributions reflect the relative reactivity of one C–H bond in each hydrocarbon

Hydrocarbon	Metal fragment		
	$Cp^*Ir(PMe_3)$	$Cp^*Rh(PMe_3)$	$Tp^*Rh(CNR)^b$
Cyclohexane	1.0	1.0	1.0
Cyclopropane	2.63	10.4	18.4
Cyclopentane	1.6	1.8	1.7
Cycloheptane	–	0.14	–
Cyclooctane	0.09	0.06	–
Cyclodecane	0.23	–	–
Benzene	4.0	19.5	70
Pentane (1°)	–	15	14.9
(2°)	–	0	0
Hexane (1°)	2.7^a	5.9	–
(2°)	1.0^a	0	–
Propane (1°)	1.5^a	2.6	15
(2°)	1.0^a	0	0
Neopentane	1.14	–	–

[a] Relative intramolecular selectivity only. Values not relative to cyclohexane
[b] R=neopentyl.

Graham has reported that the fragment [Cp*Ir(CO)] shows a 4:1 kinetic preference for cleavage of a benzene C–H bond over a cyclohexane C–H bond. Unfortunately, no other kinetic selectivities were reported for this complex [5].

The above systems have also been examined in some detail for their thermodynamic selectivity towards C–H bond activation. As mentioned above, irradiation of Cp*Ir(PMe_3)H_2 in pentane gives a mixture of primary and secondary C–H activation products. Upon heating to 110°C, however, the primary activation product is observed to increase at the expense of the secondary activation products (Eq. 4). This observation indicates that the initial product ratio indeed represented a kinetic selectivity, and that the *n*-pentyl product is thermodynamically preferred by several kJ mol^{-1}, since no secondary products were seen at equilbrium [22].

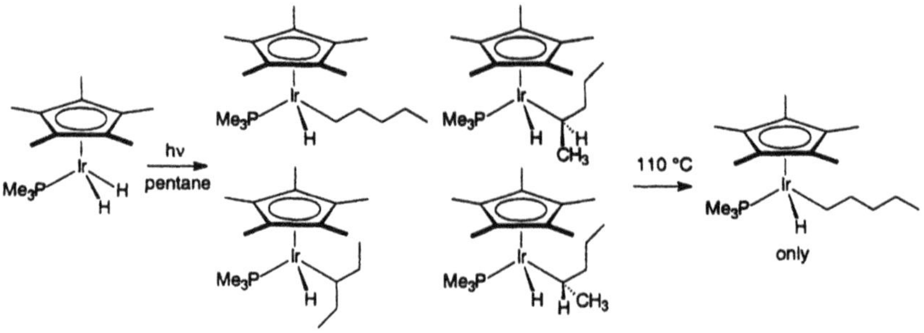

(4)

For complexes of the type Cp*Ir(PMe_3)(R)H, equilibration between two possible alkane activation products was found to occur at 140°C (Eq. 5). The mixture of alkane solvents used and the relative quantities of the two C–H insertion products was then used to calculate K_{eq} and ΔG° (Eq. 6). The results of these equilibrations are shown in Table 2. Examination of this table shows a strong preference for activation of primary over secondary C–H bonds, and a preference for less hindered primary bonds over more hindered primary bonds. This latter preference can be extreme as seen in the equilibration among the methyl groups of 2,2-dimethylbutane, which shows only activation of the C–H bonds on C4 [23].

Table 2. Thermodynamic selectivities for pairs of hydrocarbons R_1H and R_2H (Eq. 5)

M	R_1	R_2	K_{eq}	ΔG° (kJ mol^{-1})
Cp*Ir(PMe_3)	*c*-hexyl	*n*-pentyl	10.8	-8.4
Cp*Ir(PMe_3)	-$CH_2CHMeCHMe_2$	*n*-pentyl	3.5	-4.2
Cp*Ir(PMe_3)	*c*-pentyl	$CH_2CHMeCHMe_2$	1.5	-1.3
Cp*Ir(PMe_3)	*c*-hexyl	*c*-pentyl	2.0	-2.5
Cp*Ir(PMe_3)	-$CH_2CMe_2CH_2CH_3$	-$CH_2CH_2CMe_3$	>20	<-10
Tp'Rh(CNR)[a]	*c*-hexyl	phenyl	2×10^8	-47.7
Tp'Rh(CNR)[a]	*c*-hexyl	methyl	340	-14.6
Tp'Rh(CNR)[a]	*n*-pentyl	methyl	12	-6.3

[a]R=neopentyl.

$$Cp^*Ir(PMe_3)(R_1)H + R_2H \rightleftharpoons Cp^*Ir(PMe_3)(R_2)H + R_1H \tag{5}$$

$$\Delta G^\circ = -RT\ln K_{eq} = -RT\ln\left(\frac{[Cp^*Ir(PMe_3)(R_2)H][R_1H]}{[Cp^*Ir(PMe_3)(R_1)H][R_2H]}\right) \tag{6}$$

The $Cp^*Ir(PMe_3)(R)H$ complexes have also been investigated in calorimetric experiments designed to obtain absolute iridium-carbon bond strengths. These experiments involved measuring the enthalpies of reaction of $Cp^*Ir(PMe_3)H_2$, $Cp^*Ir(PMe_3)(Ph)H$, and $Cp^*Ir(PMe_3)(c\text{-hexyl})H$ with HCl, and provided the Ir-X bond strengths listed in Table 3. Independent determination of the iridium-hydride bond strength in $Cp^*Ir(PMe_3)H_2$ using photoacoustic calorimetry provided a value of 304.7±18 kJ mol^{-1} (72.9±4.3 kcal mol^{-1}), in excellent agreement with the solution calorimetric value. The large difference in Ir-Ph vs. Ir-*c*-hexyl bond strengths (125 kJ mol^{-1}=30 kcal mol^{-1}) compared with the corresponding difference in carbon-hydrogen bond strengths (73 kJ mol^{-1}=17.5 kcal mol^{-1}) should be noted, and accounts for the strong thermodynamic preference for benzene activation [24].

Thermodynamic studies with the analogous rhodium system, $Cp^*Rh(PMe_3)(R)H$, were more difficult in that the alkyl hydride complexes were unstable at ambient temperature. It was possible to determine using kinetic techniques, however, that the complex $Cp^*Rh(PMe_3)(Ph)H$ was some 36 kJ mol^{-1} (8.7 kcal mol^{-1}) more stable than the complex $Cp^*Rh(PMe_3)(n\text{-propyl})H$ ($K_{eq}=4.6\times10^{-7}$) [25]. As mentioned earlier, Graham reported one example of thermodynamic selectivity of Tp'Rh(CO) for methane vs. cyclohexane (K_{eq}=190), and also demonstrated that benzene activation was strongly preferred over cyclohexane activation, but no further studies have been reported [12].

Another system for which thermodynamic data have been obtained in some detail is the Tp'Rh(CNneopentyl)(R)H system studied by Jones. Here, the relative thermodynamic stabilities of a number of adducts were obtained by measuring both the competitive kinetic selectivity for two types of C–H bond ($\Delta\Delta G^\ddagger$ in Fig. 2) as well as the barrier for reductive elimination of free alkane from each adduct ($\Delta G_{1\ddagger}$ and $\Delta G_{2\ddagger}$ in Fig. 2). The free energies for the latter were obtained from kinetic studies of the reductive elimination of hydrocarbon in benzene. A summary of the ΔG° values, calculated equilibrium constants, and relative metal-carbon bond strengths are given in Table 4 [26]. For $D_{C\text{-}H}$ for benzene, see ref.

Table 3. Absolute iridium-X bond strengths in $Cp^*Ir(PMe_3)X_2$ complexes

X	$D_{Ir\text{-}X}$ (kJ/mol)
H	310
Cl	377
Br	318
I	267
C_6H_5	337
C_6H_{11}	212

Table 4. Selectivities and thermodynamics for the formation of Tp'Rh(CNneopentyl)(R)H ($kJ\ mol^{-1}$) at 23°C

R	D(C–H)[a]	k_{rel} (RH)[b]	ΔG^{oc}	K_{eq}[d]	D_{rel} (M-R)[e]
phenyl	474 (2)[26]	1	0	1	0
$HCCHCMe_3$	458 (3)[26]	10	22.2	1.3×10^{-4}	39
methyl	438 (0.4)	2.3	3.0	1.6×10^{-6}	69
n-pentyl	409.6	4.7	39.1	1.4×10^{-7}	104
c-pentyl	403 (2)	24.8	47.9	3.9×10^{-9}	120
c-hexyl	401 (4)	35.2	47.7	4.2×10^{-9}	121
mesityl	370 (6)	1.3	27.8	1.3×10^{-5}	133
i-butenyl	358 (6)	4.3	32.7	1.8×10^{-6}	149

[a] Unless otherwise noted, bond strength values were obtained from *The CRC Handbook of Chemistry and Physics* [131].
[b] $k_{rel}=k_{oa}(C_6H_6)/k_{oa}(RH)$ for oxidative addition of a molecule of substrate measured at –15°C.
[c] Relative to Tp'Rh (CNneopentyl) (Ph)H.
[d] K_{eq} is for Tp'Rh (CNneopentyl) (Ph)H+RH=Tp'Rh (CNneopentyl) (R)H+PhH.
[e] Bond strengths relative to $D_{Rh\text{-}Ph}$ in Tp'Rh(CNneopentyl)(Ph)H. $D_{rel}=D_{Rh\text{-}Ph}-D_{Rh\text{-}R}$.

[27] $D_{C\text{-}H}$ for *t*-butylethylene is based on the latest value of $D_{C\text{-}H}$ for ethylene. See Ervin et al. (1990) [27]. As seen with iridium, the differences in metal-carbon bond strengths are substantially larger than the corresponding differences in carbon-hydrogen bond strengths, leading to a substantial thermodynamic preference for the activation of hydrocarbons with stronger C–H bonds [17].

2.4 C–H Bond Activation in Matrices

Many of the above complexes have been examined by photolysis in inert and reactive gas matrices. These experiments, in general, provide evidence for the photochemical generation of the 16-electron coordinatively unsaturated intermediates, their weak interaction with inert gas atoms or methane, and in several cases their eventual reaction with methane by C–H activation. The applicability of this method to a particular system depends upon the volatility of the metal complex precursor, as the species must go into the gas phase during deposition in the matrix. Several examples are given below.

Rest and Graham reported in 1984 that the metal carbonyl complexes $CpRh(CO)_2$, $CpIr(CO)_2$, and $Cp^*Ir(CO)_2$ can be deposited in methane matrices at 12 K and irradiated to give the corresponding methane oxidative addition products [28]. In addition, the dihydride $CpIr(CO)H_2$ could be irradiated in a methane/argon matrix to generate $CpIr(CO)(CH_3)H$ by an alternative route [29]. While the dicarbonyl compounds were not efficient producers of the coordinatively unsaturated intermediate, Perutz found that $CpRh(CO)(C_2H_4)$ lost ethylene easily upon photolysis, generating CpRh(CO)(CH_3)H. A competitive photodissociation of CO was also observed, generating the fragment [CpRh (C_2H_4)] which did not activate methane [30].

Perutz also examined the photochemistry of CpRh(PMe_3)H_2 and CpIr(PMe_3) H_2 in Ar, CH_4, N_2 and CO/Ar matrices. These experiments provide strong evidence for the formation of the reactive fragment [CpM(PMe_3)] in argon matrices, and adducts of the type CpM(PMe_3)(L) in N_2 and CO/Ar matrices. The methane addition product CpM(PMe_3)(CH_3)H formed in methane matrices [31].

2.5 Evidence for Alkane Sigma Complexes

The activation of C–H bonds via oxidative addition was initially thought to occur by way of a three-centered transition state involving the metal, carbon, and hydrogen atoms. The discovery of stable η^2-dihydrogen adducts, however, suggested that the analogous species formed by interaction of a C–H sigma bond with a metal center might be a species with some intermediate stability [32]. The observation of agostic metal complexes, where an intramolecular version of this type of interaction can be seen, provides strong impetus for such species as intermediates along the reaction pathway for alkane oxidative addition [33].

2.5.1 *Observation of Intermediates by Transient Absorption*

Perutz reported the time resolved infrared (TRIR) study of CpRh$(CO)_2$ in cyclohexane solution. A species was observed with a lifetime of ~15 ms, assigned as CpRh(CO)(*c*-hexyl)H. If CO is present (1.5 atm), the intermediate decays with a half-life of 1.7 ms. Similar observations were made if CpRh(CO)(C_2H_4) was used to prepare the reactive intermediate. Laser flash photolysis experiments show the formation of the hydrocarbon activation adduct within 400 ns of the flash, but did not provide evidence for an intermediate prior to its formation [34].

Bergman and Moore have similarly investigated the photochemistry of Cp*Rh$(CO)_2$ in liquid Xe and Kr solution by transient TRIR spectroscopy. Monocarbonyl species are initially observed and assigned as Cp*Rh(CO)Xe and Cp*Rh(CO)Kr, the latter reacting much more rapidly with CO. The species Cp*Rh(CO)Kr was observed to react with cyclohexane to give Cp*Rh(CO)(*c*-hexyl)H, and the rate was found to be dependent on the cyclohexane concentration. Due to the asymptotic dependence of the observed rate constant for this reaction, and its dependence on the *nature* of the alkane (i.e., C_6H_{12} vs. C_6D_{12}), the authors proposed the reversible formation of an alkane sigma complex prior to oxidative addition of the C–H bond (Eq. 7), but the alkane sigma complex could not be distinguished spectroscopically from the krypton complex [35]. For the equilibrium K_{eq}, ΔH=−4.2±0.4 kJ mol^{-1} (−1.0±0.1 kcal mol^{-1}) and ΔS=33±4 J

$mol^{-1} K^{-1}$ (8±1 cal $mol^{-1} K^{-1}$), which indicates little thermodynamic preference for krypton vs. cyclohexane binding to the Rh^{I} fragment. For the second step where C–H bond cleavage occurs, a very low activation energy of 4.8 kcal mol^{-1} was obtained. The authors also looked at the reaction of the xenon complex Cp*Rh(CO)Xe with CO and found the substitution to follow bimolecular kinetics [36]. Later work with perdeutero-neopentane at 165 K in liquid krypton showed a 1-cm^{-1} difference between the ν_{CO} stretch of Cp*Rh(CO)Kr and Cp*Rh(CO)[$C(CD_3)_4$]. This slight difference in absorption allowed for the observation of the growth and decay of the alkane sigma complex, the conversion to Cp*Rh(CO)[$CD_2C(CD_3)_3$]D occurring with a half-life of 430 μs [37].

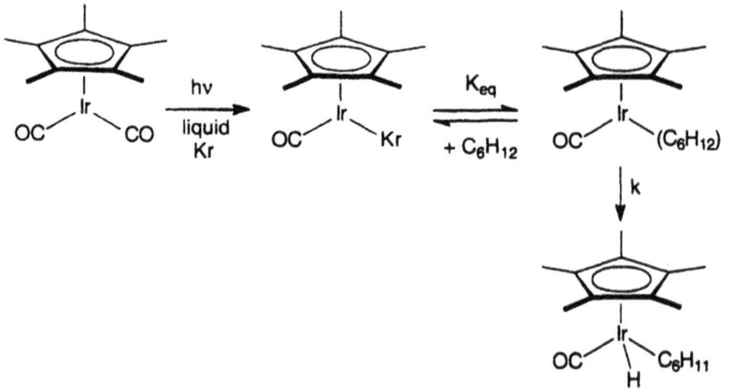

(7)

The Cp analog of the above system was examined in the gas phase by TRIR, since $CpRh(CO)_2$ is sufficiently volatile to observe intermediates spectroscopically (vapor pressure=~400 mtorr at 20°C). Irradiation in the presence of neopentane (80 mtorr) shows the initial formation of the gaseous species [CpRh(CO)] and its conversion to CpRh(CO)(neopentyl)H. An alkane complex was proposed as an intermediate, but was not detected spectroscopically [38].

More recent studies have appeared focusing on the tris-(3,5-dimethylpyrazolyl)borate-rhodiumdicarbonyl complex using femtosecond transient absorption spectroscopy. Irradiation of Tp'$Rh(CO)_2$ in cyclohexane and examination of the TRIR spectrum shows the formation and decay of two intermediates prior to the formation of the oxidative addition product Tp'Rh(CO)(*c*-hexyl)H. These intermediates have lifetimes of 200 ps and 230 ns, respectively. Based upon the carbonyl stretching frequencies of the intermediates (1972 and 1990 cm^{-1}, respectively), the first species is assigned as the alkane sigma complex η^3-Tp'Rh(CO)(σ-C_6H_{12}) and the second species is assigned as the complex η^2-Tp'Rh(CO)(σ-C_6H_{12}), where the Tp' ligand has changed hapticity by dissociation of a pyrazole ring, as shown in Scheme 1. Activation of the C–H bond is followed by the rapid recoordination of the pyrazole ring to give η^3-TpRh(CO)(*c*-hexyl)H [39]. Examination of the photochemistry of the bis-(3,5-dimethylpyrazolyl)borate rhodium dicarbonyl complex showed an IR peak at 1992 cm^{-1} in cyclohexane, which was assigned as η^2-Bp'Rh(CO)(σ-C_6H_{12}). The similarity of the

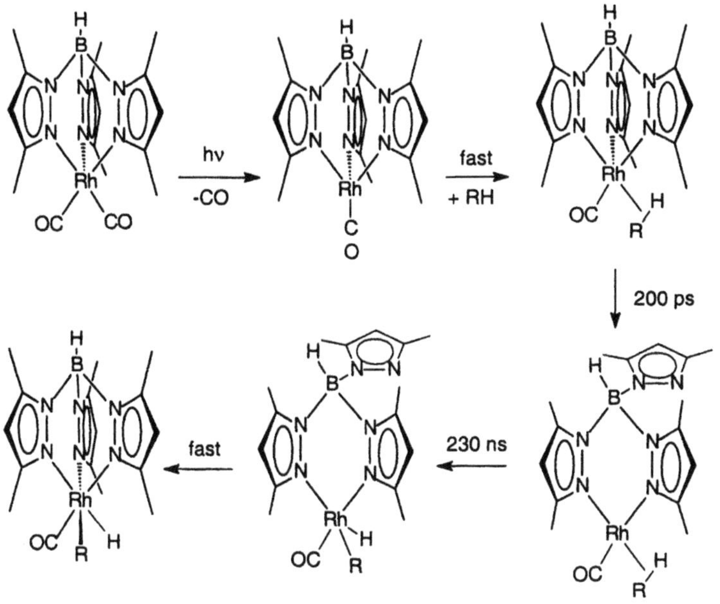

Scheme 1.

CO stretch to that in the second intermediate observed in the Tp' case provides further evidence for the assignment of the intermediate structures [40].

2.5.2
Indirect Detection in Solution

Several lines of evidence pointed towards the existence of alkane complexes even before their direct observation in matrices and by transient absorption spectroscopy. In Bergman's studies of $Cp^*Ir(PMe_3)(R)H$ complexes, the complex $Cp^*Rh(PMe_3)$(*c*-hexyl)D was prepared. Upon heating to 130°C, this complex was observed to scramble deuterium between the hydride and cyclohexyl-C1 positions as indicated in Scheme 2 [23].

Norton reported the synthesis of $Cp_2W(CH_3)D$ and its reductive elimination of methane. This complex was found to undergo H/D scrambling between the methyl group and the hydride positions, but was complicated by an intermolecular contribution to the scrambling [41]. Related studies by Jones with $Cp^*Rh(PMe_3)(CH_3)D$ showed scrambling between the methyl group hydrogen and metal deuteride position [42]. Studies of the similar intramolecular H/D scrambling in Tp'Rh(CNneopentyl)(CH_3)D and Tp'Rh(CNneopentyl)(CD_3)H provided both kinetic and equilibrium data for the isomerization [43].

Perhaps the most information packed experiment was performed by Bergman with the multiply labeled compound $Cp^*Rh(PMe_3)(^{13}CH_2CH_3)D$. This compound was observed to equilibrate with the isomer in which the deuteride

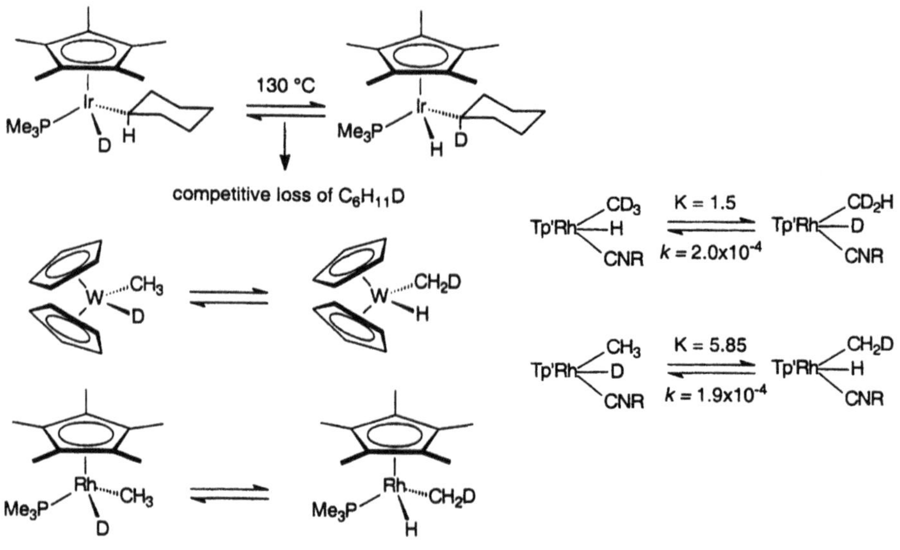

130 °C

competitive loss of $C_6H_{11}D$

$K = 1.5$

$k = 2.0 \times 10^{-4}$

$K = 5.85$

$k = 1.9 \times 10^{-4}$

Scheme 2.

ligand had scrambled into the α-ethyl position at –80°C. Upon warming to –30°C, scrambling of the point of attachment of the ethyl group changed from α to β, with the deuterium remaining attached to the labeled carbon, as shown in Eq. 8. Loss of ethane is competitive with the latter rearrangement. These scramblings were proposed to occur by way of ethane σ-complexes in which interconversion between interaction with C–H bonds on the same methyl group occurred more facilely than with C–H bonds on an adjacent methyl group [44].

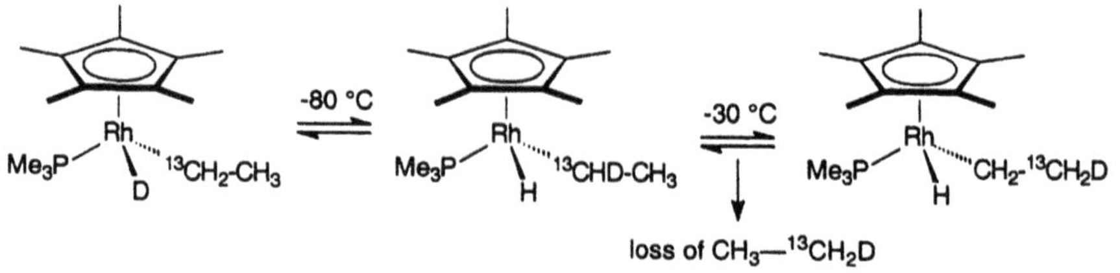

-80 °C

-30 °C

loss of CH_3—$^{13}CH_2D$

(8)

Interactions of non-reactive unsaturated metal complexes with alkanes is documented in the studies by Rayner with the fragment [$W(CO)_5$]. TRIR was used to look at the equilibrium between the bound and ligand-free complex. These data allowed determination of the actual enthalpy for coordination of the alkane to the unsaturated metal center. As seen in Table 5, there is a substantial interaction between most alkanes and the [$W(CO)_5$] fragment, with methane being the most weakly bound (no complexation was observed) [45].

George and Poliakoff have reported that photolysis of $CpRe(CO)_3$ in heptane solution produces a new dicarbonyl species that back reacts with added CO in a bimolecular reaction. The adduct was observed in the TRIR and was character-

Table 5. Binding energies of alkanes in $W(CO)_5$(alkane) complexes

Alkane	ΔH° (kJ mol^{-1})	Alkane	ΔH° (kJ mol^{-1})
CH_4	<21	n-C_5H_{12}	44
C_2H_6	31	n-C_6H_{14}	45
C_3H_8	34	c-C_3H_6	34
n-C_4H_{10}	38	c-C_5H_{10}	43
i-C_4H_{10}	36	c-C_6H_{12}	48

ized as the sigma alkane complex $CpRe(CO)_2$(σ-heptane). The complex is quite unreactive, with a second order rate constant for reaction with CO of 2000 M^{-1} s^{-1}. The corresponding xenon complex, formed by irradiation in supercritical xenon at 25°C, is similarly unreactive towards CO (k=4800 M^{-1} s^{-1}). The krypton complex, in comparison, is approximately 2000× more reactive with CO than the xenon complex [46].

The equilibrium loss of dihydrogen from $Ir(Pi\text{-}Pr_3)_2(\eta^2\text{-}H_2)(H)_2X$ complexes has been found to be strongly solvent dependent, as evidenced by the observation of vigorous effervescence when solid samples of $Ir(Pi\text{-}Pr_3)_2(\eta^2\text{-}H_2)(H)_2X$ (X=Cl, Br, or I) are dissolved in hydrocarbon solvents. Examination of the temperature dependence of the equilibria provided evidence that toluene reacts with all 3 five-coordinate species $Ir(Pi\text{-}Pr_3)_2(H)_2X$, but that only the iodide complex interacted with alkane solvents. This conclusion was made based upon the smaller ΔH and ΔS values for the iodide complex equilibrium in alkane solvents, compared with the bromide and chloride complexes (Eq. 9) [47].

$$IrL_2X(H)_2(\text{solvent}) \rightleftharpoons IrL_2X(H)_2 \overset{\pm H_2}{\rightleftharpoons} IrL_2X(H)_2(H_2) \quad (9)$$

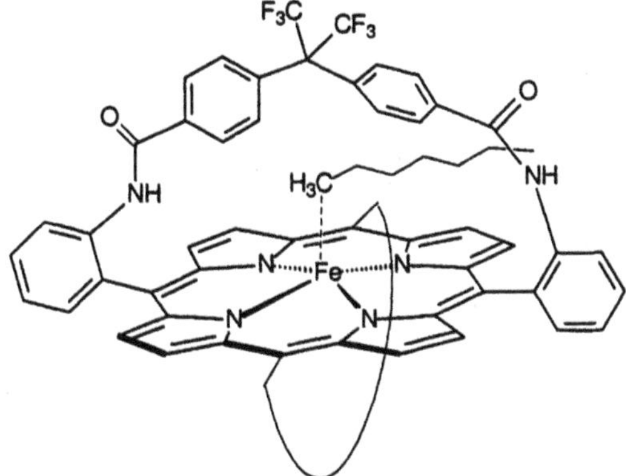

Fig. 3. Structure of a porphyrin-alkane complex

Finally, Reed has reported an interesting crystal structure of a double porphyrin A-frame iron(II) complex. This complex has a large void above the porphyrin plane, and the X-ray structure (Fig. 3) shows a disordered *n*-heptane solvent molecule trapped in the void. The heptane appears as an *n*-octane molecule with 50% occupancy of the terminal methyl groups, corresponding to a positional disorder involving a one-atom displacement along the alkane chain [48]. The iron-carbon bond distances of 2.5 and 2.8 Å are in the range commonly observed in complexes with agostic C–H interactions.

2.6 Stereochemistry of Oxidative Addition of C–H Bonds

Few reports have appeared addressing the stereochemistry of either C–H bond oxidative addition or reductive elimination. The most convincing paper was an intramolecular activation examined by Flood in which a chiral 8-ethylquinoline derivative underwent benzylic activation by $PdCl_4^{2-}$. The reaction proceeds with the net retention of configuration at carbon (Eq. 10) [49].

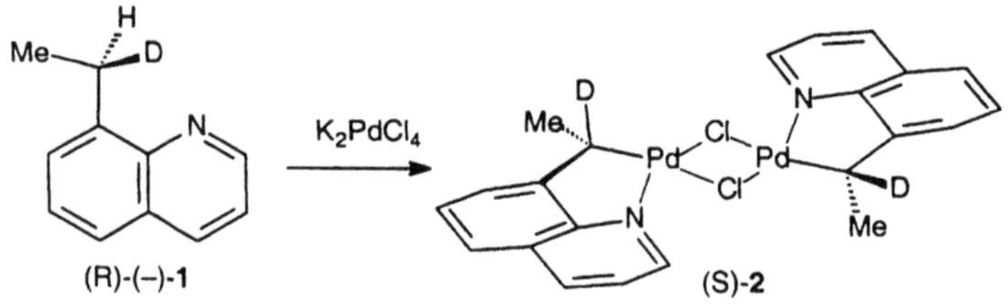

(R)-(–)-1 $\xrightarrow{K_2PdCl_4}$ (S)-2 (10)

Bergman also examined the rearrangement of a gem-dimethylcyclopropane adduct which was interpreted in terms of retention of configuration at carbon during C–H oxidative addition. Inversion at the carbon center was effected by way of isomerization to an alkane sigma complex that rearranged to a second sigma complex before reinserting into the C–H bond (Eq. 11) [50].

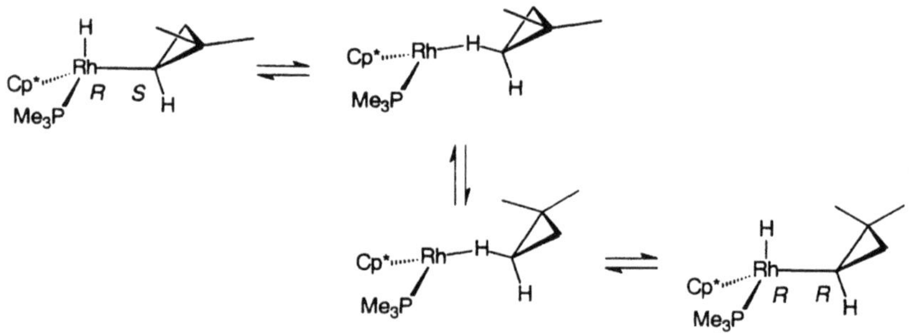

(11)

A very recent example of an enantioselective C–H activation has been reported by Bergman. A chiral metal complex was employed in several C–H activation reactions, as shown in Scheme 3. Activation of benzene results in the formation of a 1:1 mixture of possible diasteriomers. Activation of cyclohexane, however, gives only a single diasteriomer. Subsequent thermolysis of the sample at 150°C

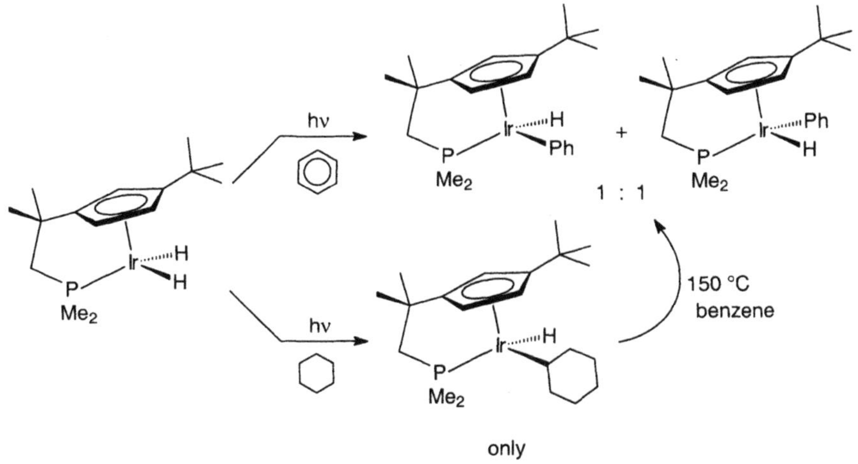

Scheme 3.

in benzene then leads to formation of a mixture of both possible phenyl hydride diastereomers [51].

2.7 Theoretical Treatment of C–H Activation

Over the past 15 years, many theoretical treatments of C–H activation have appeared. Early work by Hoffmann addressed qualitative orbital approaches to C–H activation by CpML fragments [52]. More quantitative approaches have appeared recently for the addition of methane to the [CpRh(CO)] fragment [53–56]. These more recent calculations provide support for the presence of methane σ-complexes along the reaction coordinate for methane oxidative addition, and confirm the weak nature of the interaction between the metal center and the C–H sigma bond (~20 kJ mol^{-1}). A more detailed comparison of these results is beyond the scope of this chapter.

3 Alkane Activation by Ir^{III} and Pt^{II} Complexes

One of the earliest reports of alkane C–H activation was made by Shilov in 1969 in which H/D exchange was reported between methane and a D_2O/CH_3COOD solvent in the presence of K_2PtCl_4 [57]. While the mechanistic details of this exchange were not entirely clear, the work stood as an isolated example of alkane activation for many years. Alkane activation by platinum was not reported again until 1986, when Whitesides found that $(Cy_2PCH_2CH_2PCy_2)Pt(neopentyl)H$ lost neopentane and activated a variety of alkanes at ~50°C (Eq. 12) [58, 59]. These reactions are believed to proceed by way of an initial reductive elimination to generate a Pt^0 intermediate that then oxidatively adds to another C–H bond. In this sense, the complex is similar to those discussed in the previous section.

$$\text{(Cy}_2\text{P)}_2\text{Pt(CH}_2\text{CMe}_3\text{)(H)} \xrightarrow{-\text{CMe}_4} [\text{(Cy}_2\text{P)}_2\text{Pt}] \xrightarrow{+\text{RH}} \text{(Cy}_2\text{P)}_2\text{Pt(R)(H)} \quad (12)$$

$R = Me, c\text{-}C_5H_{10}, c\text{-}C_6H_{12}, Ph$

Subsequent work with a Pt^{II} complex containing a labile triflate ligand showed evidence for the alkyl/aryl exchange labeled as type (c) in Fig. 1. While mechanistic studies were limited, the observation of little positional selectivity in the reaction with toluene argued in favor of the oxidative addition pathway, i.e., via a Pt^{IV} intermediate (Eq. 13) [60].

$$\text{(Me}_3\text{P)}_2\text{Pt(CH}_2\text{CMe}_3\text{)(OTf)} \xrightarrow[-\text{OTf}^-]{+\text{ArylH}} [\text{(Me}_3\text{P)}_2\text{Pt(H)(Aryl)(CH}_2\text{CMe}_3\text{)}]^+ \xrightarrow[+\text{OTf}^-]{-\text{CMe}_4} \text{(Me}_3\text{P)}_2\text{Pt(Aryl)(OTf)} \quad (13)$$

Horvath has reported conditions under which Pt^{II} is used to catalyze the conversion of methane to methyl chloride. The reaction conditions employed are indicated below, and avoid the hydrolysis of the methyl chloride to methanol. While the total quantity of methyl chloride formed is less than the amount of platinum initially present, the system is catalytic in Pt^{II} (Eq. 14), with Pt^{IV} serving as a stoichiometric oxidant and Cl_2 stabilizing the system against precipitation of Pt^0 [61].

$$\underset{400\ \text{psi}}{CH_4} + \underset{82\ \text{psi}}{Cl_2} + \underset{4.9\ \text{mmol}}{Na_2PtCl_6} \xrightarrow[\substack{Na_2PtCl_4 \\ 0.8\ \text{mmol}}]{H_2O,\ 100\ ^\circ C} \underset{2.3\ \text{mmol}}{CH_3Cl} \quad (14)$$

Several interesting reports on the use of Pt^{II} compounds for alkane activation have appeared by Bercaw and Labinger. One of the first of these reports was the oxidation of the methyl group of *p*-toluenesulfonic acid to the corresponding alcohol and ultimately aldehyde using Pt^{IV} as the oxidant. The reaction is very clean, producing only traces of other products. Two possible mechanisms were presented, both involving C–H bond activation by Pt^{II} and based upon Shilov's original proposal (Scheme 4). In one instance, the Pt^{II}-alkyl complex is oxidized to a Pt^{IV}-alkyl complex that then undergoes nucleophilic attack by OH^- or Cl^-. In the second instance, the nucleophile directly attacks the Pt^{II}-alkyl adduct to produce Pt^0. For the first mechanism, it was suggested that the reaction with $PtCl_6^{2-}$ might occur by electron transfer, rather than alkyl transfer, which would alter the ligands present in the subsequent steps, but the net transformation would be the same. Oxidation of ethanol gave a more complicated mixture of products, with both α and β hydroxylation, β-chlorination and overoxidation to acetic acid all being observed [62].

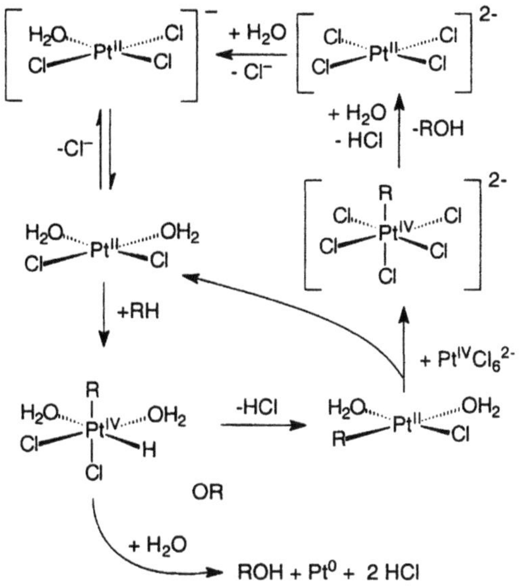

Scheme 4.

Labinger and Bercaw next examined the reductive elimination of methane from several Pt^{II}-methyl derivatives in an effort to understand the microscopic reverse of methane oxidation. Protonation of (tmeda)PtMeCl, (tmeda)Pt(CH_2Ph)Cl, (tmeda)$PtMe_2$ and *trans*-(PEt_3)$_2$Pt(CH_3)Cl (tmeda=$Me_2NCH_2CH_2NMe_2$) at low temperature in some cases led to observable Pt^{IV} intermediates, and eventually led to methane (or toluene) formation. For example, (tmeda)$PtMe_2$ reacts with HCl at –78°C to give the Pt^{IV} complex (tmeda)$PtMe_2HCl$, which loses methane upon warming to ambient temperature. In CH_3OD solvent, however, deuterium exchange into the coordinated methyl groups is observed at –40°C. This observation implies reversible formation of a sigma-methane complex. The results of all of these studies can be summarized in terms of a general mechanism as shown in Scheme 5. While the identity of the ligands around Pt can vary (halide, water, amine), the same types of intermediates account for all of the observed reactions with acids [63].

Finally, a recent report by Bercaw and Labinger demonstrates the oxidative addition of methane to a Pt^{II}-alkyl complex. Treatment of (tmeda)$PtMe_2$ with $HBAr_F$ (=HB(3,5-$C_6H_3(CF_3)_2)_4$) in perfluoropyridine leads to the production of [(tmeda)PtMe(C_5F_5N)][BAr_F]. Upon exposure to 30 atm $^{13}CH_4$, methyl group exchange is observed between the complex and free methane. Furthermore, upon heating [(tmeda)PtMe(C_5F_5N)][BAr_F] with C_6D_6, the methane isotopomers CH_4, CH_3D, CH_2D_2, and CHD_3 are observed over several days. These observations are interpreted in terms of the oxidative addition of a second C–H bond to give an alkyl aryl hydride Pt^{IV} complex and the formation of a methane sigma complex (Scheme 6) [64].

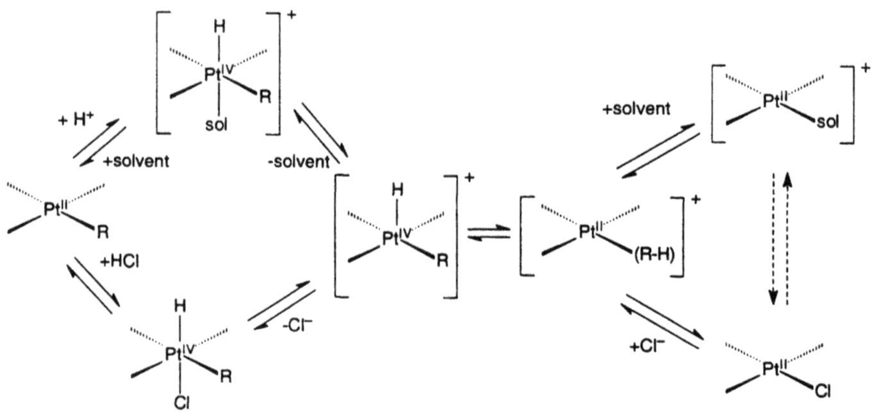

Scheme 5.

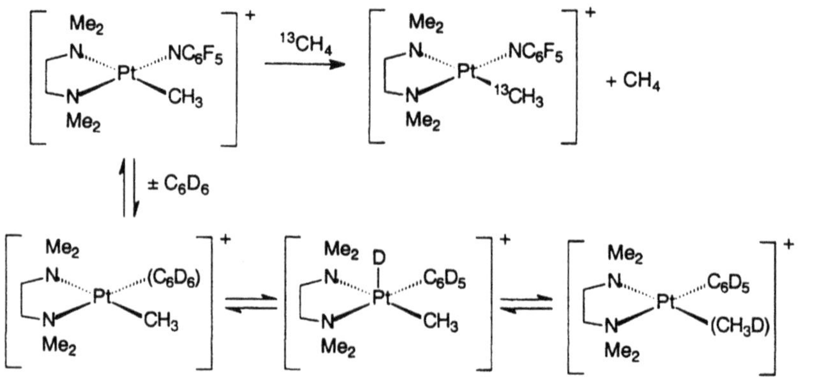

Scheme 6.

Another interesting system that appears to react by way of the path in Fig. 1c is Cp*Ir(PMe$_3$)(CH$_3$)OTf, studied by Bergman. This complex can be prepared by the disproportionation of Cp*Ir(PMe$_3$)OTf$_2$ with Cp*Ir(PMe$_3$)Me$_2$ or by reaction of Cp*Ir(PMe$_3$)Me$_2$ with one equivalent of triflic acid, and has been structurally characterized by X-ray diffraction. Reaction of the complex with $^{13}CH_4$ in dichloromethane solution gives the labeled product Cp*Ir(PMe$_3$)($^{13}CH_3$)OTf. Aromatic C–H bonds also underwent exchange with the methyl group, but cyclohexane and neopentane proved unreactive (Scheme 7) [65]. Reaction with ethane gave the ethylene hydride complex [Cp*Ir(PMe$_3$)(C$_2$H$_4$)H][OTf], via exchange and β-elimination, and reaction with diethyl ether gave the analogous ethyl vinyl ether complex. Reaction with THF gave a carbene hydride complex following methane loss [66].

Reaction of Cp*Ir(PMe$_3$)(CH$_3$)OTf with NaBAr$_F$ in dichloromethane leads to the precipitation of NaOTf and the formation of [Cp*Ir(PMe$_3$)(CH$_3$)(CH$_2$Cl$_2$)]

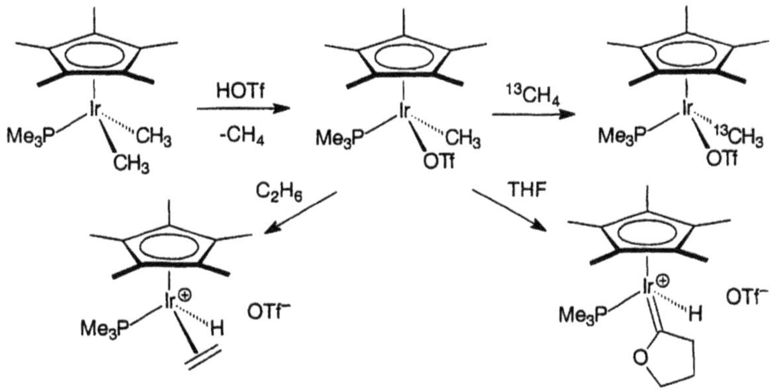

Scheme 7.

[BAr$_F$]. This cation is more reactive than Cp*Ir(PMe$_3$)(CH$_3$)OTf and reacts with terminal C–H bonds of alkanes such as pentane and methylcyclohexane to eliminate methane and give olefin hydride complexes, similar to the reaction with ethane shown in Scheme 7 [67].

One mechanism for these exchanges has been recently proposed by Chen based upon a combination of gas phase, solution phase, and computational studies. The mechanism involves cyclometallation of the PMe$_3$ ligand to give an Ir^V intermediate, followed by methane loss, followed by oxidative addition of an alkane C–H bond with opening of the metallacycle [68]. Theoretical studies by Su and Chu, however, indicate a simple mechanism involving oxidative addition of an alkane C–H bond to [Cp*Ir(PMe$_3$)(CH$_3$)]$^+$ to give an Ir^V intermediate of the type [Cp*Ir(PMe$_3$)(CH$_3$)(R)H]$^+$ [69]. A paper by Bergman also refutes the intermediacy of a phosphine metallacycle as reaction of Cp*Ir(PMe$_3$)(CH$_3$)OTf with C$_6$D$_6$ does not show any incorporation of deuterium into the phosphine ligand methyl groups [70].

One other example of alkane oxidative addition to a higher oxidation state late transition metal has been reported by Goldberg. Reaction of the trispyrazolylborate complex K[η^2-Tp'PtMe$_2$] with B(C$_6$F$_5$)$_3$ leads to the abstraction of a methyl anion and the formation of a transient species that adds to the C–H bonds of benzene, pentane, or cyclohexane (Eq. 15). This result provides the first example of the intermolecular addition of a C–H bond to a Pt^{II} species to give a stable Pt^{IV} product [71]. Earlier work by Templeton had demonstrated that the trispyrazolylborateplatinumdialkylhydride product would be stable [72].

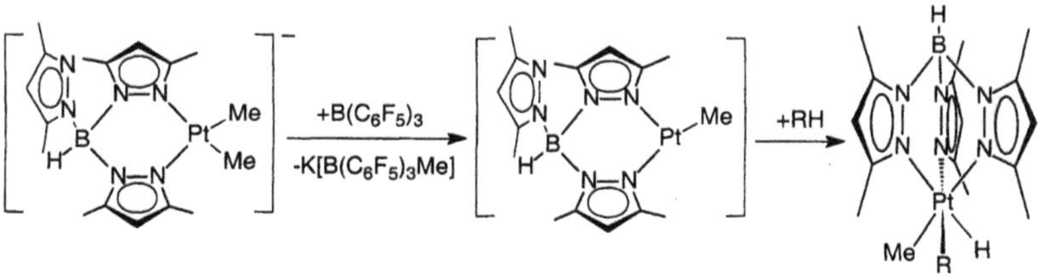

(15)

4
Electrophilic Alkane Activation

While all of the electrophilic complexes described in Sect. 3 are believed to react via oxidative addition/reductive elimination pathways, several metal complexes are believed to react via a concerted four-center mechanism as shown at the top of Fig. 1c. All of these compounds have one feature in common, namely, no *d*-electrons. Consequently, oxidative addition pathways are deemed too high in energy to be feasible. One of the earliest reports was by Watson using the lutetium complex $Cp^*{}_2LuCH_3$. Exchange with $^{13}CH_4$ occurs at 70°C in cyclohexane solution, with k_2=4.7×10^{-4} M^{-1} s^{-1} (Eq. 16). No reaction with the solvent occurs, indicating the lack of reactivity of secondary C–H bonds. Reaction with CD_4 gave $Cp^*{}_2Lu(CD_3)$. The analogous yttrium complex also undergoes methyl group exchange [73].

M—CH3 —$^{13}CH_4$→ [M···$^{13}CH_3$···H···CH_3]‡ —$-CH_4$→ M–$^{13}CH_3$

M = Lu, Y

(16)

Bercaw has reported the similar exchange of $^{13}CH_4$ with the scandocene complex $Cp^*{}_2ScMe$. The complex also reacts with benzene to give the phenyl derivative $Cp^*{}_2ScPh$ and in cyclohexane gives a head-to-tail dimer formed by attack upon the C–H bonds of two Cp* methyl groups. Reaction with hydrogen generates $[Cp^*{}_2ScH]_x$, which appears to be an extended solid. This species reacts with benzene to give $Cp^*{}_2ScPh$ in an equilibrium reaction, with K_{eq}=5.6 at 25°C. Reaction of $Cp^*{}_2ScMe$ with ethylene leads to polymer, but with propene a propenyl complex is generated (Scheme 8) [74, 75].

Marks has examined the reactivity of thorium metallacycles with hydrocarbons, where ring strain is used to provide the thermodynamic driving force for alkane activation in a reaction with methane (Eq. 17). Reaction with CD_4 shows a dramatic kinetic isotope effect, with k_H/k_D=6, which is typical of the four-centered electrophilic transition state hydrocarbon activations [76]. The metallacycle is formed by the elimination of neopentane from the bis-neopentyl derivative [77]. Reaction with cyclopropane and tetramethylsilane gave the bis-cyclopropyl product $Cp^*{}_2Th(c\text{-propyl})_2$ and the bis-TMS product $Cp^*{}_2Th(CH_2SiMe_3)_2$, respectively [78].

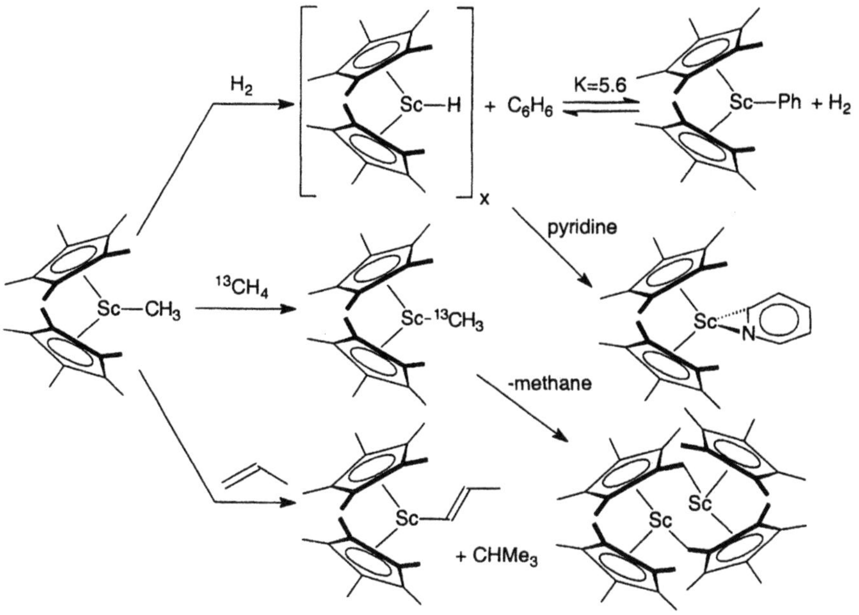

Scheme 8.

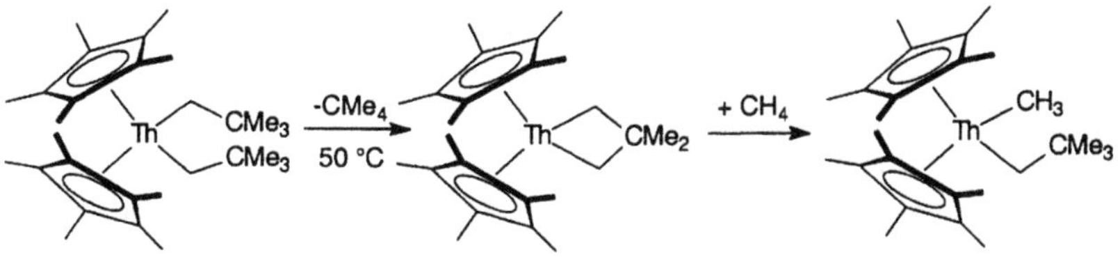

(17)

One final report of alkane activation has been reported by Moiseev. The mechanism of the reaction was not investigated, but this system might be classified as an electrophilic activation of methane, either of the Shilov type or of the concerted four-center type (Fig. 1c) where X=triflate. Reaction of methane with cobalt(III)triflate in triflic acid solution leads to the formation of methyltriflate in nearly stoichiometric quantities (90% based on Co) (Eq. 18). Carbon dioxide was also observed, but not quantified. Addition of O_2 led to catalysis (four turnovers) [79].

$$\underset{\text{10-40 atm}}{CH_4} + \underset{\text{1-5 atm}}{O_2} \xrightarrow[\text{HOTf, 100-200 °C}]{Co(OTf)_3} CH_3OTf \quad (18)$$

A wide variety of chemistry using electrophilic Pd^{II} derivatives has been investigated by Sen. This work will be reported as a separate chapter in this book.

5
Addition of C–H Bonds Across M=X Bonds

The hydrocarbon activation reactions described here all can be characterized in terms of the reaction indicated in Fig. 1d. Wolczanski described a prototypical example of this reaction in his studies of $(Bu^t{}_3SiNH)_3ZrMe$ (Scheme 9). This complex unimolecularly eliminates methane to generate a reactive imido intermediate that reacts with benzene to give the phenyl complex $(Bu^t{}_3SiNH)_3ZrPh$. Reaction of the labeled compound $(Bu^t{}_3SiNH)_3Zr(CD_3)$ with CH_4 produces the unlabeled product $(Bu^t{}_3SiNH)_3Zr(CH_3)$. The putative imido intermediate can be trapped as a THF adduct, and all alkyl and aryl derivatives can be trapped with dihydrogen [80]. On the basis of isotope effect experiments, methane sigma complexes do not appear to be involved in these reactions [81].

In addition to these exchange reactions, a number of alkane/alkane and alkane/arene exchange reactions could be studied as equilibria (benzene, toluene, cyclopropane, methane, ethane, neopentane, cyclohexane). Determination of equilibrium constants allowed calculation of $\Delta G°$ values and estimation of relative metal-carbon bond energies. Wolczanski concluded that the differences between metal-carbon bond energies and the corresponding carbon-hydrogen bond energies were essentially the same [82].

Wolczanski also investigated the chemistry of a tantalum imido system. In this system, elimination of hydrocarbon from the bis-amido imido complex occurs with difficulty at 183°C to give an amido bis-imido complex. The elimination is reversible, with the bis-imido species not being directly observed (Scheme 10). Under methane pressure, the phenyl complex loses benzene and adds methane. Neopentane, benzene, and toluene (benzylic activation) were also found to undergo activation, but not cyclohexane. The authors conclude from their equilibrium studies that the differences in metal-carbon bond strengths are approximately equal to the differences in carbon-hydrogen bond

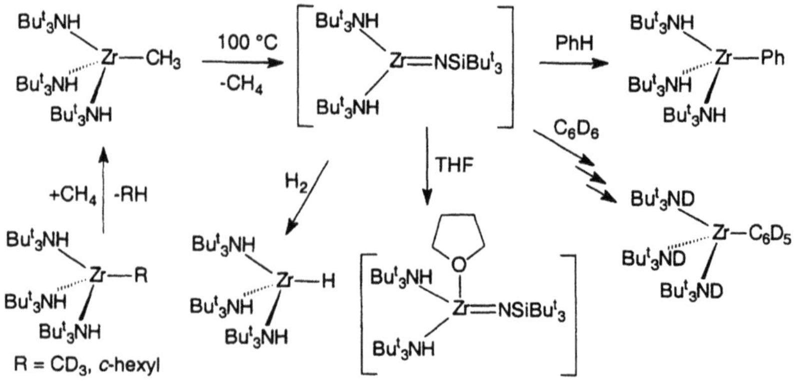

Scheme 9.

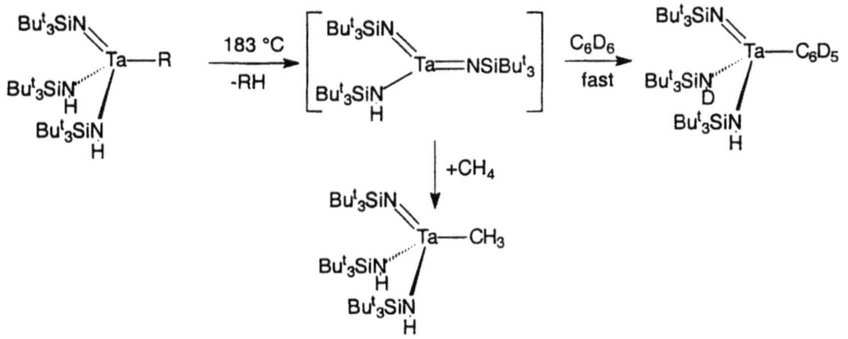

Scheme 10.

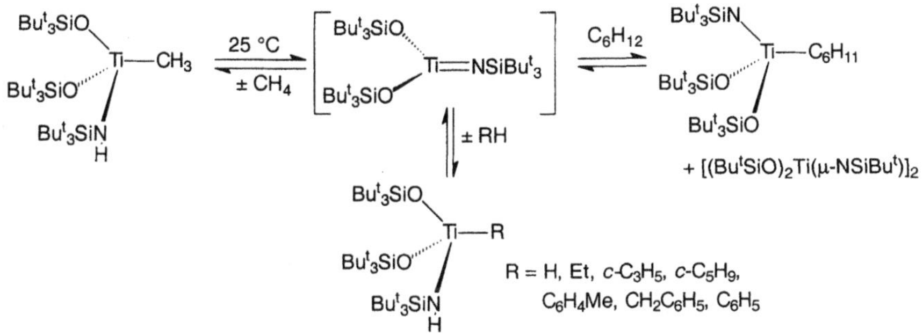

Scheme 11.

strengths, as in the zirconium complex described above. Several differences were noted, however. First, the tantalum complexes are much more stable with substantially higher kinetic barriers for hydrocarbon elimination (by ~38 kJ mol^{-1}). Second, the kinetic barrier for benzene loss was actually *lower* than that for methane loss, which does not correlate with the M-C bond strengths. The authors indicate that the reactions proceed by a late transition state with substantial N-H bond breaking [83].

Wolczanski has also examined a related bis-siloxy amido titanium complex that also reacts with C–H bonds. Once again, elimination of hydrocarbon occurred to generate an imido complex that could react with the C–H bonds of benzene, *c*-propane, *c*-pentane, neohexane, and mesitylene (Scheme 11). The intermediate imido complex could be trapped with a donor ligand such as THF, pyridine, or PMe_3. Reaction with ethylene led to the formation of an azametallacyclobutane. A large isotope effect was observed for loss of CH_3D from the N-deuterated methyl complex (k_H/k_D=13.7), implying substantial bond making character in the transition state. Once again, through equilibration, the relative stabilities of several of these titanium alkyl complexes could be determined and the relative M-C bond strengths compared. In this case, a strong correlation between D_{M-R} and D_{C-H} was observed. A plot of this correlation gives a line with a slope of 1.36, indicating that differences in metal-carbon bond strengths are substantially greater than differences in carbon-hydrogen bond strengths [84, 85]. This conclusion is in abeyance to that made from the zirconium and tantalum studies, but is similar to that made by Jones for Tp'Rh(CNneopentyl)(R)H compounds [17].

Bergman has also reported an example of C–H addition to a zirconium-nitrogen double bond. The complex $Cp_2Zr(NHR)Me$ loses methane to generate an imido complex that can either be trapped with THF or reacted with benzene (Eq. 19). No reactions with alkanes were reported [86].

$Cp_2Zr(NHR)(CH_3)$ —($-CH_4$)→ $[Cp_2Zr{=}NR]$ —($+ C_6H_6$)→ $Cp_2Zr(NHR)(Ph)$; —(THF)→ $Cp_2Zr({=}NR)(THF)$ (19)

Finally, one last report of hydrocarbon activation using an alkylidene bond has been reported by Legzdins. The reactive intermediate is generated in situ and then reacted with tetramethylsilane or cyclohexane (Eq. 20). Curiously, trimethylphosphine is required for cyclohexane activation to occur [87].

$Cp^*W(NO)(CH_2CMe_3)_2$ —(70 °C, $- CMe_4$)→ $[Cp^*W(NO)({=}CHCMe_3)]$ —($SiMe_4$)→ $Cp^*W(NO)(CH_2CMe_3)(CH_2SiMe_3)$; —($PMe_3$, cyclohexane)→ $Cp^*W(NO)(PMe_3)(C_6H_{11})$ (20)

6 Other Alkane Activations

One of the more unusual examples of hydrocarbon activation was reported by Wayland involving an example of radical homolytic cleavage of the C–H bond of methane (Fig. 1b). In this reaction, the methyl group is transferred to one porphyrin metal center and the hydrogen to a second metal center. The reaction follows termolecular kinetics, which suggests a linear transition state for the cleavage (Eq. 21) [88]. In addition to methane, only the benzylic C–H bonds of toluene proved to be reactive. The chemistry is believed to occur due to the weak Rh–Rh bond in the di-porphyrin. With R=mesityl, the Rh^{II} radical species is favored, whereas with R=3,5-xylyl, the dimer is observed as the stable species [89]. The two metalloporphyrins have been connected by a -$(CH_2)_6$- spacer to reduce the reaction to second order, resulting in more rapid rates of methane activation [90].

$$\mathrm{Rh^{II}(por)} \xrightarrow{+CH_4} \mathrm{(por)Rh{-}CH_3} + \mathrm{(por)Rh{-}H} \qquad (21)$$

R = mesityl, 3,5-xylyl

Several other examples of alkane activation have appeared, many of which occur by way of the oxidative addition pathway (Fig. 1a). Bergman reported that $CpRe(PMe_3)_3$ loses PMe_3 to generate the fragment [$CpRe(PMe_3)_2$], which reacts with benzene, cyclopropane, cyclopentane, *n*-hexane, methane, and ethylene to give C–H insertion products (Eq. 22). In the absence of a reactive hydrocarbon (such as cyclohexane), cyclometallation of the PMe_3 ligand occurs reversibly [91]. The Cp* complex was also reported to undergo similar reactions, as were the $Cp^*Re(PMe_3)_2(CO)$ and $Cp^*Re(PMe_3)(CO)_2$ complexes [92].

$$\mathrm{Cp'Re(PMe_3)_x(CO)_{3-x}} \xrightarrow{h\nu} [\mathrm{Cp'ReL_1L_2}] \xrightarrow{+\,RH} \mathrm{Cp'ReL_1L_2(R)H} \qquad (22)$$

$Cp' = C_5H_5$; $L_1=L_2=PMe_3$
$Cp' = C_5Me_5$; $L_1=CO$, $L_2=PMe_3$
$Cp' = C_5Me_5$; $L_1=L_2=PMe_3$

(cyclohexane) → $Cp'L_1Re(H)(PMe_2CH_2)$

Bergman has also found an iridium allyl hydride complex that reacts with arenes and alkanes, the allyl group being converted to an *n*-propyl group in the process. Butane and isobutane give methyl-substituted allyl derivatives under exchange with the coordinated allyl group (Eq. 23) [93].

PMe3; C_6H_6, c-C_3H_6 → $Cp^*Ir(PMe_3)(n\text{-}C_3H_7)R$
butane → methyl-substituted allyl hydride complex (23)

Field has found that the iron dihydride complex $Fe(dmpe)_2H_2$ (dmpe= $Me_2PCH_2CH_2PMe_2$) can be irradiated at low temperature to induce loss of dihydrogen. The Fe^0 fragment formed then reacts with pentane to give the *n*-pentyl hydride oxidative addition product [94]. Irradiation in liquid xenon containing methane gives the methyl hydride product [95]. These adducts are very unstable, losing alkane at ~0°C. The dihydride complex has also been investigated by laser flash photolysis, with the intermediate having a lifetime of a few milliseconds in hydrocarbon solvents [96] (Eq. 24).

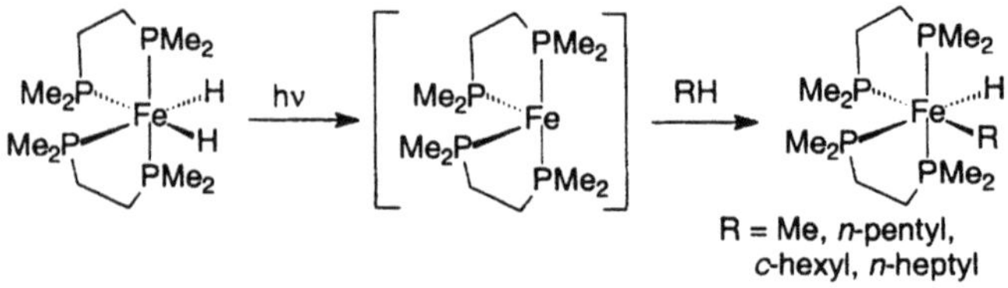

(24)

R = Me, *n*-pentyl, *c*-hexyl, *n*-heptyl

Graham has reported that irradiation of (η^6-C_6Me_6)$Os(CO)_2$ in alkane solution leads to the formation of alkane oxidative addition complexes in competition with C_6Me_6 loss [97]. Perutz and Werner have also reported the photochemical reaction of (η^6-mesitylene)$Os(CO)H_2$ in methane matrices leading to the formation of the methane activation product (η^6-mesitylene)$Os(CO)(CH_3)H$ [98].

Evidence for alkane activation has also been seen by the observation of H/D exchange between two alkanes, an alkane and an arene, or an alkane and THF. Using $CpRe(PPh_3)_2H_2$ as the photocatalyst, thousands of turnovers have been observed. While the intermediate responsible for this catalysis was not identified, it does not appear to be $[CpRe(PPh_3)H_2]$ undergoing Re^{III}/Re^{V} oxidative addition/reductive elimination, since no deuterium incorporation was observed in the dihydride catalyst [99]. Several other metal hydrides are known to catalyze H/D exchange between alkanes and deuterated benzene, such as $Ir(PMe_3)_2H_5$ [100], $CpMo(dmpe)H_3$ [101], and $Re[P(c\text{-hexyl})_3]_2H_7$ [102].

Finally, Hartwig has a recent example of a metal-catecholborane complex that photochemically activates alkanes. The boryl group is transferred to the alkyl group in yields of 22–85% (Scheme 12) [103].

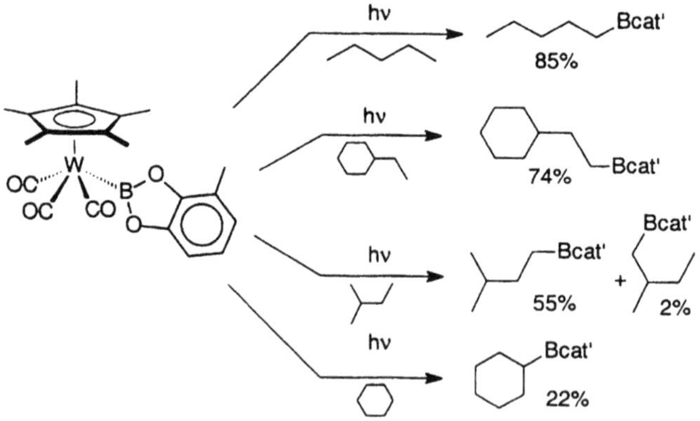

Scheme 12.

7
Arene Activation

Many compounds are known to activate arenes, so only an overview of some of the more representative or recent examples will be given here. Work by Jones showed that arene activation occurs with the electron rich 16-electron fragment $[Cp^{*}Rh(PMe_3)]$ by way of coordination to a single double bond of the arene [10, 25]. The energetics of this reaction showed that benzene activation was preferred over propane activation only slightly kinetically but by ~38 kJ mol^{-1} thermodynamically. With benzene, the η^2-complex lies about 29 kJ mol^{-1} above the oxidative addition adduct whereas with naphthalene, the η^2-complex is more stable than the C–H activation product by ~5 kJ mol^{-1} (Fig. 4) [104]. With anthracene or phenanthrene, only the η^2-complex is observed [105]. The relative stabilities of η^2-polycyclic arene complexes can be associated with the loss of resonance energy that occurs upon coordination of the metal to one double bond of the arene [106]. Evidence for similar M(aryl)H/M(η^2-arene) interconversions have been seen in Tp'Rh(CNR)(Ph)H [15] and $(Cy_2PCH_2CH_2PCy_2)$ Pt(Ph)H [107].

Several other recent examples of arene activation have also appeared. Flood reported that a triazacyclononane rhodium complex could activate arenes. These complexes are quite stable, losing methane or ethane only upon heating to 80°C (Eq. 25) [108].

1) PMe_3, 2) $NaBH_4$; then 80 °C, C_6H_6 (25)

R = H, Me
R' = Me, Et

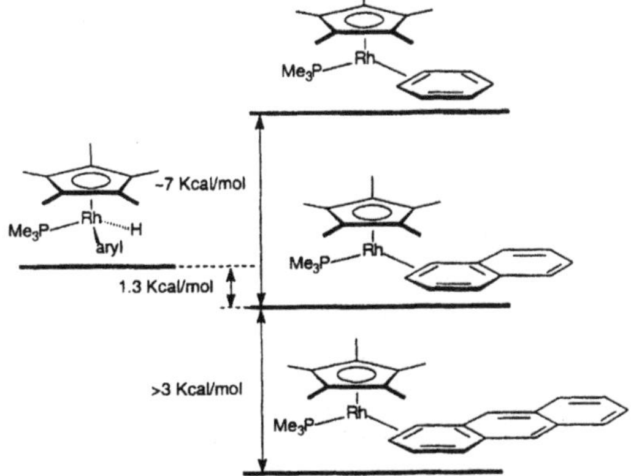

Fig. 4. Relative stabilities of arene C–H addition adducts vs. η^2-arene complexes

Flood has also reported the ability of the $Os(PMe_3)_4$(neopentyl)H to activate benzene and tetramethylsilane. The reaction is distinct from other examples that have been reported in that the mechanism does not proceed through the loss of neopentane and the formation of an $[L_4Os^0]$ intermediate, but rather via an Os^{IV} complex (Eq. 26). Evidence for this pathway includes significant incorporation of deuterium in the neopentane upon thermolysis in C_6D_6, inhibition by added PMe_3, and exchange of $P(CD_3)_3$ into the phosphines *cis* to the neopentyl and hydride groups [109, 110].

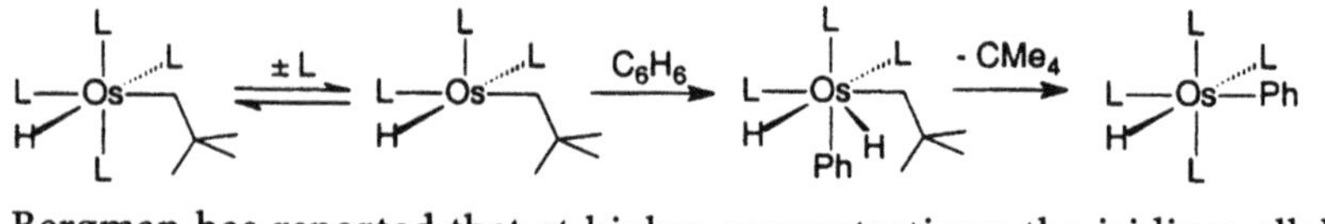

(26)

Bergman has reported that at higher concentrations, the iridium allyl complex Cp*Ir(η^3-allyl)H reacts with benzene to give a dinuclear product that has activated benzene. Furthermore, this dinuclear compound undergoes reversible benzene loss indicating that the dinuclear intermediate [Cp*Ir(η^1,η^3-allyl)IrCp*] is capable of reacting with aromatic C–H bonds (Eq. 27) [111].

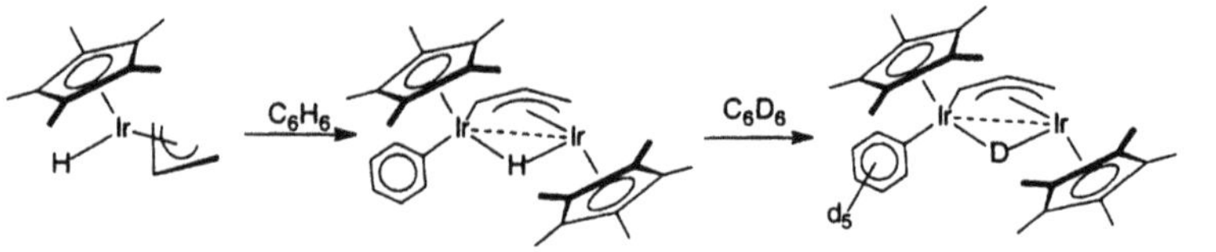

(27)

Carmona and Poveda showed that a Tp'Ir complex could activate two molecules of benzene to give a diphenyl product. Presumably, insertion of ethylene into the Ir–H bond is followed by benzene oxidative addition (via Ir^V) and loss of ethane. At this point a second benzene addition to the vinyl phenyl complex and elimination of ethylene leads to the observed product (Eq. 28). Evidence for this sequence of events comes from the trapping of these reaction intermediates with added PMe_3 [112].

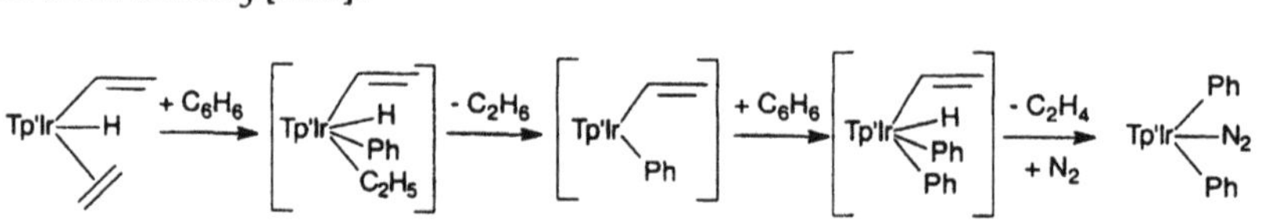

(28)

Another unusual example of arene activation was reported by Diversi. Here, an 18-electron iridium dimethyl complex is oxidized by one electron prior to reaction with benzene, and the resultant 17-electron complex readily exchanges a methyl group for an aryl group. The product oxidizes the starting material, so that the overall reaction is an example of electron transfer catalysis. The proposed mechanism is shown in Scheme 13 [113].

Sutton has reported a pentamethylcyclopentadienyl rhenium dinitrogen complex that activates benzene (Eq. 29) [114]. The intermediate involved is similar to that described above in studies by Bergman [92]. In the absence of a reactive solvent, cyclometallation was observed.

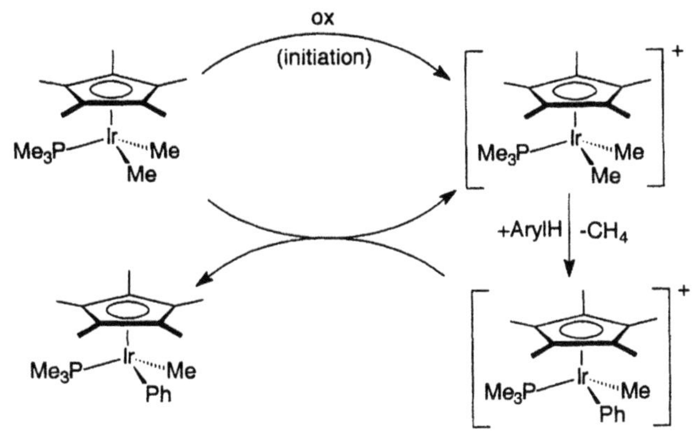

Scheme 13.

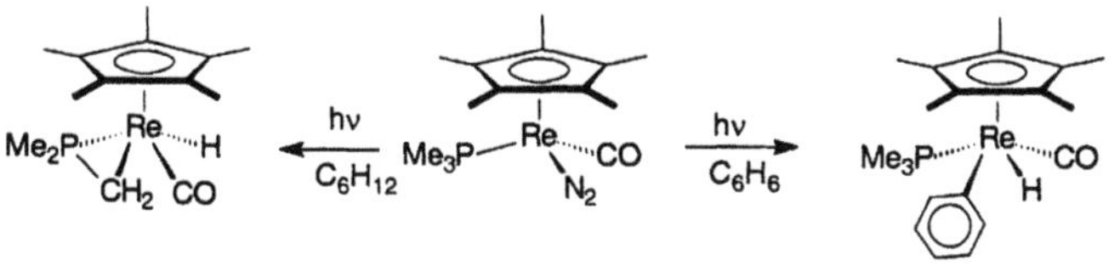

(29)

In analogy to the iron chemistry reported by Field above [94], Hartwig et al. reported that irradiation of $(Me_2PCH_2PMe_2)_2RuH_2$ in benzene led to the formation of $(Me_2PCH_2PMe_2)_2Ru(Ph)H$ [115]. Perutz studied this photochemical reaction both in the matrix and by transient absorption spectroscopy, showing that benzene activation by the Ru^0 fragment was $\sim 10^5$ times slower than reaction with H_2, CO, C_2H_4, or Et_3SiH [116].

As a final example, Brookhart reported that the cobalt complex $Cp^*Co(C_2H_4)_2$ was capable of catalyzing H/D exchange between C_6D_6 and the coordinated olefinic hydrogens at 60°C. By changing to the trimethylsilylethylene ligand, the exchange could be carried out at ambient temperature (Eq. 30) [117].

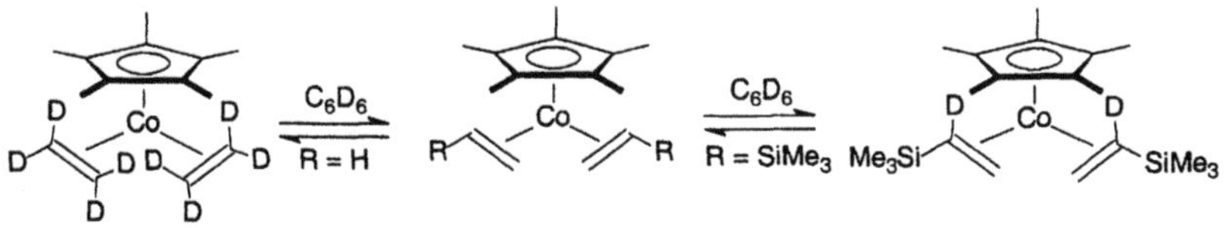

(30)

Berry has recently reported an example of catalytic arene activation and functionalization. This report involves either $Cp^*Rh(SiEt_3)_2H_2$ or (η^6-arene) $Ru(SiEt_3)_2H_2$ as catalyst and couples triethylsilane with an arene. *t*-Butylethylene serves as a hydrogen acceptor (Eq. 31). No intermediates are observed, and a carbosilane dimer is also formed as a significant fraction of the total product [118].

$$Et_3SiH + C_6H_5X \xrightarrow[\text{t-BuCH=CH}_2 \rightarrow \text{t-BuCH}_2\text{CH}_3]{Cp^*Rh(SiEt_3)_2H_2 \text{ or } (arene)Ru(SiEt_3)_2H_2} XC_6H_4SiEt_3 + Et_2HSiCHMeSiEt_3 \quad (31)$$

X = CF_3, F, H, CH_3, Cl, Br

8
Alkene Activation

Many of the complexes that are active in alkane activation have also been shown to display activity towards vinylic C–H bonds. One of the earlier examples was reported by Faller and Felkin, in which indirect evidence for vinylic activation was obtained by observing H/D exchange between *t*-butylethylene and C_6D_6 catalyzed by $Ir(PPr^i_3)_2H_5$. The terminal hydrogen *trans* to the *t*-butyl group was observed to exchange at a rate 10× that of the hydrogen *gem* to the *t*-butyl group. This observation was used to rule out H/D exchange by an insertion/β-elimination mechanism and instead favors a direct oxidative addition pathway (Eq. 32) [119].

$$\text{t-BuCH=CH}_2 \xrightarrow[C_6D_6,\ 25\ °C]{Ir(PPr^i_3)_2H_5} \text{t-BuCH=CHD} \quad (32)$$

Bergman found that the complex $Cp^*Ir(PMe_3)$(c-hexyl)H reacts with ethylene at 150°C to give a mixture of the η^2-ethylene complex and the vinyl hydride oxidative addition product. The ratio of the two products remained 1:2 over the course of the reaction (Eq. 33). Furthermore, the vinyl hydride complex was observed to rearrange intramolecularly to the ethylene complex upon heating to 170°C. These observations indicate that the reactive fragment $[Cp^*Ir(PMe_3)]$ can either coordinate to the π system of ethylene or activate the C–H bond in competitive reactions [120]. The π complex cannot be an intermediate in C–H activation, unlike the situation for benzene activation by $[Cp^*Rh(PMe_3)]$. Isotope effect studies were performed on this system using deuterated ethylenes. The observation of different kinetic isotope effects for product formation using $C_2H_2D_2$ (k_H/k_D=1.18) vs. a 1:1 mixture of C_2H_4/C_2D_4 (k_H/k_D=1.49) confirms that an intermediate different from the π-complex must be involved in the C–H activation reactions [121]. A computational study suggested the existence of two distinct transition states for C–H activation vs. η^2-coordination [122].

$$Cp^*Ir(PMe_3)(C_6H_{11})H \xrightarrow{150\ °C} Cp^*Ir(PMe_3)(CH{=}CH_2)H + Cp^*Ir(PMe_3)(\eta^2\text{-}H_2C{=}CH_2) \quad (33)$$

2 : 1 (vinyl hydride → ethylene complex: 170 °C, C_6D_6)

Perutz examined similar reactions with the C_5H_5 complex $CpIr(C_2H_4)_2$. Studies in argon matrices showed that photolysis resulted in the formation of the vinyl hydride $CpIr(C_2H_4)(CH{=}CH_2)H$. The same product could be obtained by photolysis of the bis-ethylene complex in frozen toluene and then warming the sample to 200 K, but the adduct decomposes thermally at 0°C [123]. Irradiation of $CpIr(PPh_3)(C_2H_4)$ in cold toluene resulted in the formation of $CpIr(PPh_3)(CH{=}CH_2)H$, which is stable at ambient temperature (Eq. 34). The complex rearranges to the η^2-C_2H_4 isomer upon heating to 118°C with a half-life of 27 h [124]. Perutz also looked at the complex $CpIr(CO)(C_2H_4)$, which also provided evidence for vinylic activation upon photolysis.

(34)

A related Cp complex that activates alkanes is the rhenium complex $CpRe(PMe_3)_3$. This species undergoes photochemical activation of ethylene in cyclohexane solution. The vinyl hydride product rearranges to the η^2-ethylene complex at room temperature. Irradition of the ethylene complex, however, was shown to produce the vinyl hydride product, leaving open the question as to which species is formed as the kinetic product of the reaction (Eq. 35). Independent *thermal* generation of the reactive fragment $[CpRe(PMe_3)_2]$ by loss of hexane from $CpRe(PMe_3)_2(n\text{-pentyl})H$ in the presence of ethylene confirmed that vinylic C–H activation was indeed the kinetically preferred, if not exclusive, reaction pathway [125].

(35)

It is interesting to note that in all of the above cases, the η^2-ethylene complex is thermodynamically more stable than the vinyl hydride isomer. Graham discovered an interesting example where the *reverse* thermodynamic stability could be observed. The perfluoromethyl derivative tris-(3,5-trifluoromethylpyrazolyl)-boratecarbonylethyleneiridium(I) was observed to rearrange *thermally* to the vinyl hydride isomer at 100°C (Eq. 36). By comparison, the analogous 3,5-dimethylpyrazolylborate complex was found to favor the η^2-ethylene complex, as observed in other systems, so that the effect of the perfluoromethyl groups is to strongly favor the oxidative addition product [126].

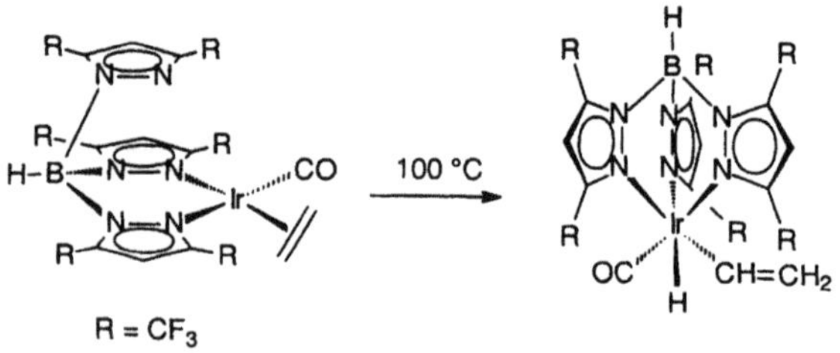

(36)

Crabtree looked at vinylic C–H activation using the unsubstituted complex trispyrazolylborate-bis-ethyleneiridium(I). Irradiation leads to the formation of the vinyl hydride complex by a mechanism that involves initial ethylene dissociation. Evidence for this pathway stems from the observed inhibition by added ethylene (Eq. 37) [127].

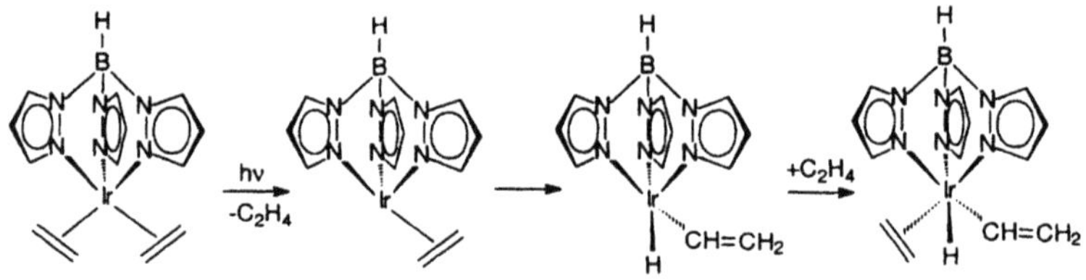

(37)

Carmona also saw ethylene activation with the tris-(3,5-dimethylpyrazolyl)borate-bis-ethyleneiridium(I) complex, but there are two main differences from Crabtree's results. First, reaction occurs thermally in solution at 60°C. Second, the C–H insertion complex undergoes further C–C coupling to give an allyl hydride product (Eq. 38) [128].

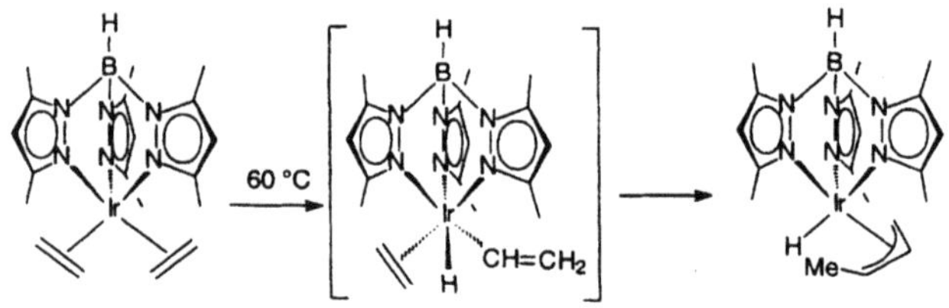

(38)

Finally, one rather different example of vinylic C–H activation has appeared using a first row transition metal. Field reported that irradiation of $Fe(dmpe)_2H_2$ in the presence of an olefin at –80°C leads to the formation of vinyl hydride products. These insertion adducts were observed with cyclopentene, ethylene, and 1-pentene. Upon warming to room temperature, the η^2-olefin complexes formed at the expense of the C–H insertion adducts (Eq. 39) [129, 130].

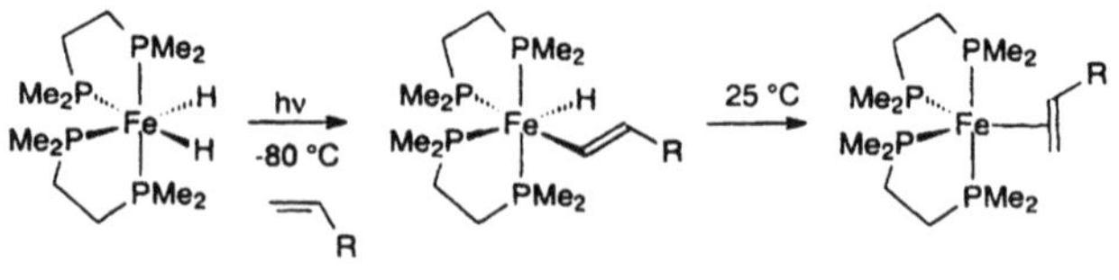

(39)

References

1. Parshall GW, Ittel SD (1992) Homogeneous catalysis, 2nd edn. John Wiley & Sons, New York
2. Crabtree RH, Mihelcic JM, Quirk JM (1979) J Am Chem Soc 101:7738
3. Abis L, Sen A, Halpern J (1978) J Am Chem Soc 100:2915
4. Janowicz AH, Bergman RG (1982) J Am Chem Soc 104:352
5. Hoyano JK, Graham WAG (1982) J Am Chem Soc 104:3723
6. Jones WD (1983) Organometallics 2:562
7. Hoyano JK, McMaster, AD, Graham WAG (1983) J Am Chem Soc 105:7190
8. Sponsler, MB, Weiller, BH, Stoutland, PO, Bergman, RG (1989) J Am Chem Soc 111:6841
9. Janowicz AH, Bergman RG (1983) J Am Chem Soc 105:3929
10. Jones WD, Feher FJ (1984) J Am Chem Soc 106:1650
11. (a) Trofimenko S (1986) Prog Inor Chem 34:115. (b) Trofimenko S (1972) Chem Rev 72:497. (c) Niedenzu K, Trofimenko S (1986) Top Curr Chem 131:1
12. Ghosh CK, Graham, WAG (1987) J Am Chem Soc 109:4726
13. Barrientos C, Ghosh CK, Graham WAG, Thomas MJ (1990) J Organomet Chem 394:C31
14. Ghosh CK, Graham WAG (1989) J Am Chem Soc 111:375
15. Jones WD, Hessell ET (1992) J Am Chem Soc 114:6087
16. Hessell ET, Jones WD (1992) Organometallics 11:1496
17. Jones WD, Hessell, ET (1993) J Am Chem Soc 115:554
18. Jones WD, Hessell ET (1991) Inorg Chem 30:778
19. Keyes MC, Young VG, Tolman WB (1996) Organometallics 15:4133
20. Bergman RG (1984) Science 223:902
21. Bergman RG (1992) Adv Chem Ser 230:211
22. Wax MJ, Stryker JM, Buchanan JM, Kovac CA, Bergman RG (1984) J Am Chem Soc 106:1121
23. Buchanan JM, Stryker JM, Bergman RG (1986) J Am Chem Soc 108:1537
24. Nolan SP, Hoff CD, Stoutland PO, Newman LJ, Buchanan JM, Bergman RG, Yang GK, Peters KS (1987) J Am Chem Soc 109:3143
25. Jones WD, Feher FJ (1989) Acc Chem Res 22:91
26. Davico GE, Bierbaum VM, DePuy CH, Ellison GB, Squires RR (1995) J Am Chem Soc 117:2590
27. Ervin KM, Gronert S, Barlow SE, Gilles MK, Harrison AG, Bierbaum VM, DePuy CH, Lineberger WC, Ellison GB (1990) J Am Chem Soc 112:5750
28. Rest AJ, Whitwell I, Graham WAG, Hoyano JK, McMaster AD (1984) J Chem Soc Chem Commun 624
29. Bloyce PE, Rest AJ, Whitwell I, Graham WAG, Holmes-Smith R (1988) J Chem Soc Chem Commun 846
30. Haddleton DM, McCamley A, Perutz RN (1988) J Am Chem Soc 110:1810
31. Partridge MG, McCamley A, Perutz RN (1994) J Chem Soc Dalton Trans 3519
32. Kubas GJ, Ryan RR, Swanson BI, Vergamini PJ, Wasserman HJ (1984) J Am Chem Soc 106:451
33. Brookhart M, Green MLH (1983) J Organomet Chem 250:395. Hall C, Perutz RN (1996) Chem Rev 96:3125
34. Belt ST, Grevels FW, Klotzbücher WE, McCamley A, Perutz RN (1989) J Am Chem Soc 111:8373
35. Weiller BH, Wasserman EP, Bergman RG, Moore CB, Pimentel GC (1989) J Am Chem Soc 111:8288
36. Weiller BH, Wasserman EP, Moore CB, Bergman RG (1993) J Am Chem Soc 115:4326
37. Bengali AA, Schultz RH, Moore CB, Bergman RG (1994) J Am Chem Soc 116:9585

38. Wasserman EP, Moore CB, Bergman RG (1992) Science 255:315
39. Lian T, Bromberg SE, Yang H, Proulz G, Bergman RG, Harris CB (1996) J Am Chem Soc 118:3769
40. Bromberg SE, Yang H, Asplund MC, Lian T, McNamara, BK, Kotz KT, Yeston JS, Wilkens M, Frei H, Bergman RG, Harris CB (1997) Science 278:260
41. Bullock RM, Headford CEL, Kegley SE, Norton JR (1985) J Am Chem Soc 107:727. Bullock RM, Headford CEL, Hennessy KM, Kegley SE, Norton JR (1989) J Am Chem Soc 111:3897
42. Jones WD, Feher FJ (1986) J Am Chem Soc 108:4814
43. Wick DD, Reynolds KA, Jones WD (1999) J Am Chem Soc, in press.
44. Periana RA, Bergman RG (1986) J Am Chem Soc 108:7332
45. Brown CE, Ishikawa Y, Hackett PA, Rayner DM (1990) J Am Chem Soc 112:2530
46. Sun, XZ, Grills DC, Nikiforov SM, Poliakoff M, George MW (1997) J Am Chem Soc 119:7521
47. Lee DW, Jensen CM (1996) J Am Chem Soc 118:8749
48. Evans DR, Drovetskaya T, Bau R, Reed CA, Boyd PDW (1997) J Am Chem Soc 119:3633
49. Holcomb HL, Nakanishi S, Flood TC (1996) Organometallics 15:4228
50. Mobley TA, Schade C, Bergman RG (1995) J Am Chem Soc 117:7822
51. Mobley TA, Bergman RG (1998) J Am Chem Soc 120:3253
52. Saillard JY, Hoffmann R (1984) J Am Chem Soc 106:2006
53. Ziegler T, Tschinke V, Fan L, Becke AD (1989) J Am Chem Soc 111:9177
54. Song J, Hall MB (1993) Organometallics 12:3118
55. Musaev DG, Morokuma K (1995) J Am Chem Soc 117:799
56. Siegbahn, PEM (1996) J Am Chem Soc 118:1487
57. Shilov AE (1984) Activation of saturated hydrocarbons by transition metal complexes. D. Reidel, Boston, and refs. therein
58. Hackett M, Ibers JA, Jernakoff P, Whitesides GM (1986) J Am Chem Soc 108:8094
59. Hackett M, Whitesides GM (1988) J Am Chem Soc 110:1449
60. Brainard RL, Nutt WR, Lee TR, Whitesides GM (1988) Organometallics 7:2379
61. Horváth IT, Cook RA, Millar JM, Kiss G (1993) Organometallics 12:8
62. Labinger JA, Herring AM, Lyon DK, Luinstra GA, Bercaw JE, Horváth IT, Eller K (1993) Organometallics 12:895
63. Stahl SS, Labinger JA, Bercaw JE (1996) J Am Chem Soc 118:5961
64. Holtcamp MW, Labinger JA, Bercaw JE (1997) J Am Chem Soc 119:848
65. Burger P, Bergman RG (1993) 115:10462
66. Luecke HF, Arndtsen BA, Burger P, Bergman RG (1996) J Am Chem Soc 118:2517
67. Arndtsen BA, Bergman RG (1995) Science 270:1970
68. Hinderling C, Feichtinger D, Plattner DA, Chen P (1997) J Am Chem Soc 119:10793
69. Su MD, Chu SY (1997) J Am Chem Soc 119:5373
70. Luecke HF, Bergman RG (1997) J Am Chem Soc 119:11538
71. Wick DD, Goldberg KI (1998) J Am Chem Soc 119:10235
72. O'Reilly SA, White PS, Templeton JL (1996) J Am Chem Soc 118:5684
73. Watson PL (1983) J Am Chem Soc 105:6491
74. Thompson ME, Bercaw JE (1984) Pure Appl Chem 56:1
75. Thompson ME, Baxter SM, Bulls AR, Burger BJ, Nolan MC, Santarsiero BD, Schaefer WP, Bercaw JE (1987) 109:203
76. Fendrick CM, Marks TJ (1984) J Am Chem Soc 106:2214
77. Bruno JW, Smith GM, Marks TJ, Fair CK, Schultz AJ, Williams JM (1986) J Am Chem Soc 108:40
78. Fendrick CM, Marks TJ (1986) J Am Chem Soc 108:425
79. Vargaftik MN, Stolarov IP, Moiseev II (1990) J Chem Soc Chem Commun 1049
80. Cummins CC, Baxter SM, Wolczanski PT (1988) J Am Chem Soc 110:8731
81. Schaller CP, Bonanno JB, Wolczanski PT (1994) J Am Chem Soc 116:4133

82. Schaller CP, Cummins CC, Wolczanski PT (1996) J Am Chem Soc 118:591
83. Schaller CP, Wolczanski PT (1993) Inorg Chem 32:131
84. Bennett JL, Wolczanski PT (1994) J Am Chem Soc 116:2179
85. Bennett JL, Wolczanski PT (1997) J Am Chem Soc 119:10696
86. Walsh PJ, Hollander FJ, Bergman RG (1988) J Am Chem Soc 110:8729
87. Tran E, Legzdins P (1997) J Am Chem Soc 119:5071
88. Sherry AE, Wayland BB (1990) J Am Chem Soc 112:1259
89. Wayland BB, Ba S, Sherry AE (1991) J Am Chem Soc 113:5305
90. Zhang XX, Parks GF, Wayland BB (1997) J Am Chem Soc 119:7938
91. Wenzel TT, Bergman RG (1986) J Am Chem Soc 108:4856
92. Bergman RG, Seidler PF, Wenzel TT (1985) J Am Chem Soc 107:4358
93. McGhee WD, Bergman RG (1988) 110:4246
94. Baker MV, Field LD (1987) 109:2825
95. Field LD, George AV, Messerle BA (1991) J Chem Soc Chem Commun 1339
96. Whittlesey MK, Mawby RJ, Osman R, Perutz RN, Field LD, Wilkinson MP, George MW (1993) J Am Chem Soc 115:8627
97. Kiel WA, Ball RG, Graham WAG (1990) J Organomet Chem 383:481
98. Brough SA, Hall C, McCamley A, Perutz RN, Stahl S, Wecker U, Werner H (1995) J Organomet Chem 504:22
99. Jones WD, Maguire JA (1986) Organometallics 5:590
100. Cameron CJ, Felkin H, Fillebeen-Khan T, Forrow NJ, Guittet E (1986) J Chem Soc Chem Commun 801
101. Grebenik PD, Green MLH, Izquierdo A (1981) J Chem Soc Chem Commun 186
102. Zeiher EHK, DeWit DG, Caulton KG (1984) J Am Chem Soc 106:7006
103. Waltz KM, Hartwig JF (1997) Science 277:211
104. Belt ST, Dong L, Duckett SB, Jones WD, Partridge MG, Perutz RN (1991) J Chem Soc Chem Commun 266
105. Jones WD, Dong L (1989) 111:8722
106. Chin RM, Dong L, Duckett SB, Partridge MG, Jones WD, Perutz RN (1993) J Am Chem Soc 115:7685
107. Hackett M, Ibers JA, Whitesides GM (1988) J Am Chem Soc 110:1436
108. Zhou R, Wang C, Hu Y, Flood TC (1997) Organometallics 16:434
109. Desrosiers PJ, Shinomoto RS, Blood TC (1986) J Am Chem Soc 108:1346
110. Desrosiers PJ, Shinomoto RS, Flood TC (1986) J Am Chem Soc 108:7964
111. McGhee WD, Hollander FJ, Bergman RG (1988) 110:8428
112. Gutiérrez E, Monge A, Nicasio MC, Poveda ML, Carmona E (1994) J Am Chem Soc 116:791
113. Diversi P, Iacoponi S, Ingrosso G, Laschi F, Lucherini A, Pinzino C, Uccello-Barretta G, Zanello P (1995) 14:3275
114. Klahn-Oliva AH, Singer RD, Sutton D (1986) 108:3107
115. Hartwig J, Andersen RA, Bergman RG (1991) Organometallics 10:1710
116. Nicasio MC, Perutz, RN, Walton PH (1997) Organometallics 16:1410
117. Lenges CP, Brookhart M, Grant BE (1997) J Organometal Chem 528:199
118. Ezbiansky K, Djurovich PI, LaForest M, Sinning DJ, Zayes R, Berry DH (1998) Organometallics 17:1455
119. Faller JW, Felkin H (1985) Organometallics 4:1488
120. Stoutland PO, Bergman RG (1985) J Am Chem Soc 107:4581
121. Stoutland PO, Bergman RG (1988) J Am Chem Soc 110:5732
122. Silvestre J, Calhorda MJ, Hoffmann R, Stoutland PO, Bergman RG (1986) Organometallics 5:1841
123. Haddleton DM, Perutz RN (1986) J Chem Soc Chem Commun 1734
124. Bell TW, Brough SA, Partridge MG, Perutz RN, Rooney AD (1993) Organometallics 12:2933

125. Wenzel TT, Bergman RG (1986) J Am Chem Soc 108:4856
126. Ghosh CK, Hoyano JK, Krentz R, Graham WAG (1989) J Am Chem Soc 111:5480
127. Tanke RS, Crabtree RH (1989) Inorg Chem 28:3444
128. Boutry O, Gutiérrez E, Monge A, Nicasio MC, Pérez PJ, Carmona E (1992) J Am Chem Soc 114:7288
129. Baker MV, Field LD (1986) J Am Chem Soc 108:7433
130. Baker MV, Field LD (1986) J Am Chem Soc 108:7436
131. Lide DR (ed) (1997–1998) CRC Handbook of chemistry and physics, 78th edn., CRC Press

Activation of C–H Bonds: Catalytic Reactions

Fumitoshi Kakiuchi* and Shinji Murai

Department of Applied Chemistry, Faculty of Engineering, Osaka University, Suita, Osaka 565-0871, Japan
E-mail: kakiuchi@chem.eng.osaka-u.ac.jp and murai@chem.eng.osaka-u.ac.jp

Direct use of the carbon-hydrogen bond in organic synthesis with the aid of the homogeneous transition metal complexes has been the subject of recent interest. This review surveys some of the recent advances in the field of the transition metal-catalyzed functionalization of carbon-hydrogen bonds.

Keywords: Catalytic, C–H bond, Transition metal, Bond formation, Olefin, Carbon monoxide

1 Introduction . . . 48

2 Catalytic Carbon-Carbon Bond Formation through Direct C–H Bond Cleavage . . . 48

2.1 Addition of C–H Bonds to Carbon-Carbon Double Bonds . . . 48
2.2 Addition of C–H Bonds to Carbon-Carbon Triple Bonds . . . 56
2.3 Coupling of C–H Bonds, Carbon Monoxide, and Olefins . . . 58
2.4 Insertion of Carbon Monoxide and Isocyanide into the C–H Bond 60

3 Catalytic Dehydrogenation of Alkanes and Arenes . . . 61

3.1 Dehydrogenation of Alkanes and Arenes . . . 61
3.2 Dehydrogenative Silylation of Alkanes and Arenes . . . 64

4 Hydroacylation with Aldehydes . . . 65

4.1 Intramolecular Hydroacylation of Olefins . . . 65
4.2 Intermolecular Hydroacylation of Olefins . . . 68
4.3 Hydroacylation of Acetylenes . . . 70

5 Addition of Active Methylene Compounds to Unsaturated Functions 72

5.1 Michael Addition and Aldol Reactions . . . 72
5.2 Addition to Carbon-Carbon Multiple Bonds . . . 73

6 Conclusion . . . 75

References . . . 76

Topics in Organometallic Chemistry, Vol. 3
Volume Editor: S. Murai

1
Introduction

One of the most valuable synthetic methods in organic synthesis is the direct use of otherwise unreactive C–H bonds with the aid of transition metal complexes. Since Kleiman and Dubeck reported in 1963 the possibility of cleavage of C–H bonds in azobenzene by Cp_2Ni complex [1], many research groups have reported the cleavage of C–H bonds by using stoichiometric amounts of transition metal complexes [2]. Over 50 review articles are now available and the fundamental features of the C–H bond cleavage reactions have been thoroughly studied [3]. In contrast, with respect to transition metal-catalyzed functionalization of the C–H bond, the chemistry still appears to be amateur and the only examples have appeared in the literature in the early 1990s.

In 1989, Jordan reported Zr-catalyzed addition of the C–H bond in α-picoline to olefin [4]. Moore and coworkers found that Ru-catalyzed three component coupling of pyridine, carbon monoxide, and olefin took place, although the use of an excess amount of one component is required [5]. Subsequently, Murai and coworkers published highly efficient and selective functionalization of C–H bonds in aromatic ketones with olefins in the presence of a ruthenium catalyst [6].

In this chapter, we will survey the transition metal-catalyzed functionalizations of C–H bonds that were published up to the end of February 1998. Those catalytic reactions involving a step of electrophilic substitution by a metal ion, such as that of benzene with $Pd(OAc)_2$, will not be dealt with.

2
Catalytic Carbon-Carbon Bond Formation through Direct C–H Bond Cleavage

2.1
Addition of C–H Bonds to Carbon-Carbon Double Bonds

Catalytic additions of carbon-hydrogen bonds to olefins would constitute one of the most efficient, economical methods for constructing a carbon-carbon framework. In 1978, Yamazaki et al. reported pioneering studies of the catalytic functionalization of C–H bonds [7]. Reaction of benzene, used as a solvent, with diphenylketene in the presence of $Rh_4(CO)_{12}$ as the catalyst gave the corresponding diphenylmethyl phenyl ketone in good yield with moderate selectivity (Eq. 1). The use of CO is essential to attain the catalytic reaction albeit with no CO incorporation. The rhodium catalyst is also effective for the dehydrogenative vinylation of benzene with ethylene to give styrenes (Eq. 2) [8, 9].

$$Ph_2C{=}C{=}O + C_6H_6 \xrightarrow[\text{CO 30 kg/cm}^2,\ 200\,^\circ\text{C, 5 h}]{Rh_4(CO)_{12}} \underset{68\%}{Ph_2CHCOPh} \qquad (1)$$

$Rh_4(CO)_{12}$, 220 °C, 7 h

30 kg/cm^2 25 kg/cm^2

CO, Ph, Ph

9170%/Rh 13500%/Rh 850%/Rh 185%/Rh

(2)

Double insertion of ethylene into aniline with the aid of rhodium(III) chloride hydrate gives the cyclization product, 2-methylquinoline (Eq. 3) [10]. Forcing reaction conditions and the use of an excess amount of aniline were required for this catalytic reaction.

Addition of sp^3 C–H bond adjacent to a nitrogen atom in dimethylamine to 1-pentene is catalyzed by tungsten amide complex to give *N*-methyl-*N*-(2-methylpentyl)amine (Eq. 4) [11]. The α C–H bonds of ether oxygen are added to *tert*-butylethylene in the presence of a catalytic amount of $IrH_5(P^iPr_3)_2$ under relatively mild reaction conditions (50°C) (Eq. 5) [12]. This reaction is formally a dehydrogenative coupling.

NH_2 + (100 kg/cm^2) — $RhCl_3$-3 H_2O, PPh_3, 200 °C → N–Et (H) (30 TON) + N (10 TON) (3)

TON = turnover numbers

Me_2NH + — $W(NMe_2)_n$, decaline, 160 °C, 14 h → Me–N(H) (7 TON) (4)

$CH_3OCH_2CH_2OCH_3$ + — $IrH_5(P^iPr_3)_2$, 50 °C, 24 h → RO (45%) + RO (44%) + RO (11%)

R = $CH_3OCH_2CH_2$

12 TON

(5)

Heteroatom directed ethylation of the benzene ring in phenol was catalyzed by ruthenium(II)-phosphite complex. The alkylation takes place at the position *ortho* to the hydroxyl group exclusively, and the corresponding 1:2 addition product is the major product (Eq. 6) [13]. Whether this reaction proceeds via the addition of an olefin to a C-metalated enolate or to an *ortho*-metalated phenol is not clear. In 1989, Jordan et al. reported α-alkylation of α-picoline with terminal olefin using a cationic zirconium catalyst (Eq. 7) [4]. For this reaction, protection of one α C–H bond of the pyridine ring with an alkyl group is required to conduct this reaction in a catalytic manner. They also reported the first example of the asymmetric α alkylation of the picoline via the C–H bond cleavage reaction using a chiral tetrahydroindenyl-zirconium complex [14]. The ee was moderate, but this result opened up the possibility of catalytic asymmetric C–H/olefin coupling.

PhOK (cat.), THF, 177 °C, 3.5 h; 95 psi; 12%, 13%, 75% (6)

H_2 1 atm, 23 °C, 25 h; 1.5 atm; > 40 TON (7)

In 1993, the first example of a highly efficient and selective C–H/olefin coupling reaction was found by these authors [6]. Reaction of aromatic ketones with olefins in the presence of a ruthenium catalyst gives the corresponding *ortho* alkylated compounds in quite high yields (Eq. 8). Various combinations of aromatic ketones and olefins can be applied to the reaction. The C–C bond formation occurred exclusively at position *ortho* to the ketone carbonyl group. The reaction would involve coordination of the carbonyl group to the ruthenium, bringing the metal closer to the *ortho* C–H bonds. The reaction was extended to the transition metal-catalyzed C–H/olefin coupling of various aromatic and olefinic compounds [15]. One of the most important findings in their studies is that the C–H bond cleavage step is not rate-determining. Rapid equilibrium exists prior to the reductive elimination step leading to C–C bond formation (Scheme 1). Functional group compatibility and the effect of substituents on the site-selection have been systematically studied in the reaction of various acetophenones [16]. Many functional groups (e.g., NMe_2, OMe, F, NEtC(O)Me, CO_2Et, and CN) are tolerant of this ruthenium-catalyzed aromatic C–H/olefin coupling. In the cases of *meta*-substituted acetophenones, two different reaction sites are present at the position *ortho* to the carbonyl group. The site selectivity is basically controlled by steric factors. Interestingly, however, the reactions of *m*-methoxy- and *m*-fluoroacetophenones with triethoxyvinylsilane take place at the much congested *ortho* positions (Eq. 9). These results suggest that the heteroatoms additionally assist in bringing the ruthenium closer to the C–H bond.

$RuH_2(CO)(PPh_3)_3$, toluene, 2 h, 135 °C (bath temp.); $Si(OEt)_3$; 93%; Ru(0) (8)

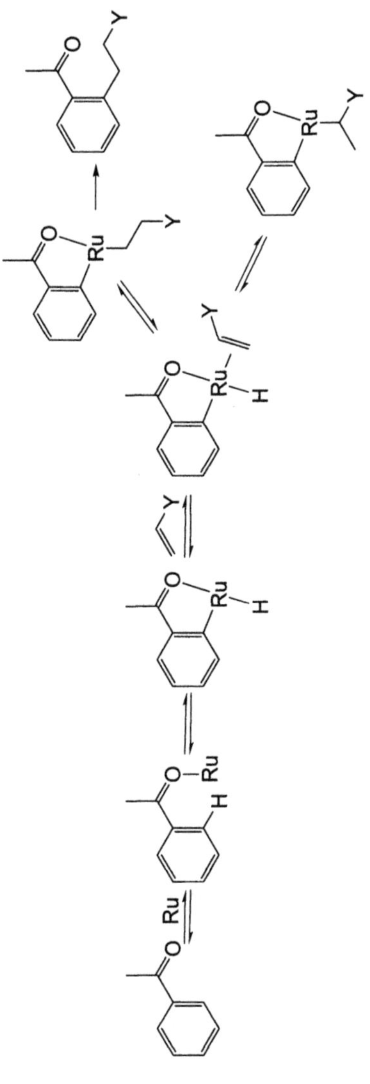

Scheme 1.

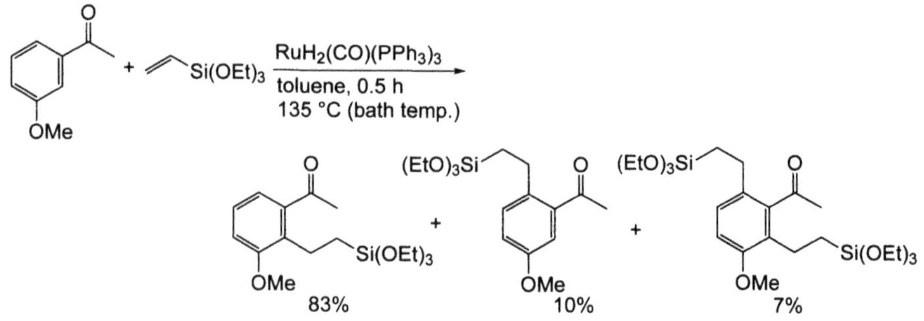

+ Si(OEt)3
RuH2(CO)(PPh3)3
toluene, 0.5 h
135 °C (bath temp.)
OMe
Si(OEt)3 (EtO)3Si (EtO)3Si Si(OEt)3
OMe 83% + OMe 10% + OMe 7%

(9)

The catalytic reactions are applicable to a variety of aromatic and heteroaromatic compounds. Examples are given in Fig. 1. The arrows in Fig. 1 show the position of the C–C bond formation.

Woodgate et al. used the ruthenium-catalyzed reaction of aromatic ketones with olefins for the synthesis of natural products (Eq. 10) [17]. They examined an alkylation of 1-(hydroxyphenyl)ethanone equivalents by using $Ru(CO)_2(PPh_3)_3$ as the catalyst. Unprotected 1-(hydroxyphenyl)ethanones, which are used as the starting material for a large number of syntheses in organic chemistry, do not react with olefin, but protection of the hydroxyl group with the silyl or alkyl group improves the reactivity of the acetophenones.

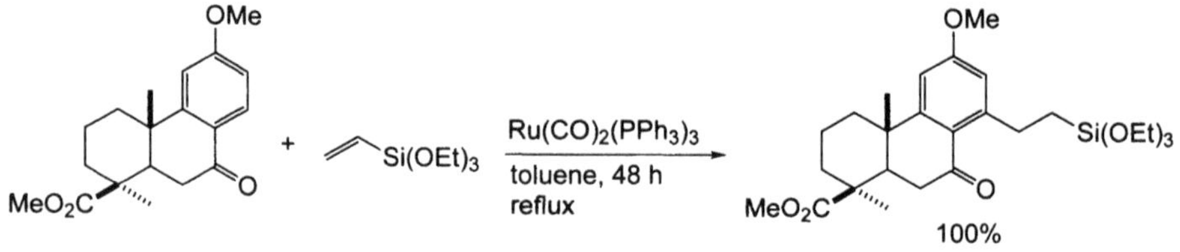

OMe
MeO2C + Si(OEt)3
Ru(CO)2(PPh3)3
toluene, 48 h
reflux
OMe Si(OEt)3
MeO2C 100%

(10)

In the case of the reaction of 3- and 4-acetylpyridines with triethoxyvinylsilane, the desired coupling takes place to give the corresponding alkylated products in high yields. The C–C bond formation also occurs at the position *ortho* to the acetyl group and the C–H bond in a more electron-deficient aromatic ring, i.e., the pyridine ring, shows a higher reactivity than that in the phenyl group (Eq. 11) [18].

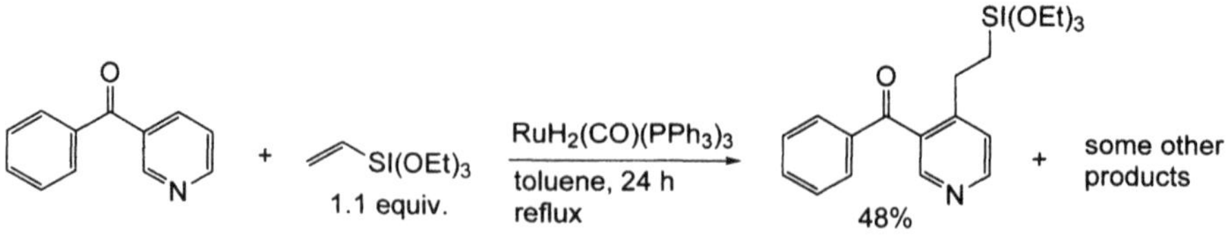

+ SI(OEt)3
1.1 equiv.
RuH2(CO)(PPh3)3
toluene, 24 h
reflux
SI(OEt)3
48% + some other products

(11)

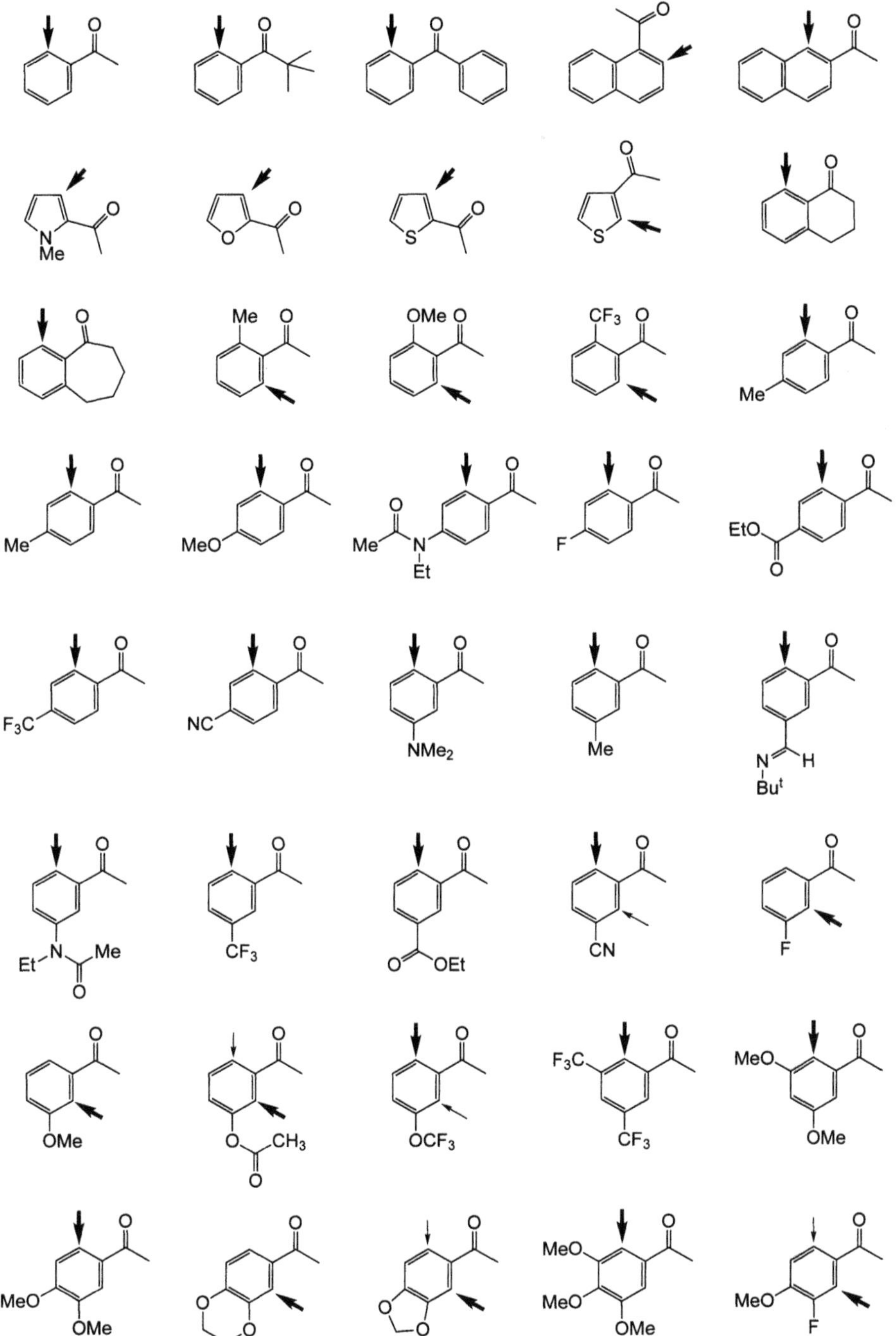

Fig. 1.

Application of C–H/olefin coupling to polymer chemistry has been accomplished by Weber's group [19]. The copolymerization of acetophenones having two free *ortho* C–H bonds and α,ω-dienes is performed using $RuH_2(CO)(PPh_3)_3$ as the catalyst (Eq. 12). The fact that this step growth polymerization gives a higher molecular weight polymer implies that each step proceeds virtually quantitatively. Especially the acetophenones with an electron-donating group, e.g., methoxy and amino groups, exhibit higher reactivities.

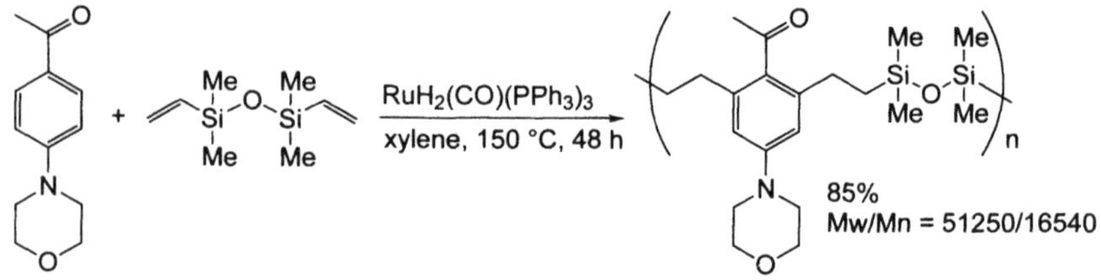

$RuH_2(CO)(PPh_3)_3$, xylene, 150 °C, 48 h; 85%; Mw/Mn = 51250/16540

(12)

The chelation-assisted C–H/olefin coupling is also applicable to the aromatic esters [20]. The esters with an electron-withdrawing group on the aromatic ring react with olefins to give the corresponding addition products in good to excellent yields (Eq. 13). The electron-withdrawing group on the aromatic ring facilitates the reductive elimination step [21].

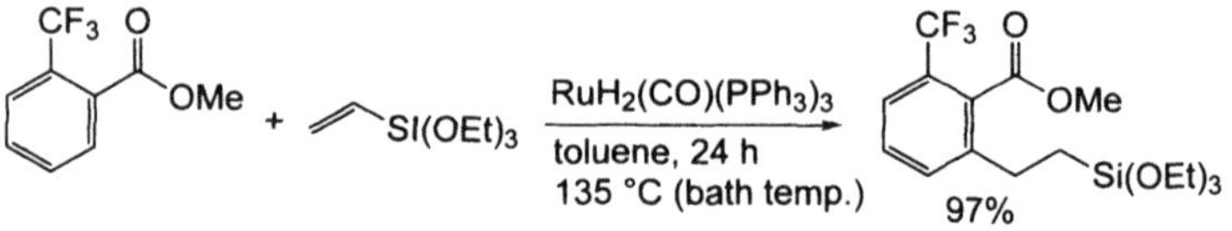

$RuH_2(CO)(PPh_3)_3$, toluene, 24 h, 135 °C (bath temp.); 97%

(13)

Nitrogen functionality can also work as the directing group. In the cases of the reactions of aromatic imines, easily derived from aromatic aldehydes or ketones with primary amine, $Ru_3(CO)_{12}$ is the superior catalyst even though this ruthenium carbonyl complex is ineffective for the reactions of aromatic ketones (Eq. 14) [22]. A by-product, i.e., a styrene derivative, was also obtained. By the use of rhodium(I) complex as the catalyst, the *ortho* C–H bonds in phenylpyridine can add to olefins [23]. Introduction of the methyl group at the 3-position of the pyridyl ring suppresses the incorporation of the second olefin into another *ortho* C–H bond. For the alkylation of 3-methyl-2-phenylpyridine, the cone angle of the phosphine ligands largely affect the reactivity rather than those of the electronic factor (Eq. 15) [24].

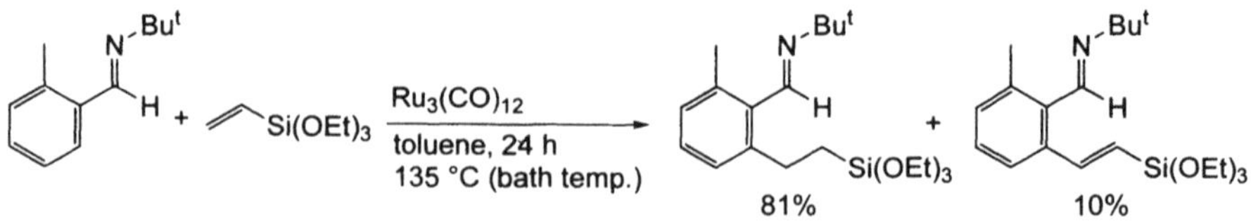

$Ru_3(CO)_{12}$, toluene, 24 h, 135 °C (bath temp.); 81% + 10%

(14)

$[RhCl(cyclooctene)_2]_2$/ 6 PCy_3, THF, 120 °C, 5 h; + $Si(OEt)_3$ → $(EtO)_3Si$ 96% (15)

Olefinic C–H bonds at the β-position in conjugate enones can be added to the carbon-carbon double bonds with the aid of the $RuH_2(CO)(PPh_3)_3$ (Eq. 16) [25]. Acylcyclohexenes exhibit high reactivities, and the presence of an oxygen atom at the allylic position of the 6-membered ring seems to increase the reactivity of the enones. Various types of olefins can be used as the C–H bond acceptor. Trost et al. have reported similar results with respect to the addition of conjugated esters to olefins using the same catalyst (Eq. 17) [26]. Various functional groups on the ester moiety are tolerated in this reaction.

Bu^t + $SI(OEt)_3$; $RuH_2(CO)(PPh_3)_3$, toluene, 0.5 h, 135 °C (bath temp.) → Bu^t, $Si(OEt)_3$ 96% (16)

OMe + $SI(OEt)_3$; $RuH_2(CO)(PPh_3)_3$, toluene, 18 h, reflux → OMe, $Si(OEt)_3$ 97% (17)

Rhodium-catalyzed alkylation of 2-isopropenylpyridine gives the addition product, in which the stereochemistry around the double bond is inverted to the thermodynamically favorable *E*-isomer (Eq. 18) [27]. Intramolecular C–H/olefin coupling provides a new entry to the carbocyclic compounds [28, 29]. Cyclization of 1-pyridyl-1,5-hexadienes giving 5-membered carbocycles is catalyzed by ruthenium or rhodium complex. This reaction also occurs even at room temperature. Deuterium-labeling experiments indicate that this reaction proceeds via the direct C–H bond cleavage pathway and the reductive elimination step seems to be the rate-determining step. This cyclization reaction proceeds in asymmetric fashion [29]. When the reaction is conducted in the presence of a monodentate chiral ferrocenyl phosphine and $[RhCl(coe)_2]_2$ (coe=cyclooctene), enantiomerically enriched carbocycles are obtained. In the case of the reaction of imidazolyl diene, the product is obtained in 82% ee at 50°C in 75% chemical yield (Eq. 19). The ee is slightly improved (87% ee) if the reaction is carried out at room temperature albeit in low chemical yield (12% yield). That a catalytic reaction involving a C–H bond cleavage proceeds at room temperature is noteworthy.

Me + ; $RhCl(PPh_3)_3$, toluene, 110 °C, 19 h → Me, H; 96% (*E* : *Z* = 93 : 7) (18)

L* = Fe OMe PPh2

[RhCl(cyclooctene)2]2/L*

THF, 50°C, 20 h

75% yields, 82 % ee

(19)

Dimerization of acrylonitrile is a cheaper way to synthesize highly valuable hexamethylenediamine [30]. In some cases of dimerization of acrylic acid esters, acrylonitriles, and acroleins, the direct C–H bond cleavage step seems to be involved in the catalytic reaction. At an early stage of catalytic dimerization of acrylonitrile, *cis*-1,4-dicyanobut-1-ene is formed as the major product, not *trans*-isomer [31]. This high *cis*-selectivity is suggested to indicate selective cleavage of C–H *cis* to CN by the metal coordinated to the nitrile group in side-on fashion.

Photolysis of methyl propionate yielded methyl 4-propionyloxybutyrate in the presence of $RhCl(CO)(PMe_3)_2$ catalyst at room temperature (Eq. 20) [32]. The C–H bond in the methyl group of the MeO moiety adds to methyl acrylate, which is presumably generated by photoassisted Rh-catalyzed dehydrogenation of propionate. Aromatic compound can also be used for this catalytic reaction, but the reaction of aliphatic compounds with methyl acrylate leads to polymerization product.

$$CH_3CH_2CO_2CH_3 \xrightarrow[\text{6 h, r.t.}]{h\nu,\ RhCl(CO)(PMe_3)_3} \underset{2035\%/Rh}{CH_3CH_2CO_2(CH_2)_3CO_2CH_3} + \underset{116\%/Rh}{CH_3O(O)C(CH_2)_4C(O)OCH_3} \quad (20)$$

2.2 Addition of C–H Bonds to Carbon-Carbon Triple Bonds

Substituted styrenes and vinylic compounds are versatile intermediates in organic synthesis, so various methods have been published in the literature [33, 34]. Among them, the Heck reaction is one of the best-studied methods for preparing these compounds [34]. However, for this reaction, the use of the halogen-carbon bond is essential for making C–C bonds. If direct addition of otherwise unreactive C–H bond to acetylenes takes place, this method will become one of the simplest methods for preparing substituted styrenes and vinylic compounds. In this section, we will describe the transition metal-catalyzed vinylation of aromatic compounds by using acetylenes.

The pioneering work of a coupling of aromatic and heteroaromatic compounds with acetylenes was reported by Yamazaki et al. in 1979 [35]. Reaction of benzene with diphenylacetylene gives triphenylethene in 45% yield. In the case of the monosubstituted benzene, the reaction of toluene proceeds at the *meta* position selectively but the reaction of anisol takes place at the *ortho* position (Eq. 21). They proposed that the site-selectivity stems from an inductive effect of an electron-negative atom. However, there is an alternative rationale more similar to the case of substituted acetophenone/olefin coupling (see Eq. 9). Competitive reaction of furan and benzene with diphenylacetylene results in an exclusive formation of the vinylfurans. Thiophene and *N*-methylpyrrol are also applicable to the vinylation reaction. These heteroaromatics are more reactive than benzene.

+ Ph—≡—Ph → $Rh_4(CO)_{12}$, 220 °C, 7 h (21)

R =	yield	*ortho*	*meta*	*para*
H	45%	-	-	-
CH_3	24%	6	65	29
OCH_3	42%	64	26	10
F	49%	70	22	8

Kisch et al. reported the cobalt- and rhodium-catalyzed additions of the C–H bond in azobenzenes to diphenylacetylene [36]. When $RhCl(PPh_3)_3$ is used as the catalyst, this coupling reaction gives 1-(arylamino)indole in good yields (Eq. 22). The kinetic study reveals a first-order dependence on $RhCl(PPh_3)_3$ and azobenzene, while a broken order of –0.3 is found for diphenylacetylene. Electron-withdrawing groups on the acetylene retard the coupling reaction. There are a couple of alternative mechanisms and these await further studies.

+ Ph—≡—Ph → $RhCl(PPh_3)_3$, 1-PrOH/HOAc 25 mL/15 μL, reflux, 24 h; 90% (22)

Ruthenium-catalyzed addition of aromatic C–H bond in aromatic ketones to acetylenes occurs site-selectively (Eq. 23) [37]. When trimethylsilyl substituted acetylenes are employed in the reaction, the desired coupling products are obtained with high regioselectivity in excellent yields. In the case of the reaction with 1-trimethylsilylpropyne, the regio- and stereochemical outcome was perfect. Therefore, only one out of possible four isomers was obtained.

+ Me—≡—$SiMe_3$ → $RuH_2(CO)(PPh_3)_3$, toluene, 3 h, 135 °C (bath temp.); 83% (23)

Rhodium-catalyzed addition of a C–H bond in benzene to acetylenes under irradiation conditions takes place, although the performance of this catalytic reaction and the selectivity of the coupling product are low [38]. A similar catalytic reaction has also been reported by Goldman et al. [39]. Catalytic coupling of benzene with phenyl acetylene gives the corresponding 1,1-disubstituted ethene (Eq. 24). For the acetylene with an alkyl group, the 1,2-addition product is the major product.

+ Ph—≡—H; hν, $RhCl(CO)(PMe_3)_2$, r.t., 19.5 h → Ph, Ph; 558%/Rh + other products (24)

2.3 Coupling of C–H Bonds, Carbon Monoxide, and Olefins

Since direct carbonylation of C–H bond leading to aldehydes is an endothermic reaction, the reaction should be conducted under irradiation conditions (see below) or by combination with further exothermic reaction of the products. Interestingly, however, in the case of the combination of the C–H bond, CO, and olefin giving ketones as the coupling product, the reaction becomes exothermic. As for this type of carbonylation reaction, Moore et al. have reported the first example of the highly selective carbonylation of aromatic ketone with the aid of ruthenium complex as the catalyst [5]. The reaction of pyridine, CO, and 1-hexene was carried out in the presence of $Ru_3(CO)_{12}$ at 150°C to give α-acylated pyridines (Eq. 25). A number of olefins can be used in this system. Terminal olefins as small as ethylene and as large as 1-eicosene afford the corresponding linear pyridyl ketones as the major products.

N (solvent) + CO + ; $Ru_3(CO)_{12}$, 150 °C, 16 h → N, O; 65% (*n* : *iso* = 93 : 7) (25)

In 1996, the present authors extended their C–H/olefin coupling to the C–H/CO/olefin coupling reaction [40]. Carbonylation of imidazole derivatives takes place in the presence of $Ru_3(CO)_{12}$ as the catalyst. The C–C bond formation occurs at the 4-position (Eq. 26). In these reactions, the linear ketones are the major products. Various functional groups, e.g., nitrile acetal, and ether, are tolerant.

Ph, N, N + CO + O, O; $Ru_3(CO)_{12}$, toluene, 160 °C, 20 h → Ph, N, N, O, O, O; 72% (*n* : *iso* = 97 : 3) (26)

When the reaction of phenylpyridine with CO and ethylene is conducted at 160°C, the benzene ring undergoes acylation at the *ortho* position (Eq. 27) [41]. In this reaction, the C–H bond in the pyridine ring does not react at all. In the case of the reaction of 2-(2'-naphthyl)pyridine, the acylation occurs exclusively at the 3'-position. This site selectivity of C–H/CO/olefin coupling is completely opposite to that of the ruthenium-catalyzed C–H/olefin coupling, i.e., reaction of 2-acetonaphthone with an olefin [15].

+ CO + ═ → ($Ru_3(CO)_{12}$; toluene; 160 °C, 20 h) 80% (27)

The similar carbonylation proceeds when the reaction of aromatic aldimine is conducted under carbonylation reaction conditions (Eq. 28) [42]. The acylated products converts to the corresponding indenone derivatives via intramolecular aldol condensation. For this reaction, ethylene, *tert*-butyl ethylene, and trimethylvinylsilane react with aromatic aldimine smoothly to provide the indenones in high yields.

+ CO + ═ → ($Ru_3(CO)_{12}$; toluene; 160 °C, 12 h) → (silica gel; 25 °C, 1 day) 82% (28)

Reaction of *N*-(2-pyridinyl)piperazines with CO and ethylene in the presence of a catalytic amount of $Rh_4(CO)_{12}$ in toluene at 160°C results in a complicated carbonylation reaction, which involves dehydrogenation and carbonylation at a C–H bond (Eq. 29) [43]. In this reaction, the carbonylation proceeds at the C–H bond α to the nitrogen atom substituted by pyridine. It is found that the reaction involves two discrete reactions: (a) dehydrogenation of the piperazine ring and (b) carbonylation at a C–H bond in the resulting olefin. An amide functionality can also serve as the directing group for carbonylation at the α C–H bond (Eq. 30) [44].

+ CO + ═ → ($Rh_4(CO)_{12}$; toluene; 160 °C, 20 h) 85% (29)

+ CO + ═ → ($Rh_4(CO)_{12}$; toluene; 160 °C, 20 h) 79% (30)

2.4
Insertion of Carbon Monoxide and Isocyanide into the C–H Bond

Several examples of transition metal-catalyzed insertions of carbon monoxide and isocyanide into the C–H bond are known. The carbonylation of a C–H bond to an aldehyde requires photoirradiation conditions. Eisenberg et al. have found iridium-[45, 46] or rhodium-catalyzed [47] photocarbonylation of benzene affording benzaldehyde, albeit with low efficiency [45–47]. They have also reported the photochemical carbonylation of benzene catalyzed by ruthenium(0) complexes [48].

Tanaka and Sakakura have developed a photochemical $RhCl(CO)(PMe_3)_2$ system [49–53] related to Eisenberg's benzene carbonylation systems. Prolonged photoirradiation results in a higher yield of the carbonylation product, but the yield of benzylalcohol formed by a reduction of benzaldehyde increased. A lower concentration of the catalyst provides a higher yield of the carbonylation product (Eq. 31) [51]. In the cases of the mono-substituted benzene, the reaction occurs at the *meta* and *para* positions, preferentially. The steric congestion seems to control the carbonylation position. They examined the effect of the wavelength in the photoassisted carbonylation of decane and found that the selectivity of the carbonylation product, i.e., C_{11} aldehydes, increased by cutting off lower wavelength (λ<325 nm) light because of the complete suppression of Norish type II reaction (Eq. 32) [52]. In the case of the photoirradiation with 295 > λ > 420 nm wave length, the carbonylation occurred preferentially at terminal carbon (86% selectivity), whereas the carbonylation took place almost equally at internal carbons (5%, 4%, 2%, and 3% selectivities).

benzene + CO —(hν, $RhCl(CO)(PMe_3)_2$, r.t., 16.5 h)→ PhCHO + $PhCH_2OH$ + PhCOOH (31)

catalyst concentration	PhCHO	$PhCH_2OH$	PhCOOH
0.7 mM	5729%/Rh	1278%/Rh	149%/Rh
7.0 mM	811%/Rh	103%/Rh	6%/Rh

$C_{10}H_{22}$ + CO —(hν, $RhCl(CO)(PMe_3)_2$, r.t., 16.5 h)→ $C_{10}H_{22}CHO$, 5 regioisomers, 610%/Rh (86% 4% 3%; 5% 2%) (32)

Several research groups have proposed the mechanism of photochemical carbonylation of benzene with $RhCl(CO)(PMe_3)_2$.[53–56]. The photoassisted carbonylation reaction has been proposed to take place through the 14-electron rhodium(I) complex by Tanaka's group [53]. Subsequently, other mechanisms have been proposed by Field [54], Goldman [55] and Ford [56]. They revealed that the primary photoprocess does not involve ligand loss leading to the 14-electron rhodium complex [55, 56]. Some theoretical calculations of the photochemical processes have also been done [57–59]. While the matter of the actual catalytically active species of this photochemical reaction remains unsettled, their photoassisted reactions open a new area of the direct functionalization of the C–H bond.

In place of carbon monoxide, isocyanides are often used as the isoelectronic compound. In 1986, Jones et al. reported that the low-valent ruthenium phosphine complex catalyzed intramolecular insertion of isocyanide into the sp^3 C–H bond under thermal conditions (Eq. 33) [60, 61]. Their finding provided a new route for synthesis of indole. An interesting feature of their reaction is that C–H bond cleavage occurs even in the presence of an excess of the trapping ligand, i.e., isocyanide.

NC; 1.5 equiv. — $RuH_2(dmpe)_2$; C_6D_6, 140 °C; 25 h → N H; 98% (33)

dmpe = P P

This reaction has been extended to photoassisted insertion of isocyanide into C–H bonds [62, 63]. When the reaction of benzene with isocyanide was carried out with the aid of $Fe(PMe_3)_2(CNCH_2Bu^t)_3$ as the catalyst under irradiation conditions, the corresponding aldimine was obtained. Under dilute reaction conditions, the reaction became catalytic and efficient (Eq. 34). Tanaka and Sakakura have also reported photoassisted $RhCl(CO)(PMe_3)_2$-catalyzed insertion of isocyanide into C–H bonds in benzene and pentane, but the turnover number of the catalyst is not high (2.45 turnovers) [64].

+ C=N$\frown$Bu^t (2 mM) — hν; $Fe(PMe_3)_2(CNCH_2Bu^t)_3$; 0.5 mM; 82% conversion → N$\frown$Bu^t, H; 5.7 TON (34)

3 Catalytic Dehydrogenation of Alkanes and Arenes

3.1 Dehydrogenation of Alkanes and Arenes

The functionalization of alkanes is a highly important subject in inorganic, organic, and organometallics chemistries. The simplest and most convenient way is dehydrogenation. In 1979, Crabtree et al. showed the possibility of dehyrogenation of alkanes by using a stoichiometric amount of iridium-phosphine complex [65]. Several years later, Baudry and Ephritikhine showed the first example with respect to the catalytic conversion of alkanes to alkenes with the aid of the homogeneous transition metal complex as the catalyst (Eq. 35) [66]. They found that rhenium polyhydride complex catalyzes dehydrogenation of cycloalkane to cycloalkene in the presence of *tert*-butylethylene as the hydrogen acceptor. The turnover number in this system was only 9. However, their result indicates the possibility of alkane dehydrogenation by a soluble transition metal catalyst.

$((p\text{-F-}C_6H_4)_3P)_2ReH_7$; 80 °C, 10 min; 9 TON (35)

Several thermal catalytic dehydrogenations of alkanes in the presence of alkenes as the hydrogen acceptor have been reported by several research groups [67–76]. A variety of iridium complexes, e.g., $[IrH_2(acetone)_2P(p\text{-}FC_6H_4)_3]SbF_6$,[67] $IrH_5(P^iPr_3)_2$ [68, 69], $[IrH_2(CF_3CO_2)(p\text{-}FC_6H_4)_2]$ [70, 71], and $[IrH_2(\eta^3\text{-}C_6H_3(CH_2PBut_2)_2\text{-}2,6)]$ [72] can catalyze the dehydrogenation of cycloalkane giving cycloalkene in the presence of *tert*-butylethylene as the hydrogen acceptor (Eq. 36).

150 °C; 82 TON/min (36)

A similar iridium complex, $[IrH_2(acetone)_2PPh_3]SbF_6$ [73], catalyzes the selective dehydrogenation of cyclohexenes to arenes. In this case, the cyclohexenes work as the substrate and also as the hydrogen acceptor.

Highly efficient transfer-dehydrogenation of alkanes was reported by Goldman [74–77], using a unique catalyst system. A high pressure (1000 psi) of hydrogen is used for the dehydrogenation reaction [75–77]. Under 1000 psi of H_2 at 100°C for 15 min, a cyclooctane solution of $RhCl(CO)(PMe_3)_2$ (0.20 mM) and norbornene (1.2 M) yielded 950 turnovers of cyclooctene (0.19 M) and norbornane (1.2 M) (Eq. 37). The proposed mechanism for this paradoxical catalytic reaction involves the addition of H_2, loss of CO, and transfer of H_2 to a sacrificial acceptor, thereby generating $RhCl(PMe_3)_2$, which is the same catalytically active species proposed in the photochemical dehydrogenation of alkanes with $RhCl(CO)(PMe_3)_2$ (Scheme 2).

$+ H_2$ (1000 psi); $Rh(PMe_3)_2Cl(CO)$; 100 °C, 15 min; 950 TON (37)

In 1990, notable advance was achieved in the dehydrogenation of alkanes by Saito et al. [78a, b] They reported the first example of efficient alkane dehydrogenation in the absence of a sacrificial hydrogen acceptor under thermal conditions with the aid of $RhCl(PPh_3)_3$-catalyst. The key of this reaction was continuous removal of molecular hydrogen from the reaction mixture. In 1993,

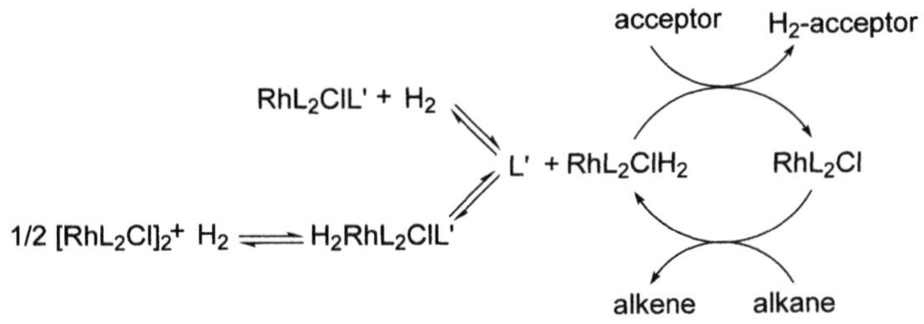

$$\frac{1}{2}\,[RhL_2Cl]_2 + H_2 \rightleftharpoons H_2RhL_2ClL' \rightleftharpoons L' + RhL_2ClH_2 \rightleftharpoons RhL_2ClL' + H_2$$

$L' + RhL_2ClH_2$ + acceptor → RhL_2Cl + H_2-acceptor; RhL_2Cl + alkane → RhL_2ClH_2 + alkene

Scheme 2.

Crabtree et al. reported the modified catalytic system for the dehydrogenation of cyclooctane with the aid of $IrH_2(CF_3CO_2)(PPh_3)_3$ catalyst. [78c] In 1997, Gupta and Goldman found a highly efficient dehydrogenation of cyclodecane to cyclodecene using a iridium complex having η^3-$C_6H_3(PBu^t_2)_2$-1,3 (PCP) ligand (Eq. 38). [78d] In this case, 360 turnover number is attained after 24 h.

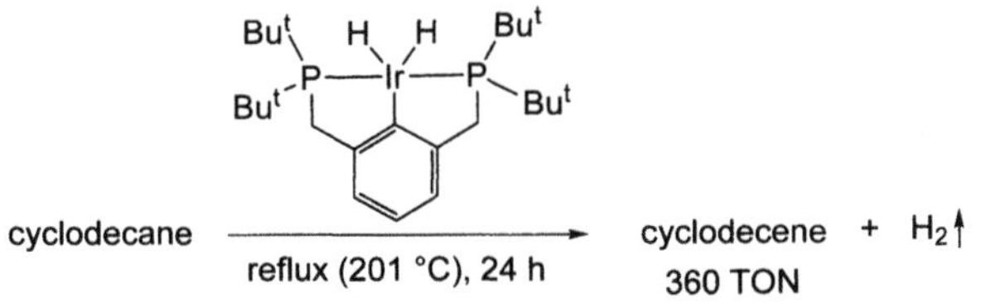

$$\text{cyclodecane} \xrightarrow[\text{reflux (201 °C), 24 h}]{(PCP)IrH_2,\ PCP = \eta^3\text{-}C_6H_3(CH_2PBu^t_2)_2\text{-}1,3} \text{cyclodecene (360 TON)} + H_2\uparrow \qquad (38)$$

Photochemical catalytic dehydrogenation of alkanes using $RhCl(CO)(PMe_3)_2$ has been reported by Saito [79]. This reaction proceeds even in the absence of the hydrogen acceptor such as *tert*-butylethylene. Tanaka and Sakakura also reported a similar type of photoassisted dehydrogenation [80–82]. In this reaction, a slow flow of nitrogen accelerated the desired reaction (Eq. 39). Therefore, the hydrogenation of the olefin was effectively suppressed. Aldehydes can also be used as the hydrogen acceptor. Reduction of aldehydes to alcohol with alkanes, which work as the hydrogen donor, occurs under photoirradiated conditions in the presence of $RhCl(CO)(PMe_3)_2$ catalyst [83]. This reaction afforded the corresponding alcohols in high yields (67–87% yields). When the reaction is conducted under an N_2 atmosphere, dehydrogenative coupling of arenes giving biaryls takes place under photochemical conditions by using $RhCl(CO)(PMe_3)_2$ as the catalyst (Eq. 40) [84]. It was proposed that free aryl radical, which might be generated from homolytic cleavage of aryl-rhodium bond, would be important in the biaryl formation step.

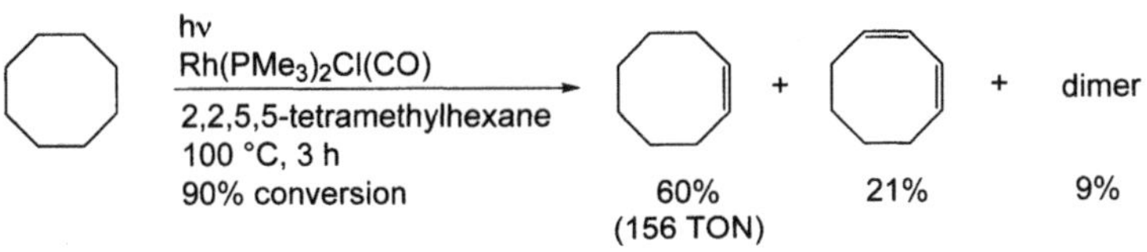

$$\text{cyclooctane} \xrightarrow[\text{2,2,5,5-tetramethylhexane, 100 °C, 3 h, 90\% conversion}]{h\nu,\ Rh(PMe_3)_2Cl(CO)} \text{cyclooctene}\ 60\%\ (156\ \text{TON}) + \text{cyclooctadiene}\ 21\% + \text{dimer}\ 9\% \qquad (39)$$

$$\text{C}_6\text{H}_6 \xrightarrow[\text{r.t., 16.5 h}]{h\nu,\ \text{RhCl(CO)(PMe}_3)_2} \text{biphenyl (360\%/Rh)} + \text{PhCHO (3\%/Rh)} + \text{PhCH}_2\text{OH (26\%/Rh)} \quad (40)$$

3.2 Dehydrogenative Silylation of Alkanes and Arenes

Dehydrogenative silylation of benzene with pentamethyldisiloxane with the aid of $IrCl(CO)(PPh_3)_2$ takes place under thermal conditions, although a prolonged reaction time (49 days) is required to obtain a relatively higher total yield of phenylated products [85]. Selective silylation of arenes with *o*-bis(dimethylsilyl)benzene is catalyzed by $Pt_2(dba)_3$ (dba=dibenzylideneacetone) complex, giving the monoarylated hydrosilanes in high yields (Eq. 41) [86]. They proposed the bis(silyl)platinum as the active catalyst species. In this dehydrogenative silylation, the reactivities of the arenes decrease in the order: anisole>chlorobenzene>benzene>toluene.

$$o\text{-C}_6\text{H}_4(\text{SiMe}_2\text{H})_2 + \text{C}_6\text{H}_6 \xrightarrow[\text{110 °C, 84 h}]{\text{Pt}_2(\text{dba})_3} o\text{-C}_6\text{H}_4(\text{SiMe}_2\text{Ph})(\text{SiMe}_2\text{H})\ \ 87\% \quad (41)$$

Self-dehydrogenative silylation of triethylsilane in the presence of *tert*-butylethylene as the hydrogen acceptor using $(\eta^5\text{-}C_5Me_5)Rh(H)_2(SiEt_3)_2$ catalyst is reported [87]. The conversion of the silane is almost quantitative and the dehydrogenative silylated product is obtained in 83% yield (Eq. 42).

$$\text{Et}_3\text{SiH} + \text{CH}_2{=}\text{CHBu}^t \xrightarrow[\text{c-C}_6\text{H}_{12},\ 150\ °\text{C, 12 h}]{(\eta^5\text{-C}_5\text{Me}_5)\text{Rh(H)}_2(\text{SiEt}_3)_2} \text{HEt}_2\text{Si–CH(Me)–SiEt}_3\ (83\%) + \text{Et}_3\text{SiCH}_2\text{CH}_2\text{CH}_2\text{Bu}^t\ (9\%) + \text{Et}_3\text{SiCH{=}CHBu}^t\ (3\%) \quad (42)$$

A similar silylation reaction of benzene with triethylsilane also took place under photoirradiation conditions [88]. Lower catalyst loading improves the selectivity of the corresponding silylbenzene (Eq. 43). This catalyst system can also be applied to the dehydrogenative silylation of arene with hexaorganodisilane.

$$\text{C}_6\text{H}_5\text{CH}_3 + \text{HSiEt}_3 \xrightarrow[\text{r.t., 16.5 h}]{h\nu,\ \text{RhCl(CO)(PMe}_3)_2} \text{CH}_3\text{C}_6\text{H}_4\text{SiEt}_3\ (161\%/\text{Rh};\ o:m:p = 4:58:38) + \text{C}_6\text{H}_5\text{CH}_2\text{SiEt}_3\ (8\%/\text{Rh}) \quad (43)$$

Ishikawa found the $Ni(PEt_3)_4$-catalyzed silylation of aromatic compounds with 3,4-benzo-1,1,2,2-tetraethyl-1,2-disilacyclobutene, providing the 1-(diethylarylsilyl)-2-(diethylsilyl)benzene in high yields (Eq. 44) [89]. In the case of the reaction of mesitylene, interestingly, the sp^3 C–H bond adds to the disilacyclobutene, albeit in low yield (28% yield) [89]. Platinum(0) complexes are also applicable to this silylation reaction as the catalyst [90].

$Ni(PEt_3)_4$, reflux, 4 h; 69% + 14% (44)

Tanaka et al. have reported platinum-catalyzed site-selective silylation of aromatic aldimines with hexamethyldisilane [91]. This silylation reaction proceeds exclusively at the position *ortho* to the imino group. In the case of the reaction of an imine having an electron-withdrawing group at the *para* position, the yield of the desired silylated product improves (Eq. 45). The key to this site selectivity is the coordination of the imino nitrogen with the platinum atom.

+ $Me_3SiSiMe_3$; 1/2 $Pt_2(dba)_3$-5/3 $P(OCH_2)_3CEt$, toluene, 160 °C, 20 h; 23% + 66% (45)

4 Hydroacylation with Aldehydes

4.1 Intramolecular Hydroacylation of Olefins

Hydroacylation has its origins in the observation by Tsuji that aldehydes are decarbonylated by $RhCl(PPh_3)_3$ complex [92]. It was proposed that the cleavage of the C–H bond in the formyl group was the initial step for the decarbonylation reaction. Several years later, intramolecular hydroacylation of enals was first observed by Sakai et al. using $RhCl(PPh_3)_3$ complex [93]. This reaction, however, required the use of a stoichiometric amount of the rhodium complex and the yield of the cyclized ketone was low (30% yield). In 1972, Miller et al. found the first example of the catalytic intramolecular hydroacylation of olefin affording a cyclic ketone (Eq. 46) [94]. Higher yields of cyclopentanone can be achieved by carrying out the catalytic reaction in ethylene-saturated chloroform.

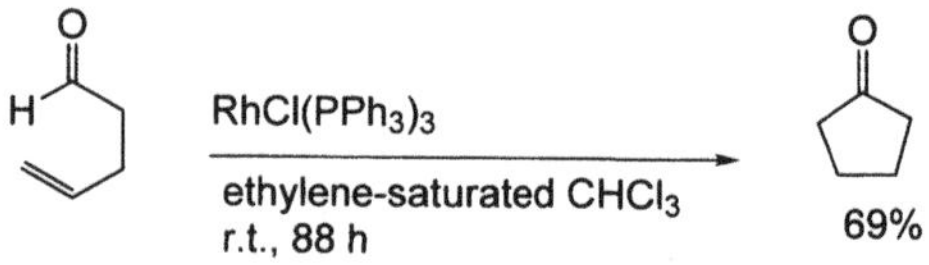

$RhCl(PPh_3)_3$, ethylene-saturated $CHCl_3$, r.t., 88 h; 69% (46)

Mechanistic studies on the intramolecular hydroacylation by using deuterium-labeling experiments have been reported by several groups [95–100]. The results of their studies showed that the addition of the Rh–H bond to a carbon-carbon double bond takes place in *syn* fashion [95, 96]. They also demonstrated that C–H bond cleavage, hydrid transfer to the double bond, and carbonyl deinsertion are all fast and reversible steps (Scheme 3) [99].

Larock et al. have extended the rhodium-catalyzed intramolecular hydroacylation of unsaturated aldehydes to a convenient method for preparing cyclopentanones [101]. Studying the reactivities of the various types of enals, they found that the alkyl substitution in either the 2 or the 5 position in 4,5-unsaturated aldehydes substantially reduces the yield of cyclic ketones. In the case of the reaction of 5-hexenals, either 5-membered ring ketones or the 6-membered ring ketones are considered to be obtained as the product, and cyclopentanone derivatives are predominantly formed. Interestingly, however, the selective formation of the 6-membered ketones was observed when intramolecular hydroacylation of 1,2-isopropylidene-3-C-ally *ribo*-pentodialdose was carried out with the aid of $RhCl(PPh_3)_3$ as the catalyst [102]. They proposed that this opposite selectivity stemmed from the ring strain of the fused 5,5,5-tricyclic ring product (Eq. 47).

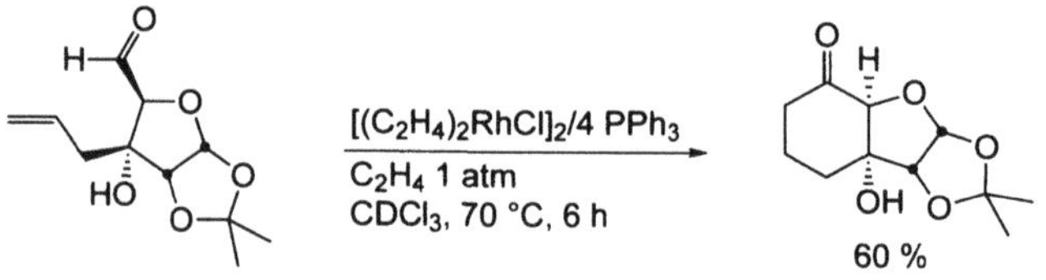

$[(C_2H_4)_2RhCl]_2$/4 PPh_3, C_2H_4 1 atm, $CDCl_3$, 70 °C, 6 h; 60 % (47)

James et al. have applied this intramolecular hydroacylation to the resolution of racemic enals using rhodium(I) complex and chiraphos [103]. In this case, 5-membered ring ketones with up to 69% ee of the optical isomer are obtained in moderate yields (15–58% yields) (Eq. 48).

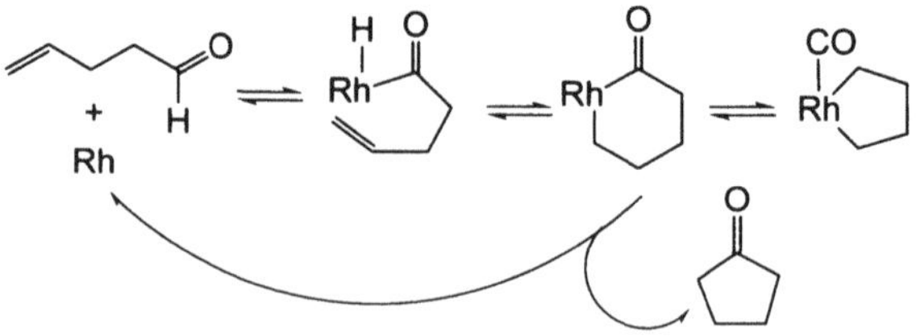

Scheme 3.

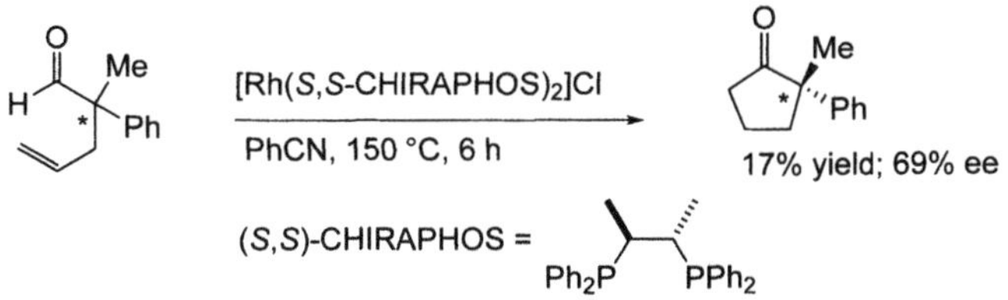

[Rh(S,S-CHIRAPHOS)$_2$]Cl
PhCN, 150 °C, 6 h
17% yield; 69% ee
(S,S)-CHIRAPHOS =
Ph$_2$P PPh$_2$

(48)

Application of this cyclization reaction to a large variety of 4-pentenals with the aid of the rhodium complex has been reported. The first example of an asymmetric cyclization of 4-pentenals via hydroacylation using a chiral rhodium diphosphine catalyst was published by Sakaki et al. in 1989 [104]. The diphosphine ligand ((1S,2S)-*trans*-1,2-bis(diphenylphosphinomethyl)cyclohexane) having a cyclohexane backbone in the chiral center shows the better asymmetric induction than DIOP ligand. Various types of enals are applicable to this asymmetric intramolecular hydroacylation reaction [105, 106]. The use of BINAP ligand as the chiral auxiliary improves the optical yield to >99% ee when 4-substituted 4-pentenals are used as the substrate (Eq. 49) [106]. Steric repulsion between the substituent at the 4-position and the substituent on the phosphine atom controls the enantiofacial selection.

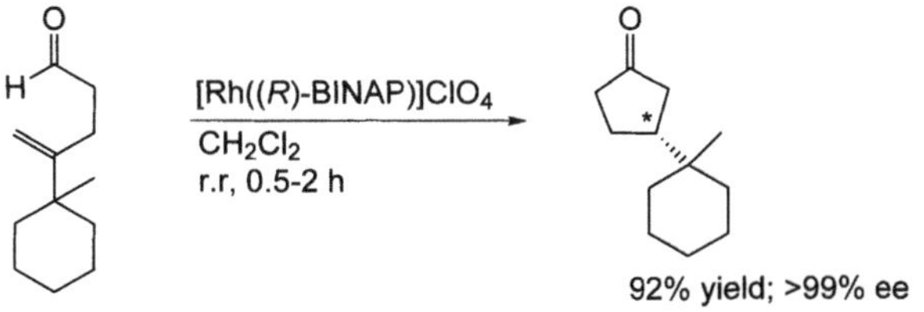

[Rh((R)-BINAP)]ClO$_4$
CH$_2$Cl$_2$
r.r, 0.5-2 h
92% yield; >99% ee

(49)

Similar asymmetric cyclization of substituted 4-pentenals using cationic chiral diphosphine rhodium complexes as the catalyst has been reported by Bosnich et al. [100, 107, 108]. They found that with the BINAP catalyst, almost complete enantioselectivity is observed for 4-pentenals bearing 4-substituted tertiary substituents, and this reaction is tolerant of a wide range of functional groups [107, 108]. They applied this high enantioselective hydroacylation reaction to efficient kinetic resolutions of 3-substituted pentanals using the BINAP catalyst, but only modest kinetic resolution was observed [100]. They concluded that the origin of the lower efficiency is that the asymmetric hydroacylation is

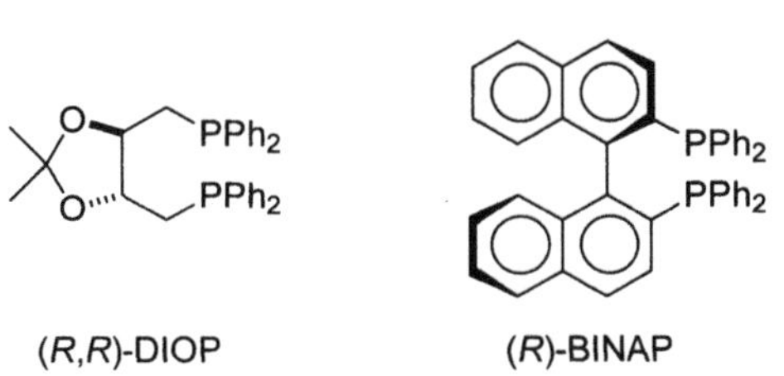

PPh$_2$
PPh$_2$
(R,R)-DIOP

PPh$_2$
PPh$_2$
(R)-BINAP

Structure 1.

not governed by a single enantioselective step, but rather that the enantioselection is controlled by a number of reversible steps, e.g., carbonyl deinsertion-insertion step, involving reaction intermediates.

Eilbracht et al. have developed rhodium- or ruthenium-catalyzed one-pot synthesis of cyclopentanones from allyl vinyl ether via tandem Claisen rearrangement and hydroacylation [109–111]. This protocol requires elevated temperature (140–220°C) and also requires alkyl or aryl substituents at the terminal position of the allylic double bond to prevent undesirable double bond migration in the intermediary formed, unsaturated aldehyde.

4.2 Intermolecular Hydroacylation of Olefins

Intramolecular transition metal-catalyzed hydroacylation reactions have opened up a new area of synthesizing cyclic ketones. This reaction can also be extended to intermolecular addition reactions. Miller et al. found the first example of an intermolecular hydroacylation of an aldehyde with an olefin giving ketones, when they were studying the mechanism of the rhodium-catalyzed intramolecular cyclization of 4-pentenal using ethylene-saturated chloroform as the solvent (Eqs. 46, 50) [112].

$Rh(C_2H_4)_2(acac)$; $CHCl_3$, r.t.; 48 h; 84% yield (50)

When the reaction of propionaldehyde with ethylene is conducted in the presence of $RuCl_2(PPh_3)_3$ as the catalyst without solvent at 210°C for 18 h, 3-pentanone was obtained together with unreacted aldehyde and aldolization products [113, 114].

Other rhodium complex also catalyzed the addition of the C–H bond in aldehyde to olefins [115–117]. The use of paraformaldehyde results in the formation of aldehydes [115]. Marder et al. proposed the reaction mechanism of CpRh(ethylene)$_2$-catalyzed addition of C–H bond in aldehyde to ethylene by the use of isotope-labeling experiments [117]. They suggested that insertion of ethylene to the Rh-H bond must take place rapidly and reversibly, and this equilibrium must be established significantly faster than either aldehyde reductive elimination or product formation (Scheme 4).

Watanabe et al. reported that the addition of C–H bonds in aldehydes to olefins took place efficiently with the aid of $Ru_3(CO)_{12}$ under a CO atmosphere at 200°C (Eq. 51) [118]. They also reported that the same ruthenium-carbonyl complex catalyzes the addition of the C–H bonds in formic acid esters and amides to olefins (Eq. 52) [119].

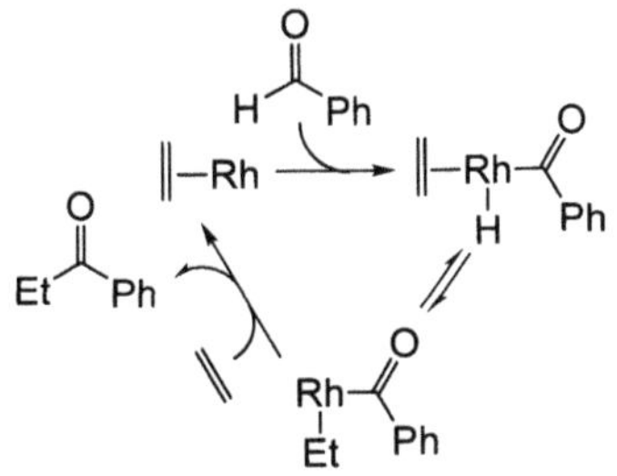

Scheme 4.

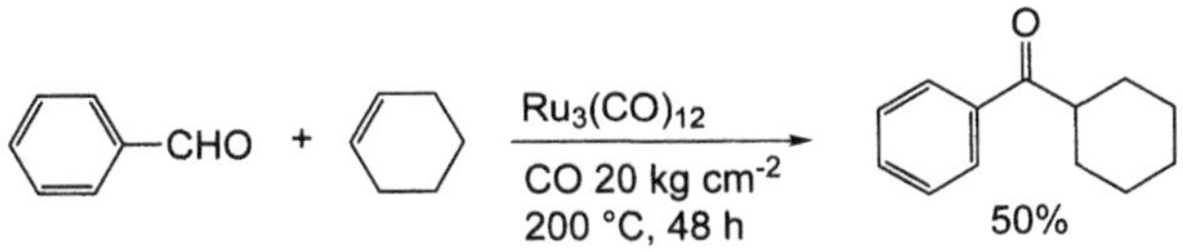

$Ru_3(CO)_{12}$, CO 20 kg cm^{-2}, 200 °C, 48 h; 50% (51)

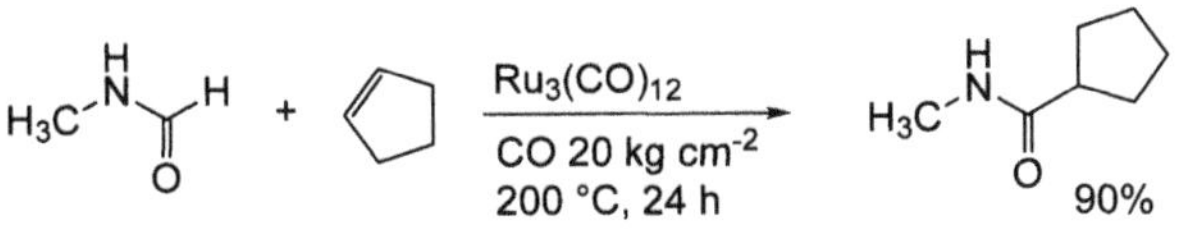

$Ru_3(CO)_{12}$, CO 20 kg cm^{-2}, 200 °C, 24 h; 90% (52)

Chelation-assisted additions of formyl C–H bonds to olefins and dienes have been reported by Jun et al. [120]. In the case of the reaction of 8-quinolinecarboxaldehyde, they proposed that the formation of the stable 5-membered metallacyclic complex [121] suppressed the undesired decarbonylation reaction (Eq. 53) [120]. The intermolecular hydroacylation of 1-alkene with 2-(diphenylphosphino)benzaldehyde by rhodium(I) catalyst has been conducted on the basis of this working hypothesis [122].

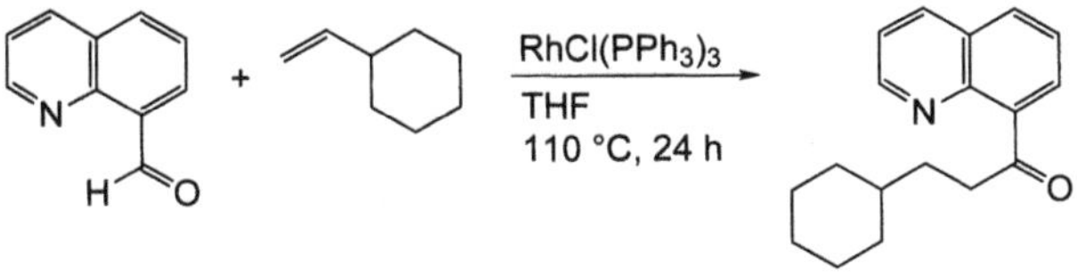

$RhCl(PPh_3)_3$, THF, 110 °C, 24 h (53)

Suggs found the first example of the addition of the C–H bond in imino group to olefin (ethylene) using $RhCl(PPh_3)_3$ (Eq. 54) [123]. The hydroiminoacylation of olefins (e.g, 1-pentene, allyl alcohol, 1,5-hexadine, and 1,5-pentadine-3-ol) can be applied to synthesis of ferrocenyl ketimines, which are masked acylferrocenes (Eq. 55) [124].

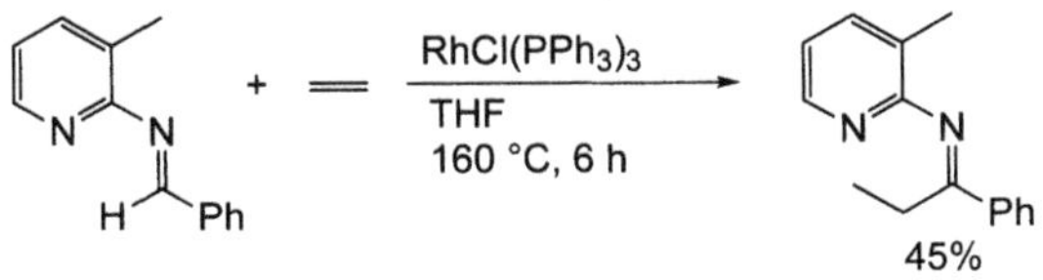

$RhCl(PPh_3)_3$, THF, 160 °C, 6 h; 45% (54)

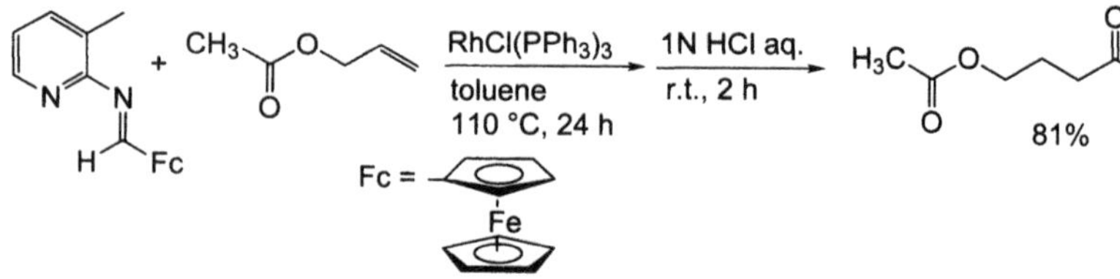

(55)

This two-step synthesis of ketones has been improved from the aldehyde into one-step synthesis with the cocatalyst system of the rhodium complex and 2-amino-3-picoline, which reacts with an aldehyde to give an aldimine in situ. The ketimine produced is easily converted to a ketone by in situ hydrolysis with H_2O, which is formed in the step of condensation of the aldehyde with the amine (Eq. 56) [125].

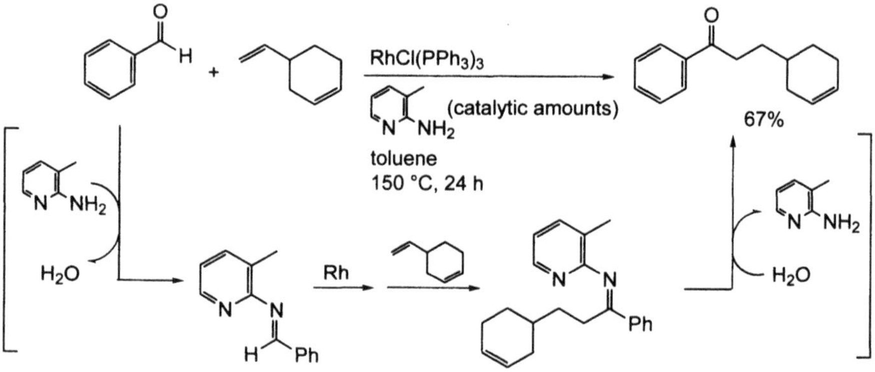

(56)

Brookhard et al. have reported the use of $(\eta^5\text{-}C_5Me_5)Co(CH_2{=}CHSiMe_3)_2$ as a catalyst precursor for both intra- and intermolecular hydroacylation of certain substrates together with a mechanistic study of these reactions [126]. Benzaldehydes having an electron-donating substituent such as the *N,N*-dimethylamino group show a higher reactivity. In the case of the use of vinyltriphenylsilane, complete conversion at 0.5% catalyst loading is observed (Eq. 57).

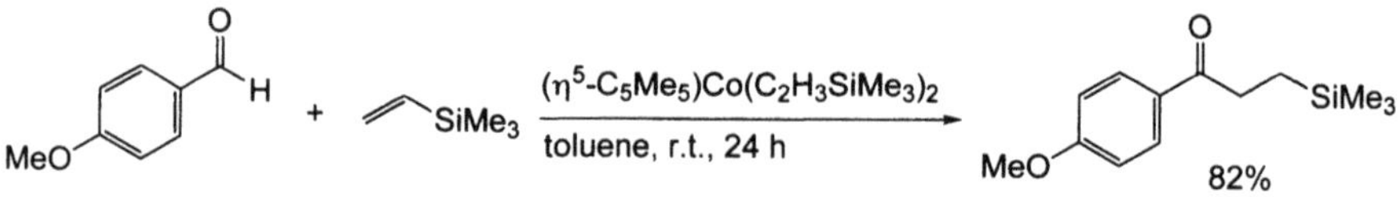

(57)

4.3 Hydroacylation of Acetylenes

The hydroacylation of acetylenes giving conjugated enones was first published in 1990 [127]. For this reaction, a nickel(0) complex is suitable for attaining high yields (Eq. 58). Both aliphatic and aromatic aldehydes are applicable to the reaction. In the case of the reaction with unsymmetrically substituted acetylenes, the regioselectivity depends upon steric bulkiness of the alkyl substituent. Therefore, the C–C bond is formed preferentially at the sterically less congested position.

^{i}PrCHO + Pr–≡–Pr → ($Ni(cod)_2$/$P(C_8H_{17})_3$, THF, 100 °C, 20 h) 93% (*E*:*Z* = 93:7) (58)

Rhodium-catalyzed reaction of a phosphine substituted aldehyde with terminal acetylenes results in the formation of hydrogenated product, which may be formed by the hydride reduction of the primarily produced enones [128]. When the reaction is conducted using sterically hindered terminal acetylene, e.g., 3,3-dimethyl-1-butyne, a good yield of the primary hydroacylation product is obtained (Eq. 59). This observation indicates that conjugated enones are apparently the primary products.

Ph_2P–C_6H_4–CHO + H–≡–Bu^t → ($[RhCl(cyclooctene)_2]_2$, THF, 90 °C, 4 h) 47% + 16% (59)

Highly efficient catalytic addition of formyl C-H bond in salicyl aldehydes to acetylenes was reported by Nomura and Miura in 1997 (Eq. 60) [129]. Combination of $[RhCl(cod)]_2$ (cod=1,5-cyclooctadiene) and diphenylphosphinoferrocene (dppf) as the cocatalyst is effective for the desired formyl C–H/olefin coupling reaction. Almost quantitative yields were attained in many runs. The key to their success is that the hydroxyl group at the position *ortho* to the formyl group directs the rhodium closer to the formyl C-H bond. It is worthy of note that propargyl alcohols and their esters, which are highly reactive towards low-valent rhodium complexes, remain intact in the coupling products.

Salicylaldehyde + H–≡–CH(OAc)C_5H_{11} → ($[RhCl(cod)]_2$/dppf, Na_2CO_3, toluene, reflux, 4 h) 70% (60)

5
Addition of Active Methylene Compounds to Unsaturated Functions

5.1
Michael Addition and Aldol Reactions

Although the catalytic reactions at more or less acidic C–H bonds are beyond the scope of this review, some should be mentioned. The addition of active methylene compounds to aldehydes and α,β-unsaturated carbonyl compounds are catalyzed by several transition metal complexes. Murahashi et al. have reported the $RuH_2(PPh_3)_4$-catalyzed addition of activated nitriles to aldehydes, ketones, and α,β-unsaturated carbonyl compounds (Eq. 61) [130]. In the case of the reaction of the nitriles with aldehydes and ketones, condensation products of the type of Knoevenagel reaction are obtained in high yields. Interestingly, the reaction of α,β-unsaturated carbonyl compounds give Michael adducts without contamination by the corresponding aldol products. They suggested the addition of C–H bond to a low-valent ruthenium as the initial step. Echavarren suggested the possibility of Michael type addition of the active methylene compounds [131]. They extended Murahashi's results to the reaction of active methylene compounds having no cyano group using ruthenium or rhodium complexes as the catalyst (Eq. 62). Similar iridium-catalyzed Knoevenagel condensations of ethyl cyanoacetate with aldehydes as well as ketones have been reported by Lin et al. [132].

EtO2C, CO2Et, NC, CO2Et, Me + methyl vinyl ketone —[$RuH_2(PPh_3)_4$, THF, -78 °C, 6 h]→ EtO2C, NC, CO2Et, CO2Et, Me + EtO2C, NC, CO2Et, CO2Et, Me

72% (97/3)

(61)

MeO2C, MeO2C + CN —[$RuH_2(PPh_3)_4$, CH_3CN, r.t., 24 h]→ MeO2C, MeO2C, CN

94%

(62)

This type of transition metal catalyzed the Michael addition of nitriles to methyl acrylate, and methyl vinyl ketone proceeds with good to high yields with the aid of $RhH(CO)(PPh_3)_3$ as the catalyst (Eq. 63) [133]. Interestingly, benzyl cyanide also shows a high reactivity with methyl vinyl ketone. In this study, the insertion of the low-valent rhodium species into the C–H bond adjacent to the cyano group has been proposed.

Ph, NC + methyl vinyl ketone —[$RhH(CO)(PPh_3)_3$, toluene, 20 °C, 3 h]→ Ph, NC, O

85%

(63)

Ito et al. reported that rhodium-catalyzed Michael addition of α-cyano carboxylates to α,β-unsaturated carbonyl compounds can be made asymmetric (Eq. 64) [134]. This Michael reaction takes place in high chemical and optical yields even in the case of low catalyst loading (0.1–1 mol%). It has been shown that *trans*-chelating chiral bisphosphines (TRAP) are more effective than *cis*-chelating bisphosphines. This catalyst system can also be applied to the asymmetric aldol reaction of 2-cyanopropionates (Eq. 65) [135]. This aldol reaction gives the corresponding adducts in high chemical and optical yields. Surprisingly, the use of formalin, i.e., aqueous solution of formaldehyde, did not affect the enantiopurity of the product much.

$RhH(CO)(PPh_3)_3$/ (*S,S*)-(*R,R*)-TRAP; benzene, 3 °C, 10 h

TRAP = 2,2''-bis[1-(diphenylphosphino)ethyl]-1,1''-biferrocene

99% yield; 86% ee (64)

$Rh(acac)(CO)_2$/ (*S,S*)-(*R,R*)-TRAP; Bu_2O, -10°C, 24 h

86% yield; 93% ee (65)

Masked formyl cyanides such as cyanohydrin alkyl ethers of formyl cyanide are also applicable to the aldol reaction. Palladium complexes, especially $Pd_2(dba)_3$-$CHCl_3$, show high catalytic activities for the additions to aldehydes (Eq. 66) [136].

$Pd_2(dba)_3$-$CHCl_3$; dppe; CH_3CN, r.t., 24 h; Ac_2O

100%

dppe = Ph_2P–CH_2CH_2–PPh_2

(66)

5.2 Addition to Carbon-Carbon Multiple Bonds

Active methylene compounds can be added to polar double bonds such as those in acrylate esters and methyl vinyl ketone as has been described in the previous section. Active methylene compounds can also be added to carbon-carbon multiple bonds in allenes and alkynes with the aid of the transition metal complexes as the catalyst. The addition of methylmalononitrile to 3-phenyl-1,2-butadiene takes place in the presence of $Pd_2(dba)_3$-$CHCl_3$ to give the corresponding addition product with *E*-stereochemistry (Eq. 67) [137 a]. The C–C bond formation occurs exclusively at the terminal position of the allenes. Trost et al. independently reported the similar results with respect to palladium-catalyzed addition of C–H bonds in active methylene compounds to allenes [137 b].

$$\mathrm{Ph(Me)C{=}C{=}CH_2} + \mathrm{MeCH(CN)_2} \xrightarrow[\text{THF, reflux, 48 h}]{\mathrm{Pd_2(dba)_3\text{-}CHCl_3/dppb}} \mathrm{Ph(Me)C{=}CHCH_2C(Me)(CN)_2}\ \ 68\% \tag{67}$$

With the allenes having electron-withdrawing groups (F, Cl, Br, CF_3 and OCF_3) on the phenyl substituents, addition takes place predominantly at the internal carbon atom. In contrast, the terminal attack is favored for the allenes having an electron-donating group (CH_3 and OCH_3) on the phenyl group [138].

In the case of the reactions of methylmalononitrile with alkoxy allenes, the carbon-carbon bond formation takes place at the position α to the oxygen atom (Eq. 68) [139]. On the contrary, in the case of a reaction of allenes having the sulfanyl group, the carbon-carbon bond formation predominantly occurs at the position γ to the sulfur atom (Eq. 68) [140]. In both cases, heteroatom substituted π-allyl palladium complexes were proposed as the intermediate. The different regioselectivities of Eq. 68 would be explained on the basis of well-accepted chemistry with respect to the heteroatom substituted allylcations and allylanions [141]. Therefore, since a carbocation is stabilized by the adjacent oxygen atom and is poorly stabilized by the adjacent sulfur atom, the nucleophilic attack onto the heteroatom substituted π-allyl ligand preferentially takes place at the position α to the oxygen atom and γ to the sulfur atom.

$$\mathrm{PhCH_2O{-}CH(C(Me)(CN)_2){-}CH{=}CH_2}\ \ 80\% \xleftarrow[\mathrm{X = O}]{\text{catalyst}} \mathrm{PhCH_2X{-}CH{=}C{=}CH_2} + \mathrm{MeCH(CN)_2} \xrightarrow[\mathrm{X = S}]{\text{catalyst}} \mathrm{PhH_2CS{-}CH{=}CH{-}CH_2C(Me)(CN)_2}\ \ 73\%\ (E/Z = 72/28)$$

catalyst = $Pd_2(dba)_3$-$CHCl_3$/dppb dppb = $Ph_2P(CH_2)_4PPh_2$

(68)

The addition of C–H bond of active methine compounds to carbon-carbon double bond in the allene moiety proceeds in intramolecular fashion in the presence of palladium catalyst, leading to the five- or six-membered carbocycles (Eq. 69) [142]. Similar intramolecular carbocyclization can be applied to the methine compounds having the acetylene moiety, leading to the five-membered *exo*-methylene cyclopentanes in good to excellent yields [143].

$$\mathrm{CH_2{=}C{=}CH(CH_2)_3CH(CN)_2} \xrightarrow[\text{THF, 70 °C, 1.5 h; dppf = Ph}_2\text{P-Fc}]{[(\eta^3\text{-}C_3H_5)PdCl]_2/\text{dppf}} \text{2-vinylcyclopentane-1,1-dicarbonitrile}\ \ 88\% \tag{69}$$

The palladium-catalyzed reaction of methyne compounds, e.g., 2-phenyl-2-cyanoacetale, with a conjugated enyne afforded the corresponding allenes in high yield (Eq. 70) [144]. The formation of allenes suggested that the C–C bond formation took place exclusively at the terminal carbon atom of the C–C double bond.

$$\text{(isopropenylacetylene)} + \text{Ph(H)C(CN)(CO}_2\text{Et)} \xrightarrow[\text{THF, 65 °C, 63 h}]{\text{Pd}_2\text{(dba)}_3\text{-CHCl}_3\text{/dppf}} \text{H}_2\text{C=C=C(CH}_3\text{)CH}_2\text{C(Ph)(CN)(CO}_2\text{Et)} \; 100\% \quad (70)$$

The reaction of active methine compounds, e.g., ethyl methylcyanoacetate and methylmalononitrile, with vinyltins using palladium catalyst gives 3-hexene derivatives (Eq. 71) [145]. Participation of two molecules of the methyne compound and two vinyl moieties of the vinyltin in this reaction is obvious, though the actual reaction pathway was not clarified.

$$\text{Bu}_2\text{Sn(CH=CH}_2\text{)}_2 + \text{Ph(H)C(CN)(CO}_2\text{Et)} \xrightarrow[\text{THF, r.t., 2 days}]{\text{Pd}_2\text{(dba)}_3\text{-CHCl}_3\text{/dppb}} \text{NC(Ph)(EtO}_2\text{C)C-CH}_2\text{CH=CHCH}_2\text{-C(Ph)(CN)(CO}_2\text{Et)} \; 81\% \quad (71)$$

When the reaction of active methine compounds with methylenecyclopropane was carried out in the presence of $Pd(PPh_3)_4$ as the catalyst, two types of ring-opening product were obtained (Eq. 72) [146]. This observation suggests that the reaction does not proceed through the trimethylenemethane-palladium intermediate, from which one ring-opening product would be formed predominantly [147].

$$\text{(methylenecyclopropane)=CH(CH}_2\text{)}_2\text{Ph} + \text{Me(H)C(CN)}_2 \xrightarrow[\text{THF, 100 °C, 2-3 days}]{\text{Pd(PPh}_3\text{)}_4} \text{(NC)}_2\text{C(Me)CH}_2\text{CH}_2\text{C(=CH}_2\text{)CH}_2\text{(CH}_2\text{)}_2\text{Ph} \; 75\% + \text{Me(NC)}_2\text{C-CH(CH}_3\text{)CH=CH(CH}_2\text{)}_2\text{Ph} \; 10\% \quad (72)$$

6 Conclusion

The use of C–H bonds is obviously one of the simplest methods in organic synthesis. From the synthetic point of view, C–H/olefin, C–H/acetylene, and C–H/CO/olefin couplings can be regarded as practical tools since these reactions exhibit high selectivity, high efficiency, and wide applicability, which are essential for practical organic synthesis. Hydroacylations of olefins and acetylenes provide unsymmetrical ketones and α,β-conjugate enones, which are highly versatile synthetic intermediates. Transition metal-catalyzed aldol and Michael addition reactions of active methylene compounds are now widely used for enantioselective and diastereoselective C–C bond formation reactions under neutral conditions.

In the past several years, the chemistry of the catalytic use of the C–H bond in organic synthesis has been rapidly expanding to various other fields, such as polymer chemistry. And in the coming decade, we look forward to fascinating new discoveries for the direct use of C–H bonds in organic synthesis.

References

1. Kleiman JP, Dubeck M (1963) J Am Chem Soc 85:1544
2. Shilov AE, Shul'pin GB (1997) Chem Rev 97:2879 and references cited therein
3. (a) Parshall GW (1970) Acc Chem Res 3:139. (b)Webster DE (1077) Adv Organomet Chem 15:147. (c) Bergman RG (1984) Science 223:902. (d) Shilov AE (1984) Activation of saturated hydrocarbons by transition metal complexes. D. Reidel, Dordrecht. (e) Crabtree RH (1985) Chem Rev 85:245. (f) Hill CL (1989) Activation and functionalization of alkanes. Wiley, New York. (g) Davies JA, Watson PL, Liebman JF, Greenberg A (1990) Selective hydrocarbon activation. VCH, New York
4. Jordan RF, Taylor DF (1989) J Am Chem Soc 111:778
5. (a) Moore EJ, Pretzer WR, O'Connell TJ, Harris J, LaBounty L, Chou L, Grimmer SS (1992) J Am Chem Soc 114:5888. (b) Moore EJ, Pretzer WR (1992) US Patent 5,081,250
6. Murai S, Kakiuchi F, Sekine S, Tanaka Y, Kamatani A, Sonoda M, Chatani N (1993) Nature 366:529
7. Hong P, Yamazaki H, Sonogashira K, Hagihara N (1978) Chem Lett 535
8. Hong P, Yamazaki H (1979) Chem Lett 1335
9. Hong P, Yamazaki H (1984) J Mol Catal 26:297
10. Diamond SE, Szalkiewicz A, Mares F (1979) J Am Chem Soc 101:490
11. Nugent WA, Ovenall DW, Holmes SJ (1983) Organometallics 2:161
12. Lin Y, Ma D, Lu X (1987) Tetrahedron Lett 28:3249
13. Lewis LN, Smith JF (1986) J Am Chem Soc 108:2728
14. Rodewald S, Jordan RF (1994) J Am Chem Soc 116:4491
15. (a) Murai S, Kakiuchi F, Sekine S, Tanaka Y, Kamatani A, Sonoda M, Chatani N (1994) Pure Appl Chem 66:1527. (b) Kakiuchi F, Sekine S, Tanaka Y, Kamatani A, Sonoda M, Chatani N, Murai S (1995) Bull Chem Soc Jpn 68:62. (c) Murai S, Chatani N, Kakiuchi F (1997) Pure Appl Chem 69:589
16. (a) Sonoda M, Kakiuchi F, Chatani N, Murai S (1995) J Organomet Chem 504:151. (b) Sonoda M, Kakiuchi F, Chatani N, Murai S (1997) Bull Chem Soc Jpn 70:3117
17. (a) Harris PWR, Woodgate PD (1996) J Organomet Chem 506:339. (b) Harris PWR, Woodgate PD (1997) J Organomet Chem 530:211
18. Grigg R, Savic V (1997) Tetrahedron Lett 38:5737
19. (a) Guo H, Tapsak MA, Weber WP (1995) Macromolecules 28:4714. (b) Guo H, Wang G, Tapsak MA, Weber WP (1995) Macromolecules 28:5686. (c) Lu P, Paulasaari JK, Weber WP (1996) Macromolecules 29:8583. (d) Wang G, Guo H, Weber WP (1996) J Organomet Chem 521:351
20. Sonoda M, Kakiuchi F, Kamatani A, Chatani N, Murai S (1996) Chem Lett 109
21. Kakiuchi F, Sonoda M, Chatani N, Murai S unpublished results
22. Kakiuchi F, Yamauchi M, Chatani N, Murai S (1996) Chem Lett 111
23. Lim Y-G, Kim YH, Kang J-B (1994) J Chem Soc Chem Commun 2267
24. Lim Y-G, Kang J-B, Kim YH (1996) J Chem Soc Parkin Trans 1 2201
25. Kakiuchi F, Tanaka Y, Sato T, Chatani N, Murai S (1995) Chem Lett 679
26. Trost BM, Imi K, Davies IW (1995) J Am Chem Soc 117:5371
27. Lim Y-G, Kang J-B, Kim YH (1996) Chem Commun 585
28. (a) Fujii N, Kakiuchi F, Chatani N, Murai S (1996) Chem Lett 939. (b) Fujii N, Kakiuchi F, Yamada A, Chatani N, Murai S (1998) Bull Chem Soc Jpn 71:285
29. Fujii N, Kakiuchi F, Yamada A, Chatani N, Murai S (1997) Chem Lett 425
30. (a) Kashiwagi K, Sugise R, Shimakawa T, Matuura T, Shirai M, Kakiuchi F, Murai S (1997) Organometallics 16:2233. (b) Murai S, Oodan K, Sugise R, Shirai M, Shimakawa T (1994) Japan Kokai Tokkyo Koho 09,531
31. (a) Rhone-Poulenc (1966) Neth Appl 6,603,115. (b) Misono A, Uchida Y, Hidai M, Kanai H (1967) J Chem Soc Chem Commun 357. (c) Misono A, Uchida Y, Hidai M, Shinohara H, Watanabe Y (1968) Bull Chem Soc Jpn 41:396. (d) Billing E, Strow CB, Pruett

RL (1968) J Chem Soc Chem Commun 1307. (e) Tsou DT, Burrington JD, Maher EA, Grasselli RK (1983) J Mol Catal 22:29
32. (a) Sakakura T, Sodeyama T, Tanaka M (1988) Chem Lett 683. (b) Sasaki K, Sakakura T, Tokunaga Y, Wada K, Tanaka M (1988) Chem Lett 685
33. (a) Collman JP, Hegedus LS, Norton JR, Finke RG (1987) Principles and applications of organotransition metal chemistry. University Science Books, California. (b) Spessard GO, Miessler GL (1997) Organometallic chemistry. Prentice-Hall, New Jersey
34. Heck RF (1985) Palladium reagents in organic syntheses. Academic Press, London
35. (a) Hong P, Cho B-R, Yamazaki H (1979) Chem Lett 339. (b) Yamazaki H, Hong P (1983) J Mol Catal 21:133. (c) Hong P, Cho B-R, Yamazaki H (1980) Chem Lett 507
36. (a) Halbritter G, Knoch F, Wolski A, Kisch H (1994) Angew Chem Int Ed Engl 33:1603. (b) Aulwurm UR, Melchinger JU, Kisch H (1995) Organometallics 14:3385. (c) Dürr U, Kisch H (1997) Synlett 1335
37. Kakiuchi F, Yamamoto Y, Chatani N, Murai S (1995) Chem Lett 681
38. Tokunaga Y, Sakakura T, Tanaka M (1989) J Mol Catal 56:305
39. Boese WT, Goldman AS (1991) Organometallics 10:782
40. Chatani N, Fukuyama T, Kakiuchi F, Murai S (1996) J Am Chem Soc 118:493
41. Chatani N, Ie Y, Kakiuchi F, Murai S (1997) J Org Chem 62:2604
42. Fukuyama T, Chatani N, Kakiuchi F, Murai S (1997) J Org Chem 62:5647
43. Ishii Y, Chatani N, Kakiuchi F, Murai S (1997) Organometallics 16:3615
44. Ishii Y, Chatani N, Kakiuchi F, Murai S (1997) Tetrahedron Lett 38:7565
45. Fisher BJ, Eisenberg R (1983) Organometallics 2:764
46. Kunin AJ, Eisenberg R (1986) J Am Chem Soc 108:535
47. Kunin AJ, Eisenberg R (1988) Organometallics 7:2124
48. Gordon EM, Eisenberg R (1988) J Mol Catal 45:57
49. Sakakura T, Tanaka M (1987) Chem Lett 249
50. Sakakura T, Tanaka M (1987) J Chem Soc Chem Commun 758
51. Sakakura T, Tanaka M (1987) Chem Lett 1113
52. Sakakura T, Sasaki K, Tokunaga Y, Wada K, Tanaka M (1988) Chem Lett 155
53. Sakakura T, Sodeyama T, Sasaki K, Wada K, Tanaka M (1990) J Am Chem Soc 112:7221
54. Boyd SE, Field LD, Partridge MG (1994) J Am Chem Soc 116:9492
55. Rosini GP, Boese WT, Goldman AS (1994) J Am Chem Soc 116:9498
56. Bridgewater JS, Lee B, Bernhard S, Schoonover JR, Ford PC (1997) Organometallics 16:5592
57. Margl P, Ziegler T, Blöchl PE (1995) J Am Chem Soc 117:12625
58. Margl P, Ziegler T, Blöchl PE (1996) J Am Chem Soc 118:5412
59. Sakaki S, Ujino Y, Sugimoto M (1996) Bull Chem Soc Jpn 69:3047
60. Jones WD, Kosar WP (1986) J Am Chem Soc 108:5640
61. Hsu GC, Kosar WP, Jones WD (1994) Organometallics 13:385
62. Jones WD, Foster GP, Putinas JM (1987) J Am Chem Soc 109:5047
63. Jones WD, Hessell ET (1990) Organometallics 9:718
64. Tanaka M, Sakakura T, Tokunaga Y, Sodeyama T (1987) Chem Lett 2373
65. Crabtree RH, Mihelcic JM, Quirk JM (1979) J Am Chem Soc 101:7738
66. Baudry D, Ephritikhine M, Felkin H, Holmes-Smith R (1983) J Chem Soc Chem Commun 788
67. Burk MJ, Crabtree RH, Parnell CP, Uriarte RJ (1984) Organometallics 3:816
68. Felkin H, Fillebeen-Khan T, Gault Y, Holmes-Smith R, Zakrzewski J (1984) Tetrahedron Lett 25:1279
69. Felkin H, Fillebeen-Khan T, Holmes-Smith R, Yingrui L (1985) Tetrahedron Lett 26:1999
70. Burk MJ, Crabtree RH, McGrath DV (1985) J Chem Soc Chem Commun 1829
71. Burk MJ, Crabtree RH (1987) J Am Chem Soc 109:8025
72. Gupta M, Hagen C, Flesher RJ, Kaska WC, Jensen CM (1996) Chem Commun 2083

73. Crabtree RH, Parnell CP (1985) Organometallics 4:519
74. Maguire JA, Boese WT, Goldman AS (1989) J Am Chem Soc 111:7088
75. Maguire JA, Goldman AS (1991) J Am Chem Soc 113:6706
76. Maguire JA, Petrillo A, Goldman AS (1992) J Am Chem Soc 114:9492
77. Wang K, Goldman ME, Emge TJ, Goldman AS (1996) J Organomet Chem 518:55
78. (a) Fujii T, Saito Y (1990) J Chem Soc Chem Commun 161. (b) Fujii T, Higashino Y, Saito Y (1993) J Chem Soc Chem Commun 517. (c) Aoki T, Crabtree RH (1993) Organometallics 12:294. (d) Xu W-W, Rosini GP, Gupta M, Jensen CM, Kaska WC, Krogh-Jespersen K, Goldman AS (1997) Chem Commun 2273
79. Nomura K, Saito Y (1988) J Chem Soc Chem Commun 161
80. Sakakura T, Sodeyama T, Tokunaga Y, Tanaka M (1988) Chem Lett 263
81. Sakakura T, Tokunaga Y, Sodeyama T, Tanaka M (1988) Chem Lett 885
82. Sakakura T, Ishida K, Tanaka M (1990) Chem Lett 585
83. Sakakura T, Abe F, Tanaka M (1990) Chem Lett 583
84. Sakakura T, Sodeyama T, Tokunaga Y, Tanaka M (1987) Chem Lett 2211
85. Gustavson WA, Epstein PS, Curtis MD (1982) Organometallics 1:884
86. Uchimaru Y, El Sayed AMM, Tanaka M (1993) Organometallics 12:2065
87. Djurovich PI, Dolich AR, Berry DH (1994) J Chem Soc Chem Commun 1897
88. Sakakura T, Tokunaga Y, Sodeyama T, Tanaka M (1987) Chem Lett 2375
89. Ishikawa M, Okazaki S, Naka A, Sakamoto H (1992) Organometallics 11:4135
90. Ishikawa M, Naka A, Ohshita J (1993) Organometallics 12:4987
91. Williams NA, Uchimaru Y, Tanaka M (1995) J Chem Soc Chem Commun 1129
92. (a) Tsuji J, Ohno K (1965) Tetrahedron Lett 3969. (b) Tsuji J, Ohno K (1969) Synthesis 157
93. Sakai K, Ide J, Oda O, Nakamura N (1972) Tetrahedron Lett 1287
94. Lochow CF, Miller RG (1976) J Am Chem Soc 98:1281
95. Campbell, Jr RE, Miller RG (1980) J Organomet Chem 186:C27
96. Campbell Jr RE, Lochow CF, Vora KP, Miller RG (1980) J Am Chem Soc 102:5824
97. Milstein D, (1982) J Chem Soc Chem Commun 1357
98. Fairlie DP, Bosnich B (1988) Organometallics 7:936
99. Fairlie DP, Bosnich B (1988) Organometallics 7:946
100. Barnhart RW, Bosnich B (1995) Organometallics 14:4343
101. Larock RC, Oertle K, Potter GF (1980) J Am Chem Soc 102:190
102. Gable KP, Benz GA (1991) Tetrahedron Lett 32:3473
103. (a) James BR, Young CG (1983) J Chem Soc Chem Commun 1215. (b) James BR, Young CG (1985) J Organomet Chem 285:321
104. Taura Y, Tanaka M, Funakoshi K, Sakai K (1989) Tetrahedron Lett 30:6349
105. Taura Y, Tanaka M, Wu X-M, Funakoshi K, Sakai K (1991) Tetrahedron 47:4879
106. Wu X-M, Funakoshi K, Sakai K (1992) Tetrahedron Lett 33:6331
107. Barnhart RW, Wang X, Noheda P, Bergens SH, Whelan J, Bosnich B (1994) J Am Chem Soc 116:1821
108. (a) Barnhart RW, Wang X, Noheda P, Bergens SH, Whelan J, Bosnich B (1994) Tetrahedron 50:4335. (b) Barnhart RW, McMorran DA, Bosnich B (1997) Chem Commun 589
109. Eilbracht P, Gersmeier A, Lennartz D, Huber T (1995) Synthesis 330
110. Dygutsch DP, Eilbracht P (1996) Tetrahedron 52:5461
111. Sattelkau T, Hollmann C, Eilbracht P (1996) Synlett 1221
112. Vora KP, Lochow CF, Miller RG (1980) J Organomet Chem 192:257
113. Isnard P, Denise B, Sneeden RPA, Cognion JM, Durual P (1982) J Organomet Chem 240:285
114. T.B. Marder, D.C Roe, and D. Milstein pointed out that these results with respect to the $RuCl_2(PPh_3)_3$-catalyzed intermolecular hydroacylation have not yet been duplicated. See footnote 4 of reference 117.
115. Okano T, Kobayashi T, Konishi H, Kiji J (1982) Tetrahedron Lett 23:4967

116. Vora KP (1983) Synth Commun 13:99
117. Marder TB, Roe DC, Milstein D (1988) Organometallics 7:1451
118. (a) Kondo T, Tsuji Y, Watanabe Y (1987) Tetrahedron Lett 28:6229. (b) Kondo T, Akazome M, Tsuji Y, Watanabe Y (1990) J Org Chem 55:1286
119. (a) Tsuji Y, Yoshii S, Ohsumi T, Kondo T, Watanabe Y (1987) J Organomet Chem 331:379. (b) Kondo T, Yoshii S, Tsuji Y, Watanabe Y (1989) J Mol Catal 50:31. (c) Kondo T, Tantayanon S, Tsuji Y, Watanabe Y (1989) Tetrahedron Lett 30:4137
120. (a) Jun C-H, Kang J-B (1993) Bull Korean Chem Soc 14:153, (1993) Chem Abstr 119:72475. (b) Jun C-H, Han J-S, Kang J-B, Kim S-I (1994) Bull Korean Chem Soc 15:204, (1994) Chem Abstr 121:133923
121. Suggs isolated and characterized the 5-membered metallacycle. See Suggs JW (1978) J Am Chem Soc 100:640
122. Lee H, Jun C-H (1995) Bull Korean Chem Soc 16:66, (1995) Chem Abstr 123:83465
123. Suggs JW (1979) J Am Chem Soc 101:489
124. (a) Jun C-H, Kang J-B, Kim J-Y (1991) Bull Korean Chem Soc 12:259, (1991) Chem Abstr 115:114722. (b) Jun C-H, Kang J-B, Kim J-Y (1993) J Organomet Chem 458:193. (c) Jun C-H, Kang J-B, Kim J-Y (1993) Tetrahedron Lett 34:6431. (d) Jun C-H, Han J-S, Kang J-B, Kim S-I (1994) J Organomet Chem 474:183
125. (a) Jun C-H, Lee H, Hong J-B (1997) J Org Chem 62:1200. (b) Jun C-H, Lee D-Y, Hong J-B (1997) Tetrahedron Lett 38:6673. (c) Jun C-H, Huh C-W, Na S-J (1998) Angew Chem Int Ed Engl 37:145
126. Lenges CP, Brookhart M (1997) J Am Chem Soc 119:3165
127. Tsuda T, Kiyoi T, Saegusa T (1990) J Org Chem 55:2554
128. Lee H, Jun C-H (1995) Bull Korean Chem Soc 16:1135, (1996) Chem Abstr 124:202420
129. Kokubo K, Matsumasa K, Miura M, Nomura M (1997) J Org Chem 62:4564
130. (a) Naota T, Taki H, Mizuno M, Murahashi S-I (1989) J Am Chem Soc 111:5954. (b) Murahashi S-I, Naota T, Taki H, Mizuno M, Takaya H, Komiya S, Mizuho Y, Oyasato N, Hiraoka M, Hirano M, Fukuoka A (1995) J Am Chem Soc 117:12436. (c) Murahashi S-I, Naota T (1996) Bull Chem Soc Jpn 69:1805
131. Gómez-Bengoa E, Cuerva JM, Mateo C, Echavarren AM (1996) J Am Chem Soc 118:8553
132. Lin Y, Zhu X, Xiang M (1993) J Organomet Chem 448:215
133. Paganelli S, Schionate A, Botteghi C (1991) Tetrahedron Lett 32:2807
134. (a) Sawamura M, Hamashima H, Ito Y (1992) J Am Chem Soc 114:8295. (b) Sawamura M, Hamashima H, Ito Y (1994) Tetrahedron 50:4439
135. Kuwano R, Miyazaki H, Ito Y (1998) Chem Commun 71
136. Nemoto H, Kubota Y, Yamamoto Y (1994) J Chem Soc Chem Commun 1665
137. (a) Yamamoto Y, Al-Masum M, Asao N (1994) J Am Chem Soc 116:6019. (b) Trost BM, Gerusz VJ (1995) J Am Chem Soc 117:5156
138. Yamamoto Y, Al-Masum M, Fujiwara N, Asao N (1995) Tetrahedron Lett 36:2811
139. Yamamoto Y, Al-Masum M (1995) Synlett 969
140. Yamamoto Y, Al-Masum M, Takeda A (1996) Chem Commun 831
141. (a) Yamamoto Y (1991) Heteroatom-stabilized allylic anions. In: Trost BM, Fleming I (eds) Comprehensive organic synthesis. Pergamon Press, Oxford, vol 2, Sect. 1.2. (b) Still WC, Macdonald TL (1973) J Org Chem 95:2715
142. Meguro M, Kamijo S, Yamamoto Y (1996) Tetrahedron Lett 37:7453
143. Tsukada N, Yamamoto Y (1997) Angew Chem Int Ed Engl 36:2477
144. Gevorgyan V, Kadowaki C, M.Salter M, Kadota I, Saito S, Yamamoto Y (1997) Tetrahedron 53:9097
145. Nakamura I, Tsukada N, Al-Masum M, Yamamoto Y (1997) Chem Comuun 1583
146. Tsukada N, Shibuya A, Nakamura I, Yamamoto Y (1997) J Am Chem Soc 119:8123
147. Trost BM, Chan DMT (1978) J Am Chem Soc 101:6432

Catalytic Activation of Methane and Ethane by Metal Compounds

Ayusman Sen

Department of Chemistry, The Pennsylvania State University, University Park, PA 16802 USA
E-mail: asen@chem.psu.edu

Methane and ethane are the most abundant and the least reactive members of the hydrocarbon family, and their selective conversion to useful chemical products is of great scientific, as well as practical, interest. This review highlights some of the recent advances in the area of low temperature, catalytic, activation and functionalization of methane and ethane. Particular emphasis has been placed on C–H and C–C activation processes leading to the formation of oxygenates.

Keywords: Methane, Ethane, C–H activation, C–C activation, Oxidation, Catalysis

1 **Introduction** 81

2 **Reactions in Strongly Acidic Media** 84

3 **Reactions in Aqueous Medium** 89

4 **Miscellaneous Radical Pathways** 90

5 **Artificial Monoxygenases** 90

References 93

1 Introduction

Methane is the most abundant and the least reactive member of the hydrocarbon family. Ethane comes second in both categories. Together, they constitute >95% of natural gas, with known reserves approaching that of petroleum [1]. A significant portion of the methane and ethane produced is not utilized because of the difficulty associated with the transportation of a flammable, low-boiling gas. Their possible use as automobile fuels is also limited by the intrinsic disadvantages of gaseous fuels, i.e., low energy content per unit volume and the hazards associated with handling and distribution. Thus, the selective conversion of methane and ethane to more useful chemical products is of great practical interest [2]. For example, three of the highest volume functionalized organics produced commercially are methanol, formaldehyde, and acetic acid, whose 1995 United States productions were 11.3×10^9, 8.1×10^9, and 4.7×10^9 lbs, respectively [3]. The current technology for the conversion of alkanes to these products involves *multistep* processes: (a) the high temperature steam reforming of alkanes to a mixture of H_2 and CO [4], (b) the high temperature conversion of the mixture of H_2 and CO to methanol [4] and, either (c) the high temperature oxidation of methanol to formaldehyde [5] or (d) the carbonylation of methanol to acetic acid [6], mainly through the "Monsanto process" [7]. Clearly, the *direct*, low temperature conversion of the lower alkanes to the above oxygenates would be far more attractive from an economical standpoint. Of particular interest would be the formation of the *same* end product(s) from different starting alkanes, thus obviating the need to separate the alkanes. For example, natural gas is principally methane with 5–10% ethane. A system that converts both methane and ethane to the same C_1 product would not require the prior separation of the alkanes. Of course, the formation of C_1 products from ethane would require the catalytic cleavage and oxidation of C–C bonds.

The lack of reactivity of methane and ethane stems from their unusually high bond energies (C–H bond energy of methane: 104 kcal/mol) and most reactions involving the homolysis of a C–H bond occur at fairly high temperatures or under photolytic conditions. Moreover, the selectivity in these reactions is usually low because of the subsequent reactions of the intermediate products, which tend to be more reactive than the alkane itself. Using methane as an example, its C–H bond energy is 10 kcal/mol higher than that in methanol. Therefore, unless methanol can be protected or removed as soon as it is formed, any oxidation procedure that involves hydrogen-atom abstraction from the substrate C–H bond would normally cause rapid overoxidation of methanol. The radical initiated chlorination of methane invariably leads to multiple chlorinations [8] (chlorination, however, is more specific in the presence of superacids [9]). In order to achieve the selective functionalization of methane and ethane, it is therefore necessary in most cases to promote a pathway that does not involve C–H bond homolysis as one of the steps. The problem is compounded by the fact that practical oxidation processes require the direct use of dioxygen as the oxidant. Because of its triplet electronic configuration, the reaction between dioxygen and alkanes most often involves unselective radical pathways [10].

In principle, the above selectivity problems can be avoided in suitably designed metal-catalyzed oxidation procedures. Transition metals, particularly those whose most stable oxidation states differ by $2e^-$, often promote nonradical pathways even in the presence of dioxygen [11]. As a bonus, metal ion catalyzed reactions usually operate at low temperatures (<200°C). The use of milder reaction conditions also avoids the loss of selectivity due to overoxidation.

While it is difficult to design a catalytic procedure for the selective functionalization of C–H bonds, it is harder still to achieve catalytic functionalization of C–C bonds even though the C–C bonds are significantly weaker than C–H bonds. Two reasons are usually cited for the general lack of C–C activation compared to corresponding C–H activation by metals [12]. First, C–C bonds are sterically less accessible to transition metal centers surrounded by ligands. Second, metal-carbon bonds tend to be weaker than metal-hydrogen bonds, again due to steric repulsions between the ligands surrounding the metal and the alkyl group bound to it. C–C cleavage is, however, commonly observed in the interaction of bare metal cations with alkanes [13]. In this case, there is no steric hindrance and the metal-carbon and metal-hydrogen bond strengths are comparable (approx. 60 kcal/mol) [13a,b, 14].

There are three basic metal-mediated alkane activation pathways [15]. The first involves the metal as an $1e^-$ oxidant (Eqs. 1, 2). From a thermodynamic standpoint, the $1e^-$ oxidation of alkanes is generally less favorable than the corresponding $2e^-$ oxidations and, therefore, require the use of either very strong oxidants or relatively high temperatures. Sometimes an auxiliary ligand on the metal may participate in the C–H bond breaking step (Eq. 2). This appears to represent nature's preferred route to alkane C–H activation. For example, it is generally accepted that in the enzyme cytochrome P-450, the species responsible for alkane C–H cleavage is a porphyrinato-Fe(V)=O complex [16]. The C–H activating species in methane monoxygenase has been less well characterized but a high-valent Fe=O species similar to that in cytochrome P-450 has been postulated [17]. The high specificity observed in enzymatic systems is presumably a result of steric restraints. More commonly, however, the organic free radicals generated will participate in a multitude of reaction pathways leading to a large number of products [10].

$$M^{N+} + R\text{-}H \rightleftharpoons M^{(N-1)+} + R\cdot + H^+ \quad (1)$$

$$M^{N+}\text{=O} + R\text{-}H \rightleftharpoons M^{(N-1)+}\text{-OH} + R\cdot \quad (2)$$

The second C–H cleavage pathway involves the oxidative addition of the C–H bond to a low-valent metal center (Eq. 3) [15]. A two-center version of the oxidative addition reaction described above has also been observed with the porphyrin Rh–Rh bonded dimer complexes [18]. In general, the presence of reactive low-valent metal species prevents the simultaneous presence of most oxidizing agents that are capable of functionalizing the bound hydrocarbyl group in the oxidative addition product. Thus, it is difficult to construct a "one pot" catalytic oxidation procedure, although other types of catalytic functionalizations, including dehydrogenation, are known for higher alkanes [15].

$$M^{N+} + R\text{-}H \rightleftharpoons M^{(N+2)+}\!\!<\!\!{}^{R}_{H} \quad (3)$$

The activation of C–H bonds by an electrophilic pathway is shown schematically (Eq. 4) and has been observed with a number of late transition metal ions [15]. The related four-center electrophilic activation by transition, lanthanide, and actinide metal centers has also been reported (Eq. 5) [15]. A driving force for the former reaction (Eq. 4) is the stabilization of the leaving group, H^+, by solvation in polar solvents. The most significant advantage of this C–H activa-

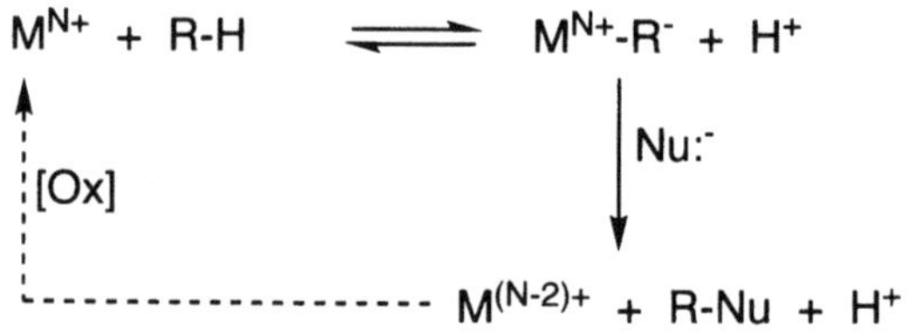

$$
\begin{array}{ccc}
M^{N+} + R\text{-}H & \rightleftharpoons & M^{N+}\text{-}R^{-} + H^{+} \\
\uparrow [Ox] & & \downarrow Nu{:}^{-} \\
\text{- - - - - - - - - -} & & M^{(N-2)+} + R\text{-}Nu + H^{+}
\end{array}
$$

($Ox = 2e^{-}$ oxidant, $Nu{:}^{-}$ = Nucleophile)

Scheme 1.

tion pathway is that the electrophilic metal center can be compatible with oxidants. Therefore, in principle, it should be possible to design a catalytic oxidation procedure that is based on an initial electrophilic C–H cleavage step, as is shown in Scheme 1.

$$M^{N+} + R\text{-}H \rightleftharpoons M^{N+}\text{-}R^{-} + H^{+} \tag{4}$$

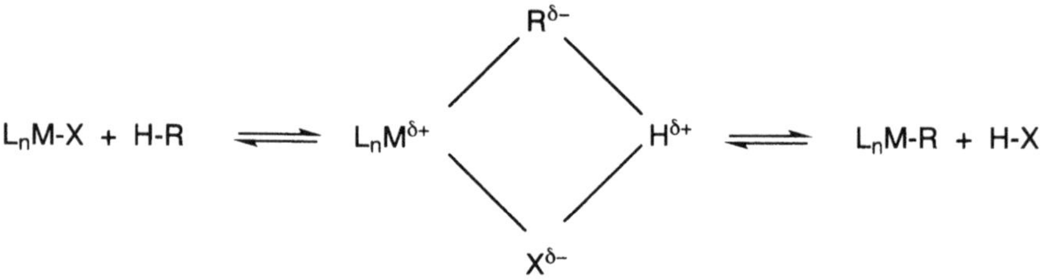

$$L_nM\text{-}X + H\text{-}R \rightleftharpoons \left[L_nM^{\delta+} \cdots R^{\delta-} \cdots H^{\delta+} \cdots X^{\delta-} \right] \rightleftharpoons L_nM\text{-}R + H\text{-}X \tag{5a}$$

$$L_nM{=}X + H\text{-}R \rightleftharpoons \left[L_nM^{\delta+} \cdots R^{\delta-} \cdots H^{\delta+} \cdots X^{\delta-} \right] \rightleftharpoons L_n\overset{R}{\overset{|}{M}}\text{—}\overset{H}{\overset{|}{X}} \tag{5b}$$

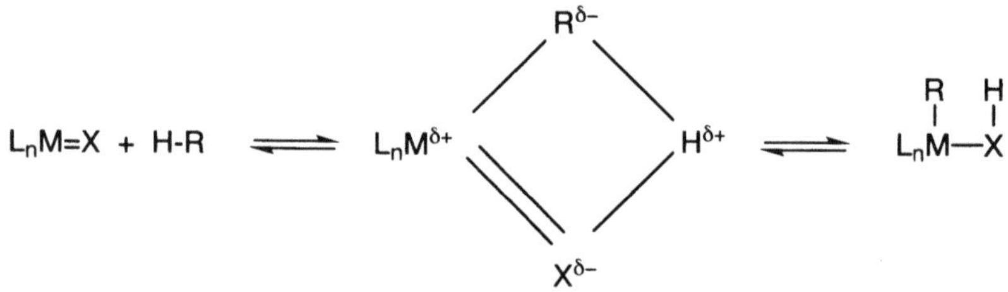

Below, we describe catalytic systems for the activation and functionalization of methane and ethane. Rather than a comprehensive review, the account highlights some of the recent advances in the area.

2
Reactions in Strongly Acidic Media

There has been a number of recent reports on metal-catalyzed electrophilic activation of methane and ethane. For two reasons much of the work in the area has been carried out in strong acids. First, the conjugate bases of strong acids are poorly coordinating, thereby enhancing the electrophilicity of the metal ion. Second, the esterification of the alcohol, the primary product of alkane oxidation, protects it from overoxidation. One impressive achievement in this area is the Hg(II) catalyzed oxidation of methane to methyl bisulfate (CH_3OSO_3H) in 100% sulfuric acid at 180°C, as described by Catalytica [19]. Both high selectivity and high conversion were achieved. The sulfuric acid served both as the solvent and the reoxidant for the metal. Although an electrophilic mechanism similar to Scheme 1 has been claimed, further studies indicate that a radical pathway, occurring at least in parallel, cannot be ruled out. For example, Sen has observed that a number of free-radical initiators, including $S_2O_8^{2-}$, also gave comparable stoichiometric yields of CH_3OSO_3H under conditions where mercury was not reoxidized by sulfuric acid [20]. The $S_2O_8^{2-}$ ion is an interesting case in point. Sen has earlier demonstrated that in water at 110°C, $SO_4^{-\bullet}$ (generated from $S_2O_8^{2-}$) abstracts a hydrogen atom from methane or ethane to form the corresponding alkyl radical, which is then converted to, inter alia, the alcohol and the bisulfate [21]. In the presence of added carbon monoxide, this radical is trapped efficiently and the resultant acyl radical is ultimately converted to a carboxylic acid. It is therefore possible that, as shown in Scheme 2 [20], one role of *all* the oxidants in sulfuric acid is to generate a methyl radical from methane by outer-sphere electron transfer followed by proton loss. The methyl radical is eventually converted to CH_3OSO_3H. In the special case of Hg(II) as the oxidant, CH_3HgOSO_3H is formed by (reversible) recombination of methyl and Hg(I) radicals. It may be noted that the proposed mechanism is similar to that suggested for some monooxygenases where both the alkyl radical and the alkyl cation, formed by electron transfers to high-valent iron-oxo species, have been implicated as intermediates [22].

The mechanistic scenario outlined above finds support in the reactivity pattern observed with ethane [20]. The radical cation formed from ethane by electron transfer would be expected to fragment some of the time by C–C cleavage (Scheme 2). The direct precedent for such a step is Olah's observation of CH_3NO_2 as the principal product in the reaction of ethane with $NO_2^+PF_6^-$ [23]. Indeed, the formation of CH_3OSO_3H was observed (up to 25% yield relative to oxidant) when ethane was contacted at 150–180°C in 98% sulfuric acid with any one of a number of radical initiators.

One possible argument against the mechanism shown in Scheme 2 is that the specific metal ions employed are not strong enough oxidants to effect a $1e^-$ oxidation of methane and ethane. However, highly electrophilic metal ions lacking donor ligands have reduction potentials significantly more positive than the corresponding ligated metal complexes. At the same time, there is a dramatic increase in oxidation potential for alkanes in strong acids [24].

The radical-initiated functionalization of methane proceeds even more readily in fuming sulfuric acid (27–33% SO_3 content by weight was employed) [25]. Thus, a variety of radical-initiators were found to convert methane to CH_3SO_3H at 90°C. For every initiator examined, the product concentration was many times the concentration of the initiator (>700 times in the case of $K_2S_2O_8$!). The

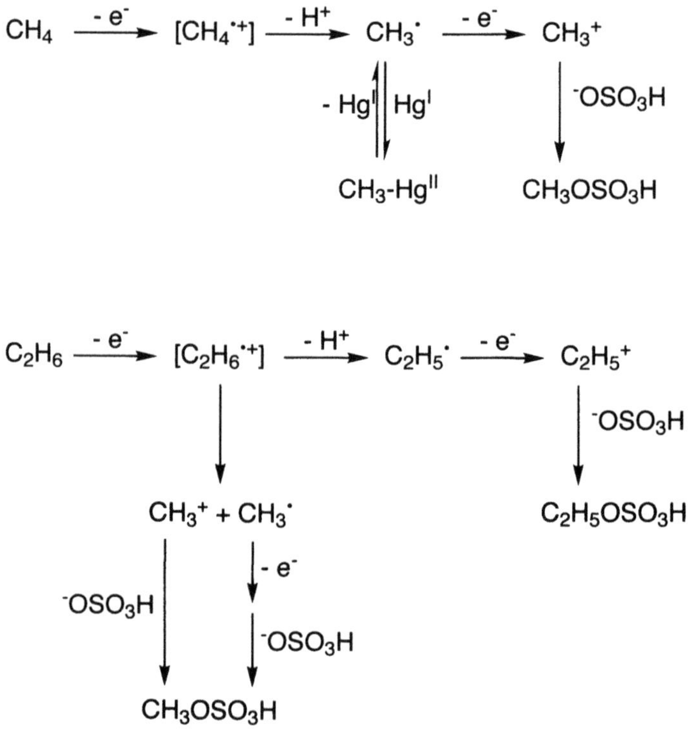

$$CH_4 \xrightarrow{-e^-} [CH_4^{\cdot +}] \xrightarrow{-H^+} CH_3^{\cdot} \xrightarrow{-e^-} CH_3^+$$

$$CH_3^{\cdot} \underset{-Hg^I}{\overset{Hg^I}{\rightleftharpoons}} CH_3\text{-}Hg^{II}$$

$$CH_3^+ \xrightarrow{^-OSO_3H} CH_3OSO_3H$$

$$C_2H_6 \xrightarrow{-e^-} [C_2H_6^{\cdot +}] \xrightarrow{-H^+} C_2H_5^{\cdot} \xrightarrow{-e^-} C_2H_5^+$$

$$C_2H_5^+ \xrightarrow{^-OSO_3H} C_2H_5OSO_3H$$

$$[C_2H_6^{\cdot +}] \longrightarrow CH_3^+ + CH_3^{\cdot}$$

$$CH_3^+ \xrightarrow{^-OSO_3H} CH_3OSO_3H$$

$$CH_3^{\cdot} \xrightarrow{-e^-} \xrightarrow{^-OSO_3H} CH_3OSO_3H$$

Scheme 2.

preference for H-atom abstraction from methane rather than the methyl group of CH_3SO_3H by the chain carrier, $CH_3SO_3^{\cdot}$, may be ascribed at least in part to its electrophilic nature (the "polar effect"). At 170°C, CH_3SO_3H was quantitatively converted to CH_3OSO_3H even in the absence of an initiator. It was also possible to directly convert methane to CH_3OSO_3H by using a radical-initiator and running the reaction at 170°C.

A significant body of work on metal-mediated electrophilic C–H activations has also been carried out in perfluorocarboxylic acids. These build upon Sen's early report on the Pd(II)-catalyzed electrophilic activation and conversion of methane to methyl ester by H_2O_2 in trifluoroacetic acid/anhydride mixture [26]. One noteworthy result in the area is the catalytic carbonylation of alkanes, including methane, in trifluoroacetic acid as reported by Fujiwara [27]. For the most part, the oxidant used was the $S_2O_8^{2-}$ ion. Three distinct catalysts, Pd(II), Pd(II)+Cu(II), and Cu(II), were employed. Mechanistic studies, including the examination of the propensity towards ring versus benzylic attack in xylene and toluene, appear to indicate that the first two catalysts activate C–H bonds through an electrophilic pathway whereas alkyl radicals are involved when Cu(II) alone is used as the catalyst. Fujiwara has also employed the copper system for the catalytic aminomethylation of alkanes, including ethane, by *tert*-amine N-oxides [27a].

Moiseev has reported the Co(II)/(III)-catalyzed oxidation of methane and ethane to alcohol derivatives by dioxygen in trifluoroacetic acid [28]. Interestingly, as in the case of oxidation in sulfuric acid, a significant amount of C–C cleavage products were obtained from ethane. A mechanism similar to that shown in Scheme 2 was proposed [28a].

In many of the oxidation reactions carried out in perfluorocarboxylic acids, the corresponding anhydride was added to rapidly esterify the alcohol derived from the alkane. However, as with sulfuric acid, the mechanism of metal-mediated C–H activations in such solvent systems should be approached with caution. For example, Sen has discovered that in the presence of a radical-initiator (e.g., H_2O_2), perfluorocarboxylic anhydrides act as oxidants towards ethane forming the mixed anhydride, $CH_3CH_2COOCOR_f$, and the ketone, $CH_3CH_2COR_f$, in varying ratios [29]. For a fixed amount of initiator, the amount of products formed increased with increasing amount of anhydride employed and was always higher than the initiator added. In particular, with $PbEt_4$ close to 500 equivs. of products were formed from ethane for every equiv. of $PbEt_4$ employed!

The mechanism of this curious reaction involves the formation of $C_2H_5^{\bullet}$ radical, which then attacks $(R_fCO)_2O$ at one of the carbonyl carbons to form an alkoxy radical (Scheme 3) [29]. The attack by the $C_2H_5^{\bullet}$ radical occurs at the most electron deficient site because of the alkyl radical's nucleophilic nature [30]. The alkoxy radical formed undergoes the well-known β-bond cleavage reaction. If the R_f-CO bond is cleaved, the product is the mixed anhydride. On the other hand, if the C(O)-O bond is broken, the ketone is produced. Both pathways produce the $R_f^{\bullet}$ radical: the first directly, and the second by formation and subsequent decarboxylation of the $R_fCO_2^{\bullet}$ radical. The $R_f^{\bullet}$ radical then continues the chain-reaction by abstracting a hydrogen from C_2H_6 forming the $C_2H_5^{\bullet}$ radical and R_fH.

Interestingly, unlike ethane, neither methane nor propane is able to participate in this reaction sequence, the former because the C–H bond of methane is too strong to undergo significant hydrogen-atom abstraction by the R_f. radical and the latter because only primary alkyl radicals are sufficiently reactive to attack $(R_fCO)_2O$. Thus, Sen's observation of Pd(II)-catalyzed conversion of methane to methanol derivative by H_2O_2 in trifluoroacetic acid/anhydride mixture was not complicated by the above reaction [26, 29].

Finally, there is a recent report on the-gas phase oxychlorination of methane using a combination of $PdCl_2$-heteropolyacids as catalysts [31]. The proposed mechanism involves an electrophilic attack on the alkane (Eq. 4). The chlorination of methane catalyzed by platinum supported on superacidic sulfated zirconia has also been reported [32].

Initiation:

$R^{\bullet} + CH_3CH_3 \longrightarrow RH + CH_3CH_2^{\bullet}$

Propagation:

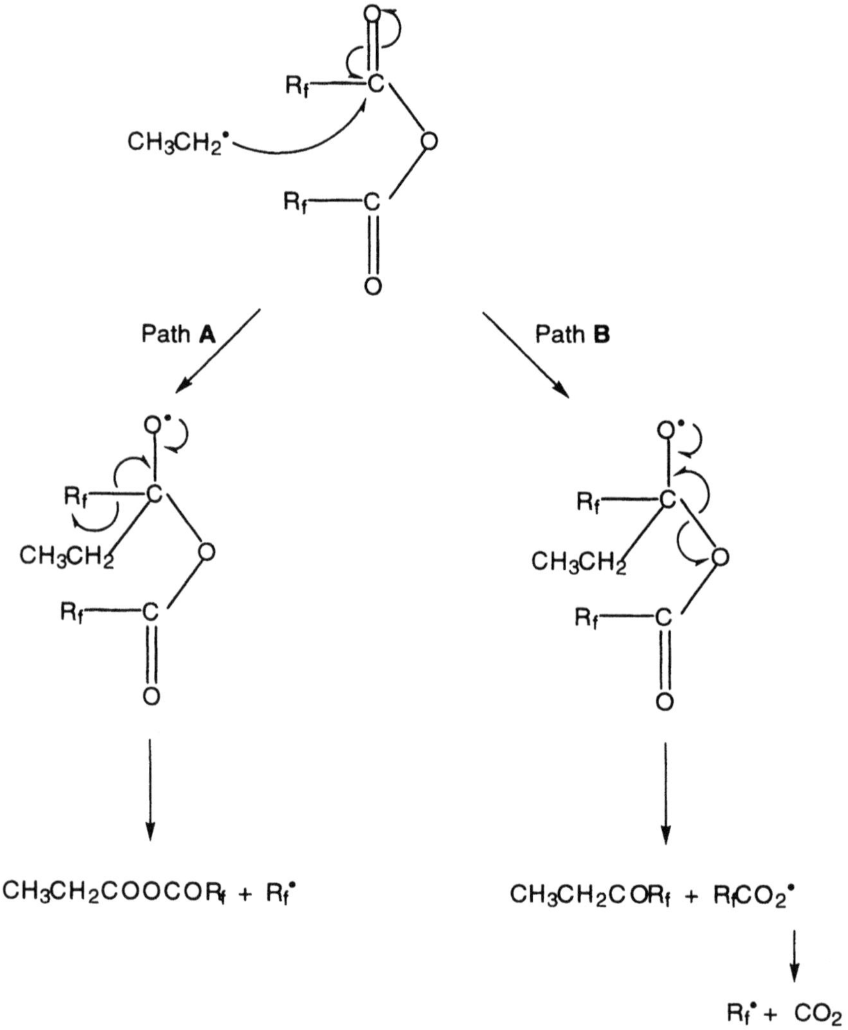

$CH_3CH_2^{\bullet}$ + $R_fC(O)OC(O)R_f$

Path **A**

$CH_3CH_2COOCOR_f + R_f^{\bullet}$

Path **B**

$CH_3CH_2COR_f + R_fCO_2^{\bullet}$

$\downarrow$

$R_f^{\bullet} + CO_2$

$CF_3^{\bullet} + CH_3CH_3 \longrightarrow CF_3H + CH_3CH_2^{\bullet}$ ($\Delta H^0 = -5$ kcal/mol at 298K)

Scheme 3.

3
Reactions in Aqueous Medium

Electrophilic C–H activations can also be effected in water. At first glance, water would appear to be particularly unpromising as a solvent for such reactions. Because of their extremely poor coordinating ability (no fully characterized alkane complex is known [33]) alkanes should not be able to compete with water for coordination sites. Moreover, the intermediate metal-alkyl species would be prone to hydrolytic decomposition. In one respect, however, water is almost an ideal medium for C–H functionalization: the O–H bond energy exceeds the corresponding C–H bond energy of even methane. Indeed, the selective oxidation of methane to methanol is carried out by methane monoxygenase in aqueous medium [17].

Shilov and his coworkers were the first to demonstrate metal-mediated alkane functionalization in water [15f]. They showed that simple Pt(II) complexes, such as $PtCl_4^{2-}$, will activate and oxidize the C–H bonds of alkanes, including methane and ethane. Sen [34], Bercaw and Labinger [35], and Horváth [36] have followed up on aspects of Shilov's work and have shown that a wide variety of substrates including methane can be functionalized with unusual selectivity through the mechanism outlined in Scheme 1. Thus, although the homolytic C–H bond energy of methane is 10 kcal/mol higher than that in methanol, a C–H bond of methanol would not be expected to be significantly more susceptible to *electrophilic* cleavage than that of methane. Indeed, Sen has observed that in water at 100°C, the rate constant for the oxidation of methane to methanol by the $PtCl_4^{2-}/PtCl_6^{2-}$ combination (the Pt(IV) species acts merely as a reoxidant for the $Pt^0 \rightarrow Pt^{II}$ step, see Scheme 1) was only one-seventh of that for methanol overoxidation by the same system [34a]. The observed similarity in rates is even more striking given the much higher binding ability of methanol to the Pt(II) center. Moving to substrates with C–H bonds somewhat weaker than that in methane resulted in actual *reversal* of commonly observed selectivity. Thus, the relative rate of C–H bond activation by the Pt(II) ion decreased in the order $H\text{-}CH_2CH_3 > H\text{-}CH_2CH_2OH > H\text{-}CH(OH)CH_3$, i.e., an order that is exactly *opposite* of that expected on the basis homolytic C–H bond energies [34a]. On a practical level, this showed that the direct conversion of ethane to ethane-1,2-diol is possible!

While it has been generally assumed that heterolytic C–H bond cleavage is involved in the Shilov system (Eq. 4), the possibility that C–H activation proceeds through an oxidative addition step (Eq. 3) resulting in the intermediacy of a Pt(IV)(alkyl)(hydride) has been raised based on studies of model systems [37].

The activation and functionalization of C–H bonds by the Pt(II) ion is particularly attractive because of the unusual regioselectivity, high oxidation level specificity, and the mildness of reaction conditions. Nevertheless, thus far it suffers from one crippling drawback: dioxygen cannot be used efficiently as the reoxidant for the Pt^0 formed from Pt(II) during substrate oxidation [38].

4
Miscellaneous Radical Pathways

Two examples of low temperature, catalytic, methane oxidation by hydrogen peroxide should be included in this section. The first involves conversion to methanol using cis-[Ru(2,9-dimethyl-1,10-phenanthroline)(solvent)$_2$](PF$_6$)$_2$ as the catalyst [39]. A ruthenium-oxo species has been proposed as the C–H activating species. In the second report, conversion of methane to methyl hydroperoxide is claimed [40]. The catalyst is a combination of [NBu$_4$]VO$_3$ and pyrazine-2-carboxylic acid. While the mechanism is uncertain, the actual oxidant is believed to be dioxygen with HO• derived from hydrogen peroxide acting as the initiator.

Finally, Crabtree has reported the gas-phase mercury photosensitized reaction of methane with ammonia to yield methylene imine as the ultimate product [41]. Higher imines are also produced if the gas-phase residence time of methylene imine is prolonged.

5
Artificial Monoxygenases

Recently, Sen has reported two catalytic systems which *simultaneously* activate dioxygen and alkane C–H and C–C bonds, resulting in the direct oxidations of alkanes. In the first system, metallic palladium was found to catalyze the oxidation of methane and ethane by dioxygen at 70–110°C in the presence of carbon monoxide [42]. In aqueous medium, formic acid was the observed oxidation product from methane while acetic acid, together with some formic acid, was formed from ethane [42a]. *No* alkane oxidation was observed in the absence of added carbon monoxide. The essential role of carbon monoxide in achieving "difficult" alkane oxidation was shown by a competition experiment between ethane and ethanol, both in the presence and absence of carbon monoxide. In the absence of added carbon monoxide, only ethanol was oxidized. When carbon monoxide was added, almost half of the products were derived from ethane. Thus, the more inert ethane was oxidized *only* in the presence of added carbon monoxide.

Studies indicated that the overall transformation encompasses three catalytic steps in tandem (Scheme 4) [42a]. The first is the water gas shift reaction involving the oxidation of carbon monoxide to carbon dioxide with the simultaneous formation of dihydrogen. It is possible to bypass this step by replacing carbon monoxide with dihydrogen. The second catalytic step involves the combination of dihydrogen with dioxygen to yield hydrogen peroxide [43] (or its equivalent). The final step involves the metal catalyzed oxidation of the substrate by hydrogen peroxide (or its equivalent).

While acetic acid was formed in good yield from ethane, the analogous formation of formic acid from methane proceeded only in low yield because of the general instability of the latter acid under the reaction conditions. Since formic acid is a much less desirable product from methane than is methanol, the possibility of halting the oxidation of methane at the methanol stage was examined.

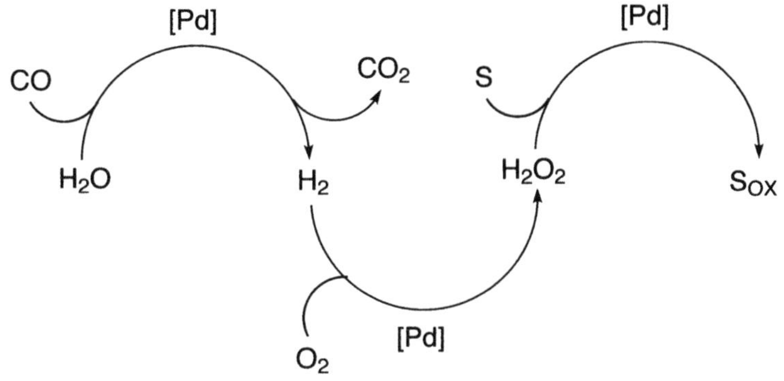

(S = substrate, S_{OX} = oxidized substrate)

Scheme 4.

Simply changing the solvent in the Pd-based catalytic system from water to a mixture of water and a perfluorocarboxylic acid (some water was necessary for the reaction, see Scheme 4) had no significant effect on product composition: formic acid was still the principal product from methane. However, the addition of copper (I) or (II) chloride to the reaction mixture had a dramatic effect. Methanol and its ester now became the preferred products, with virtually no acetic and little formic acid being formed [42b]! The activation parameters for the overall reaction determined under the condition when the rate was first-order in both methane and carbon monoxide were: $A=2\times10^4$ s^{-1}; $E_a=15.3$ kcalmol^{-1}. Since methyl trifluoroacetate is both volatile and easily hydrolyzed back to the acid and methanol, it should be possible to design a system where the acid is recycled and methanol is the end-product.

In the second (slower) system, $RhCl_3$, in the presence of several equivalents of Cl^- and I^- ions, was found to catalyze the direct functionalization of methane in the presence of carbon monoxide and dioxygen at 80–85°C [44]. The reaction proceeded in water to give acetic acid as the principal product [44a]. However, a much higher rate was observed in a 6:1 (v/v) mixture of perfluorobutyric acid and water with the products being methanol and acetic acid [44b]. It is possible to selectively form *either* methanol *or* acetic acid by a simple change in the solvent system. The ratio of alcohol derivative to the corresponding higher acid may be assumed to be a function of the relative rates of nucleophilic attack versus carbon monoxide insertion into a common Rh-alkyl bond (i.e., k_{Nu}/k_{CO}, see Scheme 5). While, to a first-order approximation, k_{CO} is likely to be independent of the solvent, k_{Nu} would depend on the nature of the nucleophile derived from the solvent. Presumably, the perfluorobutyrate ion is a better nucleophile than water since more of the alcohol derivative was formed in perfluorobutyric acid-water mixture than in pure water. This also explains why acetic acid was once again the major product when the perfluorobutyrate ion was tied up as the ester.

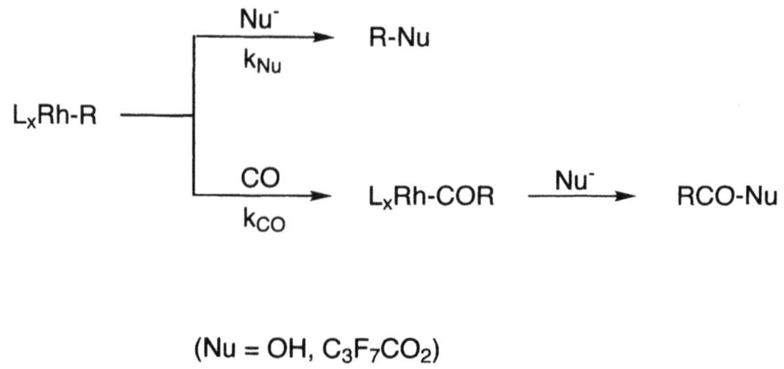

$$L_xRh\text{-}R \begin{cases} \xrightarrow[k_{Nu}]{Nu^-} R\text{-}Nu \\ \xrightarrow[k_{CO}]{CO} L_xRh\text{-}COR \xrightarrow{Nu^-} RCO\text{-}Nu \end{cases}$$

$(Nu = OH, C_3F_7CO_2)$

Scheme 5.

Consistent with the mechanistic scenario shown in Scheme 5 was also the observation that the ratio of acetic acid to methanol derivative formed from methane increased with increasing pressure of CO although the overall reaction was sharply inhibited at high CO pressures.

In addition to Sen's work on the rhodium-catalyzed oxidative carbonylation of methane, Grigoryan has also reported a similar reaction in acetic acid [45]. Predictably, the reaction rate is in-between that observed in pure water and in the perfluorocarboxylic acid-water mixture. Finally, Otsuka has reported the oxidative carbonylation of methane to acetic acid by rhodium-doped iron phosphate [46].

The Pd/Cu and the Rh-based systems show similar selectivity patterns that are, for the most part, without precedent. For example, in both cases, methane is *significantly more reactive* (at least 5 times) than methanol [42b, 44]. However, this does not take into account the increase in the C–H bond energy when methanol is converted to the ester (the following C–H bond-energy data illustrate the point: $H\text{-}CH_2OH$, 94 kcal/mol; $H\text{-}CH_2OCOC_6H_5$, 100.2 kcal/mol). For the Rh-based system, even methyl iodide was found to be less reactive than methane [44b]!

A more interesting reactivity pattern exhibited by these two systems is their preference for C–C cleavage over C–H cleavage for higher alkanes [42b, 44b]. Indeed, we are unaware of any other catalytic system that effects the oxidative cleavage of alkane C–C bonds under such mild conditions. For example, the Rh-based system converts ethane to a mixture of methanol, ethanol, and acetic acid, with the ratio of products formed through C–H versus C–C cleavage of approx. 0.6 on a per bond basis [44b]. As with methanol, control experiments indicated ethane is more reactive than ethanol. Additionally, neither ethanol nor acetic acid is the precursor to methanol. Finally, even part of the acetic acid is formed by initial C–C cleavage of ethane followed by carbonylation of the resultant C_1 fragment. For C_4 and higher alkanes, C–C cleavage products were *virtually all* that were observed; specially noteworthy was the formation of ethanol from n-butane, which indicates that vicinal diols are not the precursors to the C–C cleavage products. The above reactivity profile exhibited by the two systems, together with other observations, appears to be inconsistent with the intermediacy of *free* alkyl radicals in the oxidation process.

A curious aspect of the Pd/Cu and Rh-based systems is that, apart from their ability to simultaneously activate both dioxygen and alkane, both require a coreductant (carbon monoxide) [42, 44]. Thus, there is a striking resemblance with monooxygenases [16, 17]. In nature, while the dioxygenases utilize the dioxygen molecule more efficiently, it is the monooxygenases that carry out "difficult" oxidations, such as alkane oxidations. In the latter, one of the two oxygen atoms of dioxygen is reduced to water in a highly thermodynamically favorable reaction and the free-energy gained thereby is employed to generate a high-energy oxygen species, such as a metal-oxo complex, from the second oxygen atom (Eq. 6). Several other systems are also designed on this premise. This includes the "Gif" system [47], as well as a recently reported Eu-based system for the oxidation of methane to methanol that uses zinc as the coreductant [48]. In at least the metallic Pd-based system, the coreductant, carbon monoxide, was employed to generate dihydrogen (Eq. 7), the latter being formally equivalent to $2H^{+}+2e^{-}$ that is employed in the biological systems (Eqs. 6, 8).

$$O_2 + 2\,H^+ + 2e^- \longrightarrow H_2O + [O] \qquad (6)$$

$$CO + H_2O \longrightarrow CO_2 + H_2 \qquad (7)$$

$$O_2 + H_2 \longrightarrow H_2O + [O] \qquad (8)$$

How general is this requirement for a coreductant (e.g., CO or H_2) in achieving "difficult" catalytic hydrocarbon oxidations by dioxygen? Sen's work has provided two examples of catalytic systems that operate in this manner (i.e., as monoxygenase analogs) [42, 44]. There have been other recent publications on catalytic systems for the oxidation of hydrocarbons, including olefins and aromatics, that also call for either CO or H_2 as the coreductant [49]. While, from a practical standpoint, it is more desirable for both oxygen atoms of O_2 to be used for substrate oxidation, there appears to be no currently known catalytic system that operates as an artificial "dioxygenase" under mild conditions towards "difficult" substrates, such as those possessing unactivated primary C–H bonds.

References

1. (a) Axelrod MG, Gaffney AM, Pitchai R, Sofranko JA (1994) In: Curry-Hyde HE, Howe RF (eds) Natural gas conversion II; Elsevier, Amsterdam, p 93. (b) Masters CD, Root DH, Attanasi ED (1991) Science 253:146. (c) Starr C, Searl MF, Alpert S (1992) Science 256:981
2. Recent reviews: (a) Sen A (1996) In: Herrmann WA, Cornils B (eds) Applied homogeneous catalysis with organometallic compounds, vol 2. VCH, Weinheim, p 1081. (b) Olah GA, Molnár A (1995) Hydrocarbon chemistry. Wiley, New York. (c) Crabtree RH (1995) Chem Rev 95:987. (d) Labinger JA (1995) Fuel Process Technol 42:325. (e) Hall

TJ, Hargreaves JSJ, Huchings GJ, Joyner RW, Taylor SH (1995) Fuel Process Technol 42:151. (f) Fierro JLG (1993) Catalysis Lett 22:67. (g) Srivastava RD, Zhou P, Stiegel GJ, Rao VUS, Cinquegrane G (1992) Catalysis (London) 9:183. (h) Brown MJ, Parkynes ND (1991) Catalysis Today 8:305
3. Chemical & Engineering News April 8, 1996, p 17
4. (a) Cheng WH, Kung HH (eds) (1994) Methanol production and use. Marcel Dekker, New York. (b) Wade LE, Gengelbach RB, Trumbley JL, Hallbauer WL (1981) In: Kirk-Othmer Encyclopedia of Chemical Technology, vol 15. Wiley, New York, p 398
5. Gerberich HR, Stautzenberger AL, Hopkins, WC (1980) In: Kirk-Othmer Encyclopedia of Chemical Technology, vol 11. Wiley, New York, p 231
6. (a) Agreda VH, Zoeller JR (eds) (1993) Acetic acid and its derivatives. Marcel Dekker, New York. (b) Wagner FS (1978) In: Kirk-Othmer Encyclopedia of Chemical Technology, vol 1. Wiley, New York, p 124
7. Review: Forster D (1979) Adv Organomet Chem 17:255
8. (a) March J (1985) Advanced organic chemistry. Wiley, New York, p 620 and references therein. (b) Poutsma ML (1973) In: Kochi JK (ed) Free radicals, vol II. Wiley, New York, p 159
9. Olah G (1987) Acc Chem Res 20:422
10. Reviews: (a) Parshall GW, Ittel SD (1992) Homogeneous catalysis. Wiley, New York, p 237. (b) Howard JA (1973) In: reference 8b, p 3
11. Reviews: (a) Drago RS (1992) Coord Chem Rev 117:185. (b) Simándi LI (1992) Catalytic activation of dioxygen by metal complexes. Kluwer Academic, Dordrecht, p 74
12. (a) Crabtree RH (1992) In: Patai S, Rappoport Z (eds) The chemistry of alkanes and cycloalkanes. Wiley, New York, p 653. (b) Halpern J (1985) Inorg Chim Acta 100:41. (c) Halpern J (1982) Acc Chem Res 15:238
13. Reviews: (a) Armentrout PB (1990) In: Davies JA, Watson PL, Greenberg A, Liebman JF (eds) Selective hydrocarbon oxidation and functionalization. VCH, New York, p 467. (b) Armentrout PB, Beauchamp JL (1989) Acc Chem Res 22:315. (c) Eller K, Schwarz H (1991) Chem Rev 91:1121. (d) Schwarz H (1989) Acc Chem Res 22:282
14. Simoes JAM, Beauchamp JL (1990) Chem Rev 90:629
15. Recent reviews: (a) references 2a–d, 12a. (b) Arndtsen BA, Bergman RG, Mobley TA, Peterson TH (1995) Acc Chem Res 28:154. (c) Davies JA, Watson PL, Greenberg A, Liebman JF (eds) (1990) Selective hydrocarbon oxidation and functionalization. VCH, New York, Chaps 1–5. (d) Hill CL (ed) (1989) Activation and functionalization of alkanes. Wiley, New York. (e) Shilov AE, Shul'pin GB (1997) Chem Rev 97:2879. (f) Shilov AE (1984) Activation of saturated hydrocarbons by transition metal complexes. D. Reidel, Dordrecht
16. Reviews: (a) Groh SE, Nelson MJ (1990) In: reference 15c, p 305. (b) Valentine JS (1994) In: Bertini I, Gray HB, Lippard SJ, Valentine JS (eds) Bioinorganic chemistry. University Science Books, Mill Valley, CA, p 253. (c) Mansuy D, Battioni P (1993) In: Reedijk J (ed) Bioinorganic catalysis. Marcel Dekker, New York, p 395. (d) Omura T, Ishimura Y, Fujii-Kuriyama Y (eds) (1993) Cytochrome P-450. VCH, New York, p 17
17. Reviews: (a) Liu KE, Lippard SJ (1995) Adv Inorg Chem 42:263. (b) Feig AL, Lippard SJ (1994) Chem Rev 94:759. (c) Que L (1993) In: Reedijk J (ed) Bioinorganic catalysis. Marcel Dekker, New York, p 347
18. (a) Zhang XX, Wayland BB (1994) J Am Chem Soc 116:7897. (b) Wayland BB, Ba S, Sherry AE (1991) J Am Chem Soc 113:5305
19. Periana RA, Taube DJ, Evitt ER, Löffler DG, Wentrcek, PR, Voss G, Masuda T (1993) Science 259:340. See also: Snyder JC, Grosse AV (1950) U.S. Patent 2,493,038
20. Sen A, Benvenuto MA, Lin M, Hutson AC, Basickes N (1994) J Am Chem Soc 116:998
21. Lin M, Sen A (1992) J Chem Soc, Chem Commun 892
22. (a) Valentine AM, Wilkinson B, Liu KE, Komar-Panicucci S, Priestley ND, Williams PG, Morimoto H, Floss HG, Lippard SJ (1997) J Am Chem Soc 119:1818. (b) Newcomb M, Le Tadic-Biadatti MH, Chestney DL, Roberts ES, Hollenberg PF (1995) J Am Chem Soc 117:12085

23. Olah GA, Lin HC (1971) J Am Chem Soc 93:1259
24. Review: Fabre PL, Devynck J, Trémillon B (1982) Chem Rev 82:591
25. Basickes N, Hogan TE, Sen A (1996) J Am Chem Soc 118:13111
26. (a) Sen A (1991) Platinum Metals Rev 35:126. (b) Kao LC, Hutson AC, Sen A (1991) J Am Chem Soc 113:700
27. Reviews: (a) Fujiwara Y, Takaki K, Taniguchi Y (1996) Synlett 591. (b) Nakata K, Yamaoka Y, Miyata T, Taniguchi Y, Takaki K, Fujiwara Y (1994) J Organomet Chem 473:329
28. (a) Stolarov IP, Vargaftik MN, Shishkin DI, Moiseev II (1991) J Chem Soc, Chem Commun 938. (b) Vargaftik MN, Stolarov IP, Moiseev II (1990) J Chem Soc, Chem Commun 1049
29. Hogan T, Sen A (1997) J Am Chem Soc 119:2642
30. Motherwell WB, Crich D (1992) Free radical chain reactions in organic synthesis. Academic Press, London, p 4
31. Volkova LK, Tret'yakov VP (1995) Theoretical Experimental Chem 31:27
32. Batamack P, Bucsi I, Molnár A, Olah, GA (1994) Catal Lett 25:11
33. Review: Hall C, Perutz RN (1996) Chem Rev 96:3125
34. (a) Sen A, Benvenuto MA, Lin M, Hutson AC, Basickes N (1994) J Am Chem Soc 116:998. (b) Sen A, Lin M, Kao LC, Hutson AC (1992) J Am Chem Soc 114:6385. (c) Kao LC, Sen A (1991) J Chem Soc, Chem Commun 1242. (d) Basickes N, Sen A (1995) Polyhedron 14:197. (e) Hutson AC, Lin M, Basickes N, Sen A (1995) J Organomet Chem 504:69
35. (a) Labinger JA, Herring AM, Lyon DK, Luinstra GA, Bercaw JE, Horváth IT, Eller K (1993) Organometallics 12:895. (b) Luinstra GA, Labinger JA, Bercaw JE (1993) J Am Chem Soc 115:3004. (c) Luinstra GA, Wang L, Stahl SS, Labinger JA, Bercaw JE (1994) Organometallics 13:755. (d) Luinstra GA, Wang L, Stahl SS, Labinger JA, Bercaw JE (1995) J Organomet Chem 504:75
36. Horváth IT, Cook RA, Millar JM, Kiss G (1993) Organometallics 12:8
37. (a) Wick DD, Goldberg KI (1997) J Am Chem Soc 119:10235. (b) Stahl SS, Labinger JA, Bercaw JE (1996) J Am Chem Soc 118:5961. (c) Zamashchikov VV, Popov VG, Rudakov ES (1994) Kinet Katal 35:436. (d) Theoretical study: Siegbahn PEM, Crabtree RH (1996) J Am Chem Soc 118:4442
38. (a) Freund MS, Labinger JA, Lewis NS, Bercaw JE (1994) J Mol Catal 87:L11. (b) Geletii Yu V, Shilov AE (1983) Kinet Katal 24:486
39. Goldstein AS, Drago RS (1991) J Chem Soc, Chem Commun 21
40. Nizova GV, Süss-Fink G, Shul'pin GB (1997) J Chem Soc, Chem Commun 397
41. (a) Michos M, Krajnik J, Sassano C, Crabtree RH (1993) Angew Chem, Int Ed Engl 32:1491. (b) Krajnik J, Michos M, Crabtree RH (1993) New J Chem 17:805
42. (a) Lin M, Sen A (1992) J Am Chem Soc 114:7307. (b) Lin M, Hogan TE, Sen A (1997) J Am Chem Soc 119:6048
43. Gosser LW (1987) U.S. Patent 4,681,751
44. (a) Lin M, Sen A (1994) Nature 368:613. (b) Lin M, Hogan TE, Sen A (1996) J Am Chem Soc 118:4574
45. Chepaikin EG, Boiko GN, Bezruchenko AP, Lescheva AA, Grigoryan EA (1997) Doklady Phys Chem 353:120
46. Wang Y, Katagiri M, Otsuka K (1997) J Chem Soc, Chem Commun 1187
47. Barton DHR, Doller D (1992) Acc Chem Res 25:504
48. Yamanaka I, Soma M, Otsuka K (1996) Chemistry Lett 565
49. Representative examples: (a) Tabushi I (1988) Coord Chem Rev 86:1. (b) Otake M (1995) Chemtech 36. (c) Miyake T, Hamada M, Sasaki Y, Oguri M (1995) Appl Catal A: General 131:33. (d) Teranishi T, Toshima N (1995) J Chem Soc, Dalton Trans 979. (e) Wang Y, Otsuka K (1995) J Catal 155:256

Cleavage of Carbon–Carbon Single Bonds by Transition Metals

Masahiro Murakami* and Yoshihiko Ito

Department of Synthetic Chemistry and Biological Chemistry
Kyoto University, Yoshida, Kyoto 606-8501, Japan
E-mail: murakami@sbchem.kyoto-u.ac.jp

Cleavage of carbon-carbon bonds by transition metals under homogeneous conditions has recently received much scientific and technological interest. In this review, an overview of this field is presented. The first part deals with stoichiometric reactions involving carbon-carbon bond breaking. The second part features catalytic reactions, especially those related to organic synthesis.

Keywords: Activation, C–C bond, Transition metal, Cleavage, Oxidative addition, β-carbon elimination, Directionality, Homogeneous

1 Introduction 97

2 Stoichiometric Reactions Involving C–C Bond Cleavage 98

2.1 Insertion of a Transition Metal into a C–C Bond 98
2.1.1 Utilization of Ring Strain 99
2.1.2 Utilization of a Carbonyl Functionality 104
2.1.3 Utilization of a Pincer-Type Chelating Ligand 107
2.1.4 Utilization of Aromatization 109
2.1.5 Miscellaneous Types of Metal Insertion 110
2.2 β-Carbon Elimination 111
2.3 Miscellaneous Stoichiometric Reactions 118

3 Catalytic Reactions Involving C–C Bond Cleavage 119

4 Perspective 126

References 127

1 Introduction

A wide variety of organic functionalities can be activated by transition metal complexes. However, carbon-carbon single bonds are arguably one of the least reactive "functional" groups. The robust nature of carbon-carbon single bonds presents a fundamental challenge to organometallic chemists. Furthermore, petroleum plays a significant role in our daily lives, making this challenge also of technological interest, since the selective activation of carbon-carbon bonds is crucial for petroleum refining and transformation.

This review will focus upon the subset of interesting reactions involving the apparent cleavage of carbon-carbon single bonds promoted by transition metals under homogeneous conditions. An excellent previous review should also be referred to [1]. Reactions in the presence of heterogeneous catalysts or in the vapor phase with naked metal ions [2] are outside the scope of this review. Reactions breaking carbon-carbon double bonds, such as olefin metathesis, are not included herein.

2 Stoichiometric Reactions Involving C–C Bond Cleavage

2.1 Insertion of a Transition Metal into a C–C Bond

Oxidative addition of a C–C bond to a transition metal or, in other words, the insertion of a transition metal into a C–C bond provides a direct method for C–C bond cleavage. Reactions involving this elementary step, however, are still difficult to achieve due to the inertness of C–C σ-bonds. The lack of reactivity of carbon-carbon single bonds can be attributed to their thermodynamic stability as well as kinetic inertness. A carbon-carbon single bond is thermodynamically stable; the dissociation energy is around 356 kJ/mol (85 kcal/mol). Through oxidative addition of a C–C bond onto a transition metal, less stable M–C bonds [around 293 kJ/mol (70 kcal/mol)] are formed at the expense of a more stable C–C bond. Another reason for the difficulty of breaking a C–C single bond is the constrained directionality of its σ-orbital. A comparison of the interactions of metal orbitals with non-polarized bonds like C–C single and double bonds and C–H single bonds is shown in Fig. 1. The π-orbital of a C–C double bond is oriented sideways, making its interaction with a metal orbital facile. Although the σ-orbital connecting a hydrogen and a carbon atom lies along the bond axis, the constituent 1s orbital of the hydrogen atom is spherical. It has no other substituents except the bonded carbon, making an end-on approach to a metal sterically viable. In fact, Crabtree et al. proposed that the trajectory of oxidative addition of a C–H bond to a metal begins with an end-on approach prior to side-on coordination [3]. In contrast, the σ-orbital of a carbon-carbon single bond possesses high directionality, constrained straightway along the bond axis. Moreover, there are several substituents on both ends. The interaction of this directionally and sterically constrained orbital with metal orbitals is much more difficult than that of a C–C double bond or even of a C–H single bond. This kinetic barrier renders the C–C bond considerably inert.

Thus, the breaking of C–C bonds by the insertion of soluble transition metal complexes has been a challenging issue in the field of organometallic chemistry.

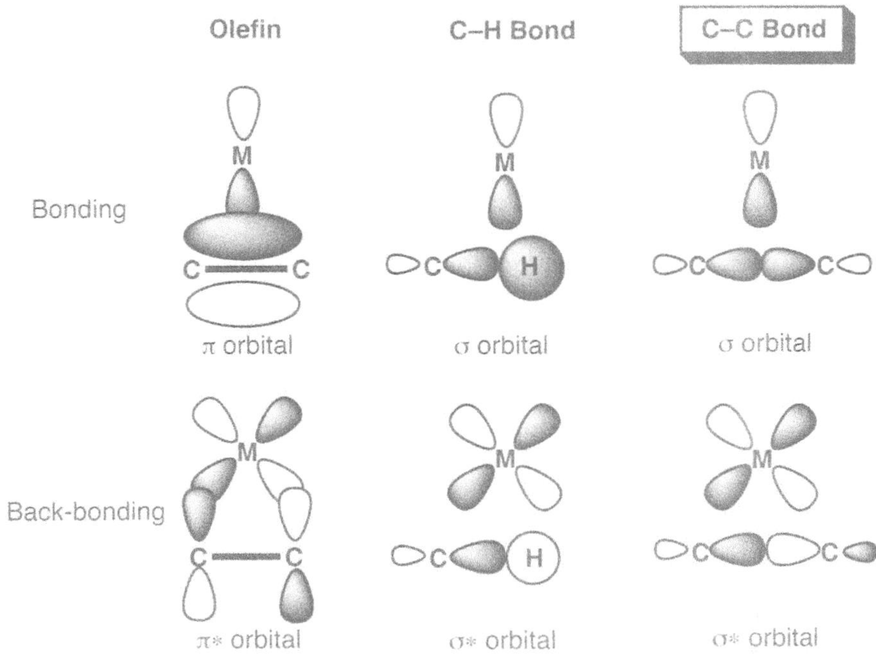

Fig. 1.

In the latter part of this century, a number of strategies have been devised to achieve C–C bond activation. The stoichiometric reactions mentioned in this chapter are organized according to the strategies employed.

The term "oxidative addition" is recognized as a process in which a low-valent metal inserts into an X–Y bond to form an X–M–Y species. It should be noted, however, that most if not all cases presented in this review imply no mechanistic feature of "oxidative addition." The real mechanistic trajectory of the reaction can be quite different from a "frontal assault," that is, a three-centered one in which a C–C•••M triangular bridged complex is an intermediate stage leading directly to a C–M–C complex.

2.1.1
Utilization of Ring Strain

2.1.1.1
Cleavage of Three-Membered Rings

The use of cyclopropane as a substrate for C–C bond activation is advantageous kinetically as well as thermodynamically. Formation of an adduct complex is thermodynamically driven by relief of the structural strain of the three-membered ring. In addition, the orbitals connecting the carbon atoms are bent outward from the internuclear axis. These orbitals are ready for interaction with metal orbitals; thus their kinetic accessibilities are also increased. In 1955, the insertion of a transition metal into a C–C single bond was first reported by Tipper, who observed the formation of platinacyclobutane **1** by the reaction of $PtCl_2$ with cyclopropane [4]. The structure of the product was later unambiguously confirmed [5, 6]. It was proposed that $PtCl_2$ acted as an electrophile based on the relative reactivities of substituted cyclopropanes [7]; cyclopropanes substituted with more electron-donating groups reacted faster.

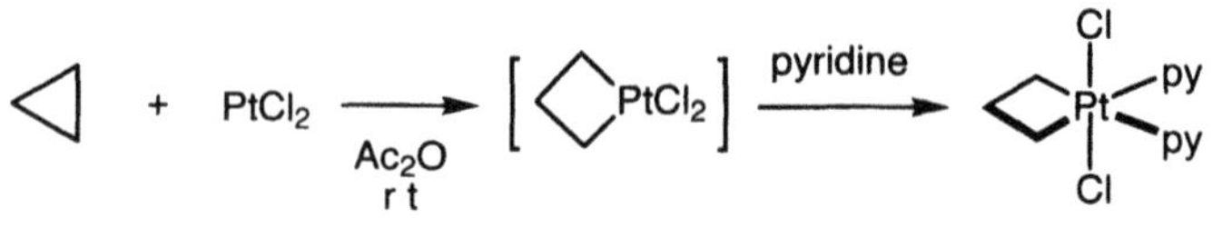

(1)

On the other hand, zerovalent platinum and palladium can be inserted into a C–C single bond between the two tertiary carbon atoms of 1,1,2,2-tetracyanocyclopropane [8, 9]. Notably, 1,1,2,2-tetracyanocyclopropane is much more susceptible to insertion than 1,2-dicyanocyclopropane. Also of interest is that these zerovalent metals attack the most positively charged carbon atoms rather than the most sterically accessible.

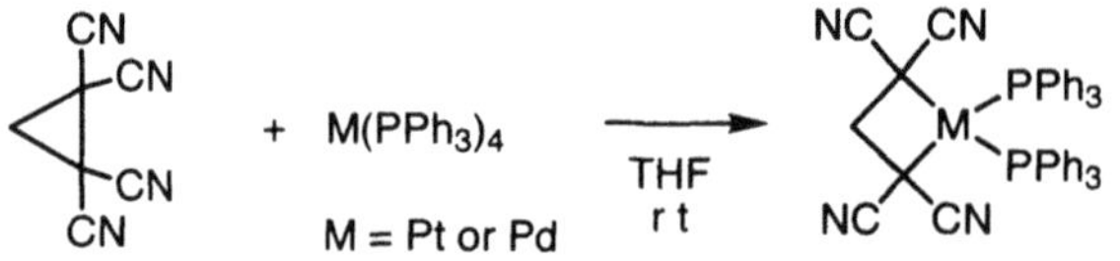

(2)

The reaction of cyclopropane with $[Rh(CO)_2Cl]_2$ results in the formation of 1-rhodacyclopentan-2-one **2** [10–12]. Oxidative addition onto rhodium is followed by insertion of the ligand carbonyl group. An analogous acylrhodium complex was obtained from the reaction of quadricyclene with $[Rh(CO)_2Cl]_2$ [13].

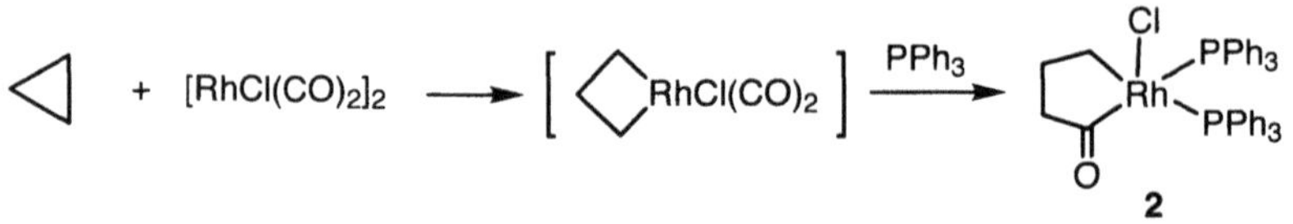

(3)

Bergman et al. presented an important mechanistic study of the oxidative addition of cyclopropane [14]. The reaction of cyclopropane with coordinatively unsaturated rhodium complex **3** at –60°C results in C–H insertion. No C–C bond cleavage was observed at that temperature. Upon raising the temperature to –20°C, (cyclopropyl)(hydride)rhodium complex **4** undergoes direct rearrangement to rhodacyclobutane **5**. The C–H insertion product is kinetically favored, and the C–C insertion product is thermodynamically favored. The kinetic preference for C–H insertion clearly demonstrates the greater steric accessibility of the C–H bond compared with the C–C bond, as mentioned above. Evidence suggesting that the formation of a rhodacyclopentane proceeded by an analogous rearrangement of a (cyclobutyl)(hydride)rhodium complex was also obtained [14].

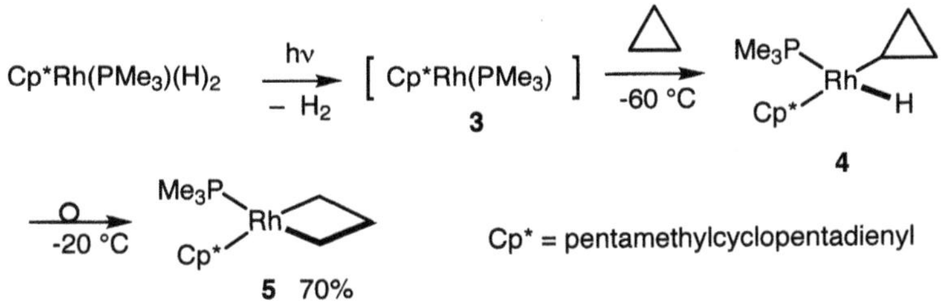

Cp*Rh(PMe3)(H)2 —hν, – H2→ [Cp*Rh(PMe3)] 3 —cyclopropane, -60 °C→ 4 —-20 °C→ 5 70% (4)

Cp* = pentamethylcyclopentadienyl

A number of investigations have appeared documenting the metal-mediated cleavage of a variety of cyclopropane derivatives, particularly those which are more reactive than ordinary cyclopropanes due to additional structural strain, unsaturation and/or substituents. In a pioneering study, methylene cyclopropane was shown to react with $Fe_2(CO)_9$ to furnish trimethylenemethane complex **6** [15].

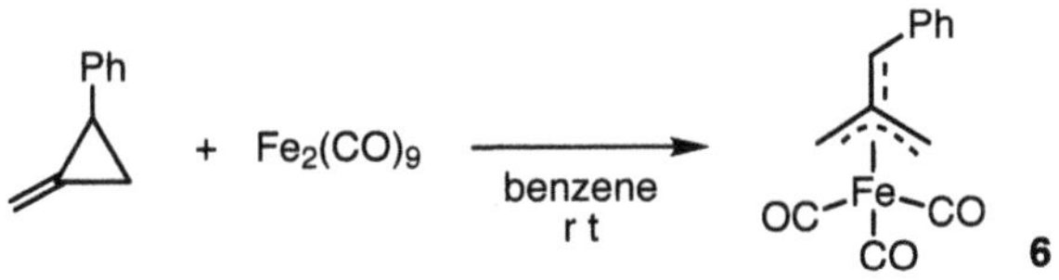

+ Fe2(CO)9 —benzene, r t→ 6 (5)

Cleavage of a C–C bond of methoxy- or siloxycyclopropane is mediated by $[Pt(CH_2{=}CH_2)Cl_2]_2$ to give a β-platinum substituted ketone through concomitant R^1-O bond breaking [16, 17].

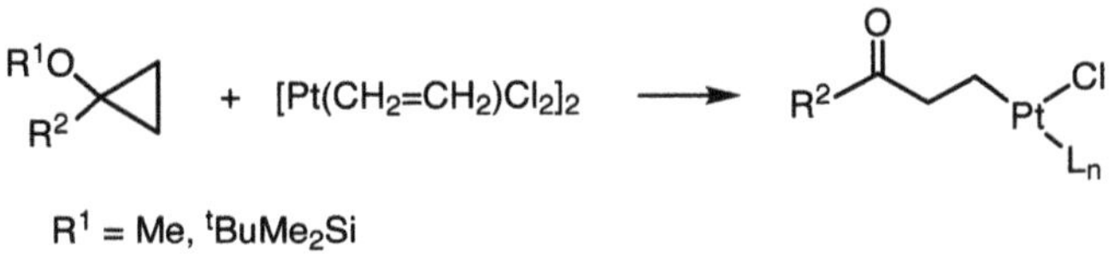

+ [Pt(CH2=CH2)Cl2]2 → (6)

R1 = Me, tBuMe2Si

Perfluorocyclopropene reacts with $Pt(CH_2{=}CH_2)(PPh_3)_2$ to afford tetrafluoroplatinacyclobutene complex 7 [18].

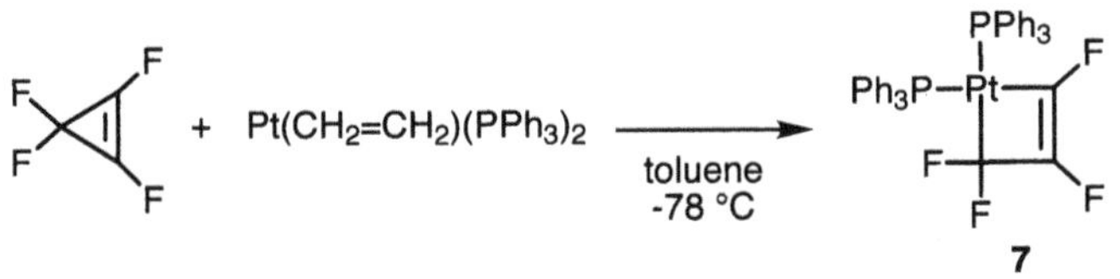

+ Pt(CH2=CH2)(PPh3)2 —toluene, -78 °C→ 7 (7)

Other examples of C–C bond cleavage of cyclopropane derivatives like cyclopropene, benzocyclopropane, bicyclo[1.1.0]butane, bicyclo[2.1.0]pentane, methylenecyclopropane, and vinylcyclopropane have been reviewed comprehensively [19–23].

2.1.1.2
Cleavage of Four-Membered Rings

A prototypical reaction involving oxidative addition of a four-membered ring to a transition metal is seen in the $Cr(CO)_6$ mediated transformation of biphenylene to 9-fluorenone [24]. The product can be viewed as arising from the insertion of chromium into the central C–C bond bridging the two aromatic rings, subsequent carbonylation, and reductive elimination.

+ $Cr(CO)_6$ → 225 °C; 88% (8)

The cleavage of biphenylene was unambiguously identified by Eisch using a nickel(0) complex [25]. The ease of the insertion depends largely on the ligands attached to nickel, with more basic ligands being preferred. The cleaved complex **8** undergoes a facile dimerization to yield the coupled complex **9** upon standing at 18–25°C.

+ $Ni(PEt_3)_4$ → 0 °C; **8** → 18 °C, - PEt_3; **9** (9)

An iridium(I) complex also cleaves the aryl-aryl bond of biphenylene to afford a stable five-coordinated iridium(III) complex [26].

+ $[Ir(cod)Cl]_2$ → CH_2Cl_2, 90 °C, 2 h → PPh_3, 20 °C, 5 h; 88% (10)

A coordinatively unsaturated rhodium(I) complex generated from **10** reacts with biphenylene to give C–H inserted complex **11** as the kinetic product. Complex **11** is then thermally converted to the C–C inserted complex **12** [27]. This result may be relevant to the reaction of cyclopropane shown in Eq. 4 in that a C–H insertion product is kinetically favored, and a C–C insertion product is thermodynamically favored. Complex **12** fails to react with dihydrogen.

+ Cp*Rh(PMe$_3$)(Ph)(H) **10** — 65 °C, 12 h → **11** — 85 °C, 5 d → **12** quant (11)

The rhodium and cobalt complexes **13** and **14**, which contain labile ethylene ligands, also cleave the central C_{sp2}–C_{sp2} bond of biphenylene to afford dinuclear complexes **15** and **16**, respectively [28]. Treatment of the cobalt complex **16** with carbon monoxide furnishes fluorenone. However, both complexes are resistant to hydrogenation.

+ Cp*M(CH$_2$=CH$_2$)$_2$ (**13** M = Rh; **14** M = Co) — 83 °C, 8 h → **15** M = Rh; **16** M = Co (12)

16 + CO — 120 °C, 1 h → fluorenone + Cp*Co(CO)$_2$

Oxidative addition of a four-membered ring to a transition metal was observed with cubane, which consists of only sp3 carbon atoms but is severely strained [29]. On treatment with PPh$_3$, carbonyl insertion ensues to afford a cyclic ketone.

+ [RhCl(CO)$_2$]$_2$ — CHCl$_3$ → [Rh(CO)L$_n$ complex] — PPh$_3$ → ketone (13)

In an analogous manner to 1,1,2,2-tetracyanocyclopropane, 1,1,2,2-tetracyanocyclobutane also undergoes zerovalent platinum-mediated cleavage at the sterically congested single bond between the positively charged carbon atoms to afford a metallacyclopentane [30]. Of note is that 1,1,2,2-tetracyanocyclobutane is more reactive than 1,2-dicyanocyclopropane. The extent of charge disproportion rather than strain energy is dominant in this case.

$$\text{EtO-cyclobutane(CN)}_4 + Pt(CH_2{=}CH_2)(PPh_3)_2 \xrightarrow[\text{rt, overnight}]{Et_2O\text{–}THF} \text{platinacyclopentane}(PPh_3)_2 \quad (14)$$

2.1.2
Utilization of a Carbonyl Functionality

The C–C single bond between a carbonyl carbon and the α-carbon is relatively weaker than other C–C single bonds. Moreover, it is likely that a carbonyl group kinetically facilitates insertion of a transition metal into the α C–C bond. In 1965, transfer of a carbonyl group from a solvent such as cyclohexanone to rhodium was reported [31]. More explicit cleavage was identified with highly strained unsaturated ketones. Platinum(0) undergoes insertion regioselectively into cyclopropenone, giving platinacyclobutenone [32, 33]. An olefin complex is intermediate to the ring-opened complex.

$$\text{Me-cyclopropenone} + Pt(CH_2{=}CH_2)(PPh_3)_2 \xrightarrow[-65\,°C]{CDCl_3} \text{olefin complex} \xrightarrow{-30\,°C} \text{platinacyclobutenone} \quad (15)$$

In contrast, when allowed to react with $Rh(CO)(PPh_3)_2(OTf)$ at 60–65°C, cyclopropenone forms cationic complex **17**, in which cyclopropenone is bound to rhodium through the oxygen atom instead of the C–C double bond. Further heating at 60–65°C leads to the formation of the metal carbonyl insertion product **18** [34]. In refluxing benzene, **18** decomposes to diphenylacetylene and $Rh(CO)(PPh_3)_2(OTf)$.

$$\text{Ph}_2\text{-cyclopropenone} + Rh(CO)(PPh_3)_2(OTf) \xrightarrow[60\text{–}65\,°C]{\text{benzene}} \mathbf{17} \xrightarrow{60\text{–}65\,°C} \mathbf{18} \xrightarrow[\text{reflux}]{\text{benzene}} Ph\text{–}C{\equiv}C\text{–}Ph + Rh(CO)(PPh_3)_2(OTf) \quad (16)$$

Diketones have also been targets for C–C bond cleaving reactions. Benzil reacts with $Pt(PPh_3)_4$ to afford $(PhCO)_2Pt(PPh_3)_2$ [35]. Treatment of benzocyclobutane-1,2-dione with $Pt(PPh_3)_4$ causes unsymmetrical cleavage of the four-membered ring [36]. Cyclobutenediones undergo analogous cleavage [37].

+ $Pt(PPh_3)_4$ → (benzene, r t) → Ph_3P, PPh_3, Pt (17)

Unlike platinum(0), rhodium, cobalt, and iron afford symmetrical complexes resulting from insertion between two carbonyl groups of benzocyclobutane-1,2-dione [38, 39].

+ $Fe(CO)_5$, $RhCl(PPh_3)_3$, $CoCl(PPh_3)_3$, $CoCp(CO)_2$ → M–L_n (18)

M-L_n = $Fe(CO)_4$, $RhCl(PPh_3)_2$, $CoCl(PPh_3)_2$, CoCp(CO)

Cyclobutenone also oxidatively adds onto rhodium(I) or cobalt(I) to afford a metallacycle [40].

EtO + $RhCl(PPh_3)_3$ → (60 °C, 5 h) → EtO, PPh_3, Rh–Cl, PPh_3 (19)

Dialkynyl ketones react with $RhCl(PPh_3)_3$ affording conjugated diynes together with $Rh(CO)Cl(PPh_3)_2$. Insertion of rhodium between the carbonyl carbon and the α sp carbon is likely to occur prior to the decarbonylation [41].

$$\text{Ph-C}\equiv\text{C-C(=O)-C}\equiv\text{C-Ph} + RhCl(PPh_3)_3 \xrightarrow{\text{xylene reflux, 2 h}} \text{Ph-C}\equiv\text{C-C}\equiv\text{C-Ph} + Rh(CO)Cl(PPh_3)_2 \quad (20)$$

Suggs and Jun exploited 8-quinolyl alkyl ketones **19–22** as substrates for C–C bond cleavage, as shown in Eq. 21. The formation of complexes **23–26** by insertion between the carbonyl carbon and the α-carbon is favored due to the general preference for five-membered chelate rings observed for cyclometalated complexes [42]. No deuterium is lost in the reaction of 8-quinolyl alkyl ketone **20**, which has both benzylic positions deuterated. Interestingly, the chirality of the α-carbon is retained during the insertion process as exemplified by the reaction of **21** [43]. It was proposed that a tetrahedral intermediate arising from direct attack of rhodium on the carbonyl carbon occurs en route to C–C bond cleavage without the intervention of a C–H activation process.

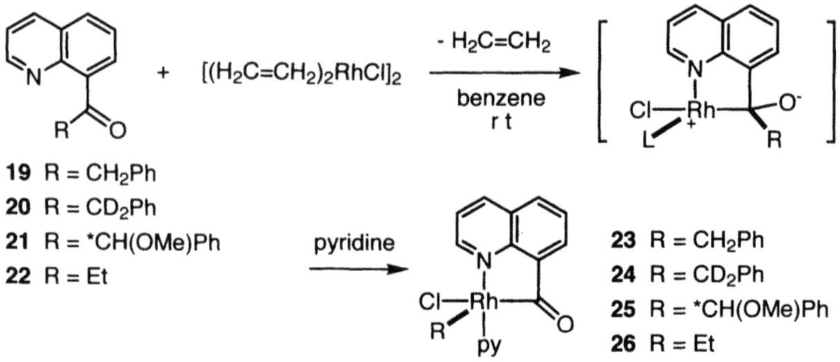

+ $[(H_2C{=}CH_2)_2RhCl]_2$ → ($-H_2C{=}CH_2$; benzene, r t)

19 R = CH_2Ph
20 R = CD_2Ph
21 R = *CH(OMe)Ph
22 R = Et

pyridine →

23 R = CH_2Ph
24 R = CD_2Ph
25 R = *CH(OMe)Ph
26 R = Et

(21)

σ,η^3-Allyl rhodium complex **27**, derived from a ring-opening reaction of vinylcyclopropane with a rhodium(I) complex, also cleaves the C–C bond of 8-quinolyl alkyl ketone **19** with regeneration of vinylcyclopropane [44]. It is remarkable that oxidative addition of the 8-quinolyl alkyl ketone is preferred over that of a strained vinylcyclopropane.

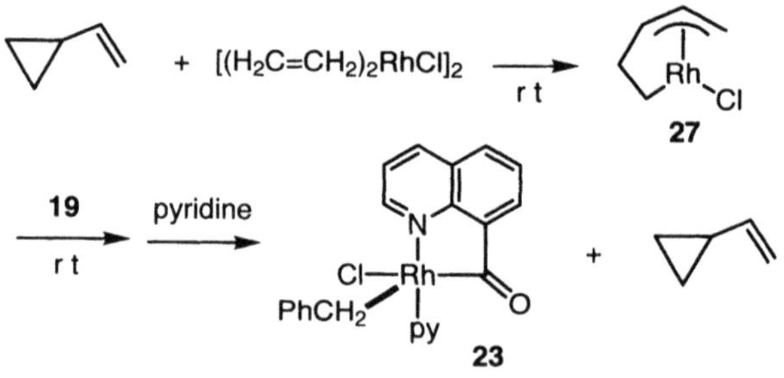

+ $[(H_2C{=}CH_2)_2RhCl]_2$ → (r t) **27**

19 (r t) → pyridine → **23** +

(22)

Exchange of the pyridine ligand of the C–C cleaved complex **26** with soft ligands such as phosphines, phosphites, or CO brings about reductive elimination to regenerate the 8-quinolyl ketone **22** [45].

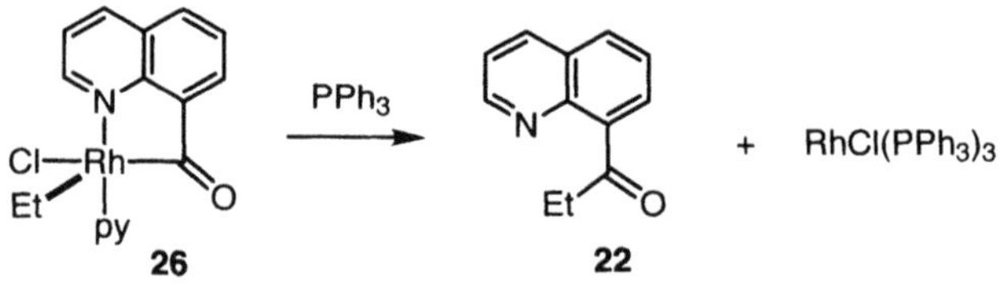

26 → (PPh_3) **22** + $RhCl(PPh_3)_3$ (23)

Ordinary cyclic mono-ketones were also decarbonylated by the action of a stoichiometric amount of $RhCl(PPh_3)_3$ [46, 47]. Treatment of cyclobutanone with $(Ph_3P)_3RhCl$ results in decarbonylation to afford the corresponding cyclopropane together with the rhodium carbonyl complex **28**. Insertion of rhodium into the α-bond, extrusion of the carbonyl group, and subsequent reductive elimination accomplishes decarbonylation. The production of a more strained three-membered ring compound is possible because the highly stable rhodium carbonyl complex **28** is formed concomitantly.

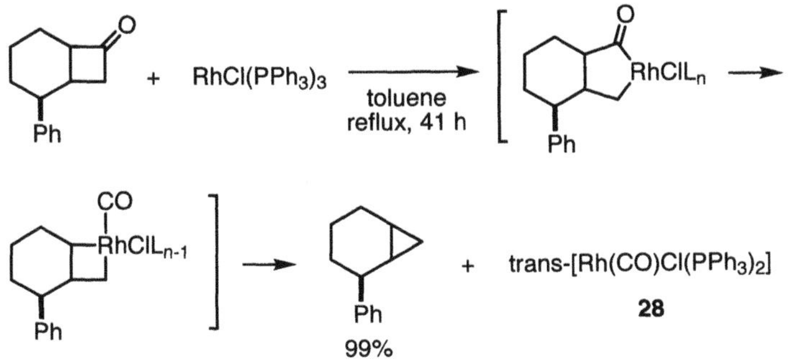

O
Ph + $RhCl(PPh_3)_3$ → toluene reflux, 41 h → [O RhClL$_n$ Ph] →
CO RhClL$_{n-1}$ Ph] → Ph 99% + trans-$[Rh(CO)Cl(PPh_3)_2]$ **28**

(24)

Cyclopentanone **29**, which is much less strained than the cyclobutanone analog, undergoes gradual decarbonylation to give cyclobutane under analogous conditions. Decarbonylation of cyclododecanone does happen although it requires a higher temperature [46, 47].

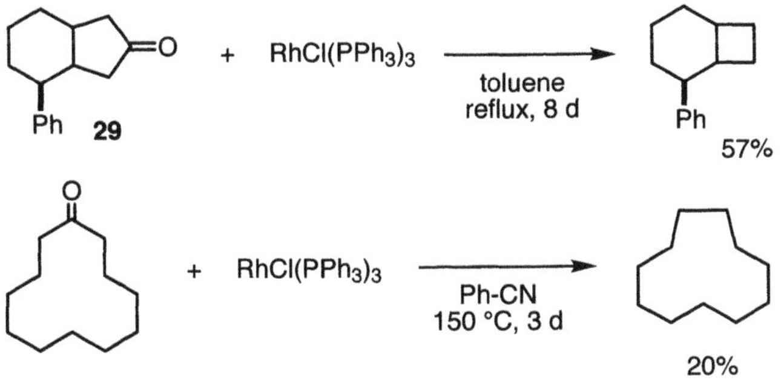

=O Ph **29** + $RhCl(PPh_3)_3$ → toluene reflux, 8 d → Ph 57%
O + $RhCl(PPh_3)_3$ → Ph-CN 150 °C, 3 d → 20%

(25)

2.1.3
Utilization of a Pincer-Type Chelating Ligand

The presence of coordinating functionalities in a target molecule induces "precoordination," whereby a transition metal is brought into the close proximity of a specific C–C bond, thus facilitating insertion of metal. Milstein at al. designed diphosphine pincer-type chelating ligands and observed selective cleavage of an alkyl group attached to an aromatic ring. The reaction of **30** with $HRh(PPh_3)_4$ at 25°C results in C–H activation to yield the thermally stable rhodium(I) complex **31** [48]. Heating of **31** at 90°C under dihydrogen brings about cleavage of a non-strained aryl-methyl bond to furnish complex **32**, with liberation of methane. The overall result is that a methylene group is excised from **30** and transferred into the H-H bond of dihydrogen. Similar transfer of a methylene group occurs into hydrosilane (H–Si bond), disilane (Si–Si bond), and benzene (C–H bond) [49]. In the hydrogenolysis of an analogous pincer-type ethylbenzene derivative, the aryl-Et bond is selectively cleaved in the presence of the Csp3–Csp3 bond of the ethyl group, despite their similar accessibilities [50]. Platinum also cleaves the aryl-methyl bond of a pincer-type ligand to transfer the methylene group into a polar substrate HCl [51].

$$\mathbf{30} + HRh(PPh_3)_4 \xrightarrow[25\,^{\circ}C]{THF} \mathbf{31} \xrightarrow[90\,^{\circ}C]{H_2} \mathbf{32} + CH_4 \qquad (26)$$

An appropriate choice of both the starting rhodium complex and the phosphorus substituents of the pincer ligand renders it possible to isolate the intermediate C–C activated complex [52, 53]. A system consisting of $[Rh(olefin)_2Cl]_2$ and a bulky, basic phosphine ligand **33** leads to selective metal insertion into an aryl-methyl bond at room temperature. Initially, parallel formation of C–H activated complex **34** and C–C activated complex **35** is observed. Complex **34** is gradually converted to **35** C at room temperature, demonstrating that the C–C activated complex **35** is thermodynamically more stable than the C–H activated complex **34**. Furthermore, a kinetic study revealed that, if the numbers of the bonds available for activation are taken into account, metal insertion into the C–C bond is also kinetically preferred over the competing insertion into the C–H bond. Electronic perturbation of the aromatic ring by introduction of a methoxy group has no effect on the reaction rate or the product ratio, suggesting that the C–C oxidative addition proceeds directly via a three-centered nonpolar transition state similar to that postulated for C–H bond activation. Facile oxidative addition of **33** bearing tertiary butyl groups on phosphorus can be ascribed to the precise steric as well as electronic properties of the metal-ligand complex. In other words, the high electron density and availability of coordinative unsaturation on the metal promote C–C bond cleavage. Oxidative addition of **33** was also observed with iridium.

An analogous system was successfully applied to cleavage of a stronger aryl-CF_3 bond [54].

33 + $[Rh(CH_2{=}CH_2)_2Cl]_2$ → (benzene, r t) **34** + **35** (27)

quant, **34** : **35** = 1.25 : 1 → 0 : 1

A phosphine-amine pincer ligand reacts with a rhodium olefin complex more easily than diphosphine pincer ligands to give a C–C bond activated complex in minutes at room temperature. In this case, a C–H activated complex was not observed upon monitoring the reaction even at –50°C [55].

+ $[Rh(CH_2{=}CH_2)_2Cl]_2$ → (benzene, r t, 3 min) (28)

2.1.4
Utilization of Aromatization

The driving force of aromaticity can be exploited for C–C bond cleavage in pre-aromatic systems. Pentamethylcyclopentadienyl complexes are obtained from reactions of hexamethyl-(Dewar benzene) with $RhCl_3$ and $IrCl_3$ [56]. A η^4-(endo-ethylcyclopentadienyl)molybdenum complex rearranges with breaking of the Cp–Et bond to a η^5-(cyclopentadienyl)(ethyl)molybdenum complex upon generation of a vacant coordination site on the metal [57].

$MoCp[PPhMe_2)$ → ($TlBF_4$) $[MoCp(PPhMe_2)]^+$ (29)

A number of variations of this reaction have been reported with iron [58], manganese [59], and rhenium [60]. A saturated cyclic hydrocarbon molecule can be also viewed as a pre-aromatic system. The reaction of 1,1-dimethylcyclopentane with cationic iridium complex **36** in the presence of an olefin as a hydrogen scavenger affords the dehydrogenated complex **37**, which subsequently undergoes C–C bond cleavage of the ligand to give complex **38** [61].

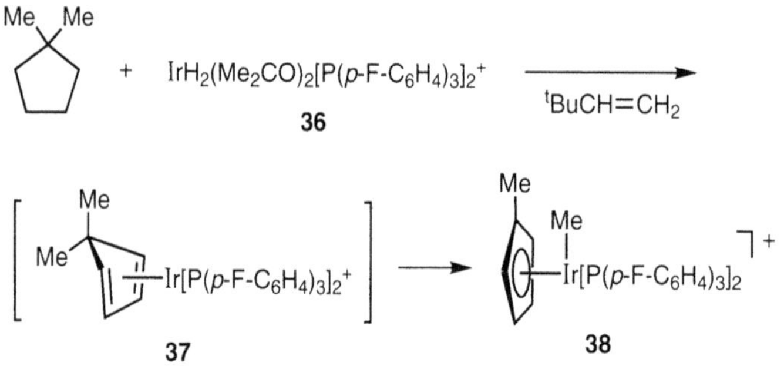

Me Me
+
IrH2(Me2CO)2[P(p-F-C6H4)3]2+
36
tBuCH=CH2
Me
Me
Ir[P(p-F-C6H4)3]2+
37
Me
Me
Ir[P(p-F-C6H4)3]2
38

(30)

Crabtree et al. also identified migration of an ethyl group on a cyclopentadienyl ring with η^4-(1,1-diethylcyclopentadienyl)iridium complex **39**, which should involve a C–C bond cleavage/formation sequence [61].

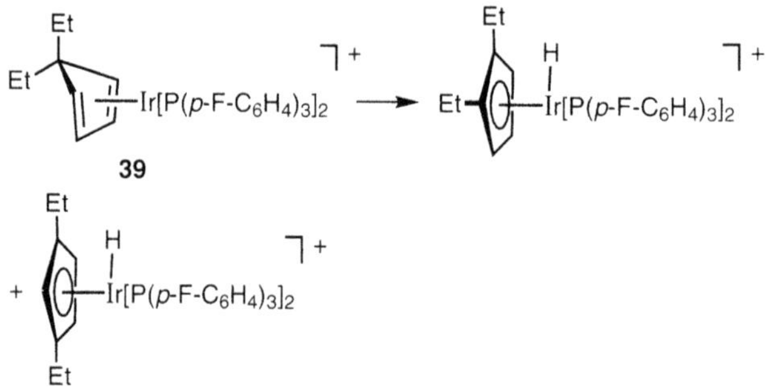

Et
Et
Ir[P(p-F-C6H4)3]2
39
Et
H
Et
Ir[P(p-F-C6H4)3]2
+
Et
H
Ir[P(p-F-C6H4)3]2
Et

(31)

2.1.5
Miscellaneous Types of Metal Insertion

Suzuki et al. reported an interesting example of a C–C bond cleavage of an ordinary non-strained diene substrate with a multimetallic complex [62]. The Csp2-Csp3 bond of cyclopentadiene is cleaved by trinuclear ruthenium pentahydride complex **40** to afford ruthenacyclohexadiene **41**, which then rearranges to 2-methylruthenacyclopentadiene complex **42**. The three metal centers cooperate for cleavage; two of the three centers act as coordination sites for the diene moiety and the third is inserted into the C–C bond.

+
Cp*
Ru
H
H
H
H
Ru
Ru
H
*Cp
Cp*
40
Cp*
Ru
Ru
Ru
*Cp
Cp*
41
Cp*
Ru
Me
Ru
Ru
*Cp
Cp*
42

Bridging hydrogens are omitted.

(32)

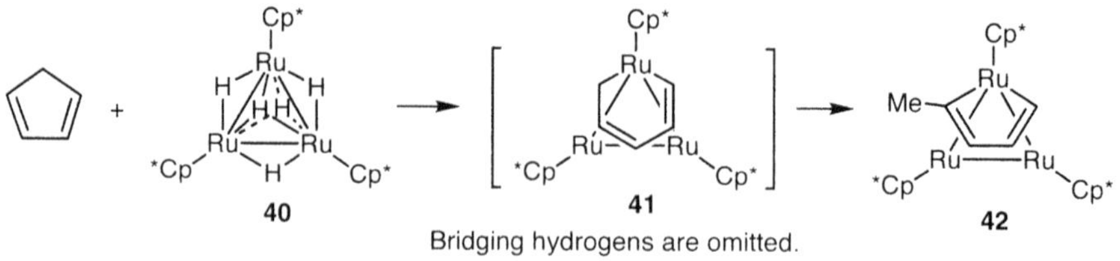

Protonation of (norbornadiene)cobalt complex **43** induces C–C bond cleavage of the norbornenyl ring to form cationic complex **44** [63, 64]. Re-protonation of the reduced complex **45** induces a second cleavage of a non-strained cyclopentene ring to give an open η^5-pentadienyl complex **46**. It is postulated that a three-center interaction of the highly electrophilic metal center with the σ-electrons of the adjacent C–C bond is involved.

Cp* Co **43** → ($HBF_4 \cdot OEt_2$, 120 °C, 1 h) → [Cp* Co]$^+$ **44** → ($LiBHEt_3$, 120 °C, 1 h) → Cp* Co **45** (33)

→ ($HBF_4 \cdot OEt_2$, 120 °C, 1 h) → [Cp* Co]$^+$ Et **46**

Reactions of curved polyaromatic hydrocarbon ligands like C_{60} with transition metals is of current interest. Oxidative addition of a strained five-membered ring of a C_{60}-derived molecule to cobalt provides a candidate complex for the inclusion of a metal into the C_{60} framework [65].

C–C bond breaking of a five-membered ring was also observed with fullerene subunit **47** [66].

47 + $Pt(CH_2{=}CH_2)(PPh_3)_2$ → (toluene, r t, 15 h, reflux, 1 h) → Pt, Ph_3P, PPh_3 (34)

2.2 *β-Carbon Elimination*

As described in Sect. 2.1, oxidative addition provides a direct approach to C–C bond cleavage. An alternative method for breaking C–C bonds employs σ alkylmetal complexes. The bond between the β- and γ-carbon atoms in these complexes can be cleaved via β-carbon elimination. Examples of C–C bond cleavage via this process are quite rare.

–M–C(α)–C(β)–C(γ)– → (β-carbon elimination) → –M–C–, C=C (35)

Watson et al. reported a leading example of β-carbon elimination observed with a well-defined metal complex [67]. Thermal decomposition of a lutetium-isobutyl complex having a vacant coordination site leads to the formation of a lutetium-methyl complex and propene by way of β-methyl elimination, the microscopic reverse of olefin insertion. A concerted four-center transition state is proposed. This study demonstrated that β-carbon elimination is an energetically accessible process, and provided a model for the chain transfer that occurs during propene oligomerization.

Cp*$_2$–Lu ⟶ (22 °C) Cp*$_2$–Lu–Me + propene (36)

❏ = vacant site

There are some recent reports wherein β-methyl elimination is directly observed with well-defined metallocene derivatives of highly Lewis-acidic early transition metals. Bridged scandocene-isobutyl complex **48** decomposes to scandocene-methyl complex **49** along with propene, which is ultimately transformed to various hydrocarbons [68].

48 ⟶ (25 °C) **49** + propene-derived hydrocarbons (37)

Reversible β-methyl elimination takes place on warming a solution of cationic (neopentyl)zirconocene complex **50** to 25°C, affording (methyl)zirconocene complex **51** and isobutene [69].

Cp$_2$–Zr (Me–B(C_6F_5)$_3$) **50** ⇌ (25 °C) Cp$_2$–Zr (Me–B(C_6F_5)$_3$)–Me **51** + isobutene (38)

Migration of more complex alkyl groups was recently reported [70]. Reversible migratory insertion/β-carbon elimination occurs between the coordinated alkyne and the bound alkyl group of alkyl-niobium(alkyne) complex **52**.

52 ⇌ (70 °C, toluene) [intermediate] ⇌ product (39)

β-Carbon elimination is now recognized as an important and predominant chain transfer step during the Ziegler-Natta polymerization of propene. Teuben et al. identified β-methyl elimination as a termination step during propene oligomerization mediated by cationic d^0 complex **53** (M=Zr, Hf) [71]. A methyl group is transferred from the growing chain to the metal during termination, regenerating the methyl complex **53**. The dominant occurrence of β-methyl elimination rather than β-hydride elimination was explained by assuming that, with intermediate **54**, steric repulsions caused by the bulky pentamethylcyclopentadienyl (Cp*) groups force the β-methyl substituent of the growing chain into the equatorial plane of the Cp^*_2 M wedge, which is the requisite position for β-elimination [72]. β-Methyl elimination was also observed in propene oligomerization with $Cp^*_2MCl_2$/methylaluminoxane systems (M=Zr, Hf) [73, 74].

Cp*2–M(Me)(□)⁺ **53** M = Zr, Hf → Cp*2–M–CH2CH(Me)Me⁺ → Cp*2–M–CH2CH(Me)CH2CH(Me)Me⁺ → **54** → Cp*2–M(Me)(□)⁺ **53** + CH2=CHCH2CH(Me)Me (40)

Cyclopolymerization of 2-methyl-1,5-hexadiene is catalyzed by a cationic zirconocene complex [75]. Isolation of methylenecyclopentane derivatives **55–57** from the low molecular weight oligomeric products provides convincing evidence for chain transfer via β-methyl elimination.

2-methyl-1,5-hexadiene —[Cp^*_2-Zr(Me)(Me-B(C_6F_5)$_3$), 22–35 °C]→ Cp^*_2-Zr⁺–[cyclopentane]$_n$–Me ; **55**, **56**, **57** (41)

For examples of β-carbon elimination in late transition metal systems, Bergman et al. identified β-methyl transfer with four-membered ruthenacycles, which is driven by the formation of π-allyl and π-oxallyl complexes. Warming the solution of oxaruthenacycle **58** to 45°C led to formation of methane and cyclic enolate complex **60** [76]. π-Oxallyl complex **59** initially arises from β-methyl elimination. Subsequent cyclometalation at the phenyl ring produces methane and **60**.

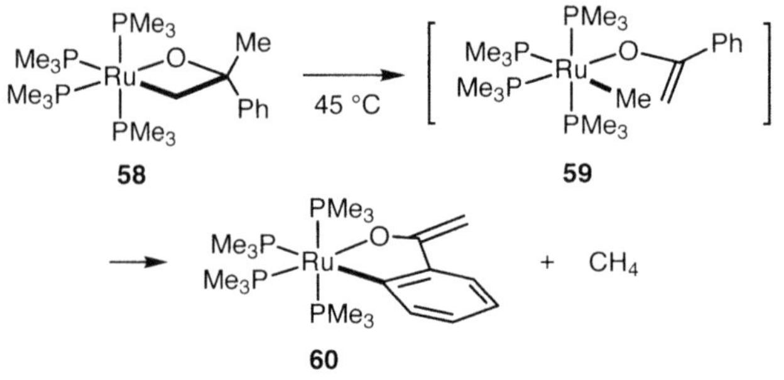

(42)

Similarly, thermolysis of ruthenacyclobutane **61** produces π-allyl complex **62** [77]. The reaction involves β-methyl transfer from the central carbon of the ligand to the metal via a formal 16-electron unsaturated intermediate. A kinetic investigation in the presence of excess phosphine revealed that the process is reversible.

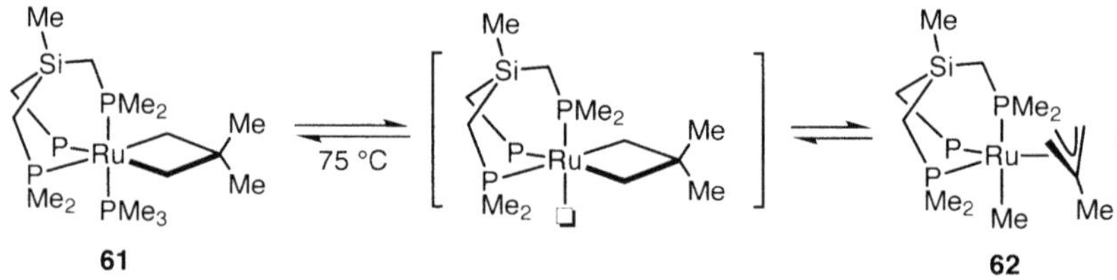

(43)

An analogous reversible process was observed in the nucleophilic addition of enolate species to the central carbon of a (π-allyl)iridium complex forming the corresponding metallacyclobutane [78].

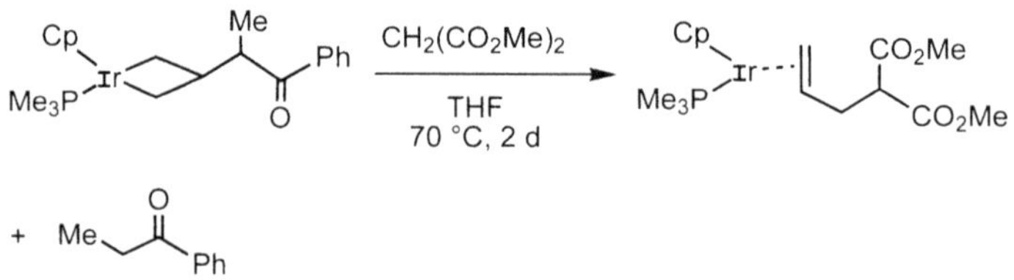

(44)

A σ,π-bonded alkyl-palladium complex undergoes β-carbon elimination on protonation of the cyclopentadiene ring [79].

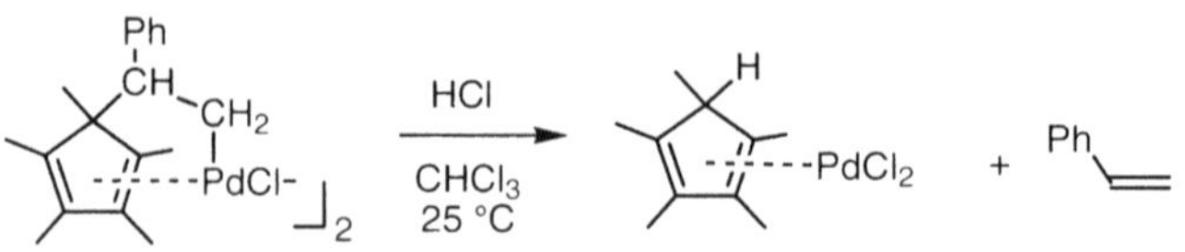

(45)

A cyclobutylmethyl-metal system provides another opportunity to observe β-carbon elimination. The ring opening process harnesses the release of the least necessitating ring strain of a four-membered ring. Scandocene hydride **63** reacts with 3-methyl-1,4-pentadiene to afford the linear π-allyl complex **65** [80]. The intermediacy of cyclobutylmethyl complex **64** which undergoes β-carbon elimination accommodates the observed rearrangement.

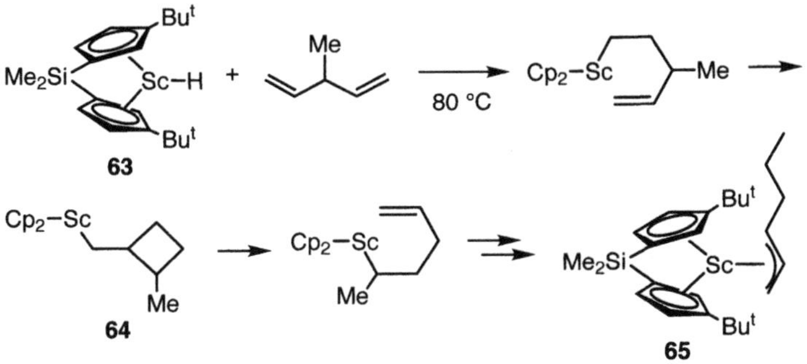

(46)

Yttrium hydride reacts with methylenecyclobutane to form pentenyl chelate complex **67** [81]. β-Carbon elimination occurs to open the ring of cyclobutylmethyl intermediate **66**.

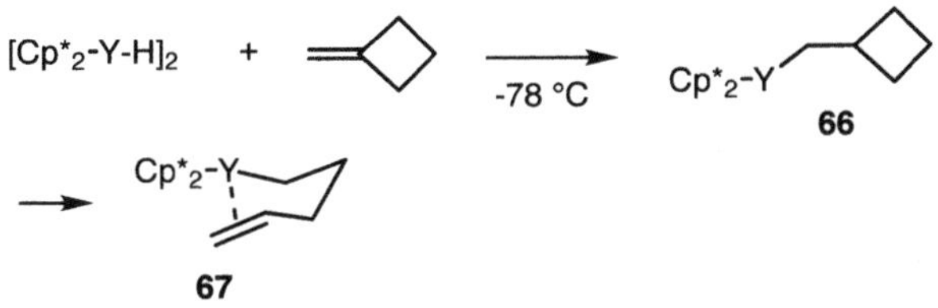

(47)

A phenylpalladium complex also causes ring-opening rearrangement of methylenecyclobutane to a π-allylpalladium complex, which arises from β-carbon elimination of an intermediate (cyclobutylmethyl)palladium complex [82].

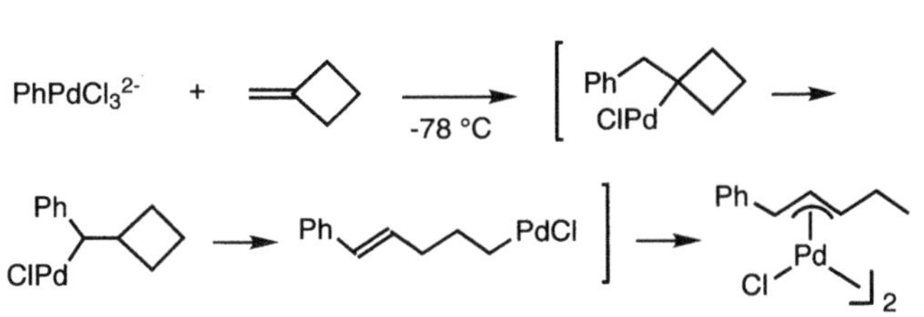

(48)

Reversible olefin insertion/β-carbon elimination occurs with a cationic pentenyl chelate platinum complex. Labeled complex **68** is reversibly converted to **70** via (cyclobutylmethyl)platinum intermediate **69** [83, 84].

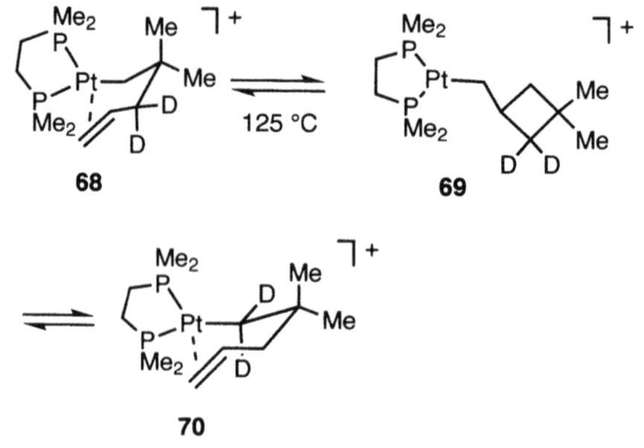

(49)

Unequivocal evidence for the ring-opening cleavage was provided by the reaction of the isolated neutral (cyclobutylmethyl)platinum complex **71** [85]. On thermal generation of a vacant coordination site, **71** undergoes β-carbon elimination to afford 2-methyl-1,4-pentadiene together with a hydride complex.

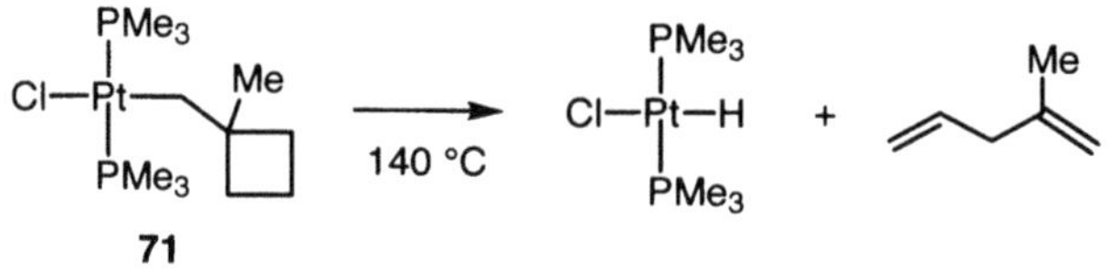

(50)

8-Quinolyl cyclobutylmethyl ketone **72** undergoes C–C bond cleavage by rhodium(I) to form the inserted intermediate **73**, as mentioned in Eq. 21. Sequential β-carbon elimination leads to ring-opening of the cyclobutane ring. Addition of pyridine and $P(OMe)_3$ induces reductive elimination to give a mixture of linear 8-quinolyl ketones **74** and **75** [86].

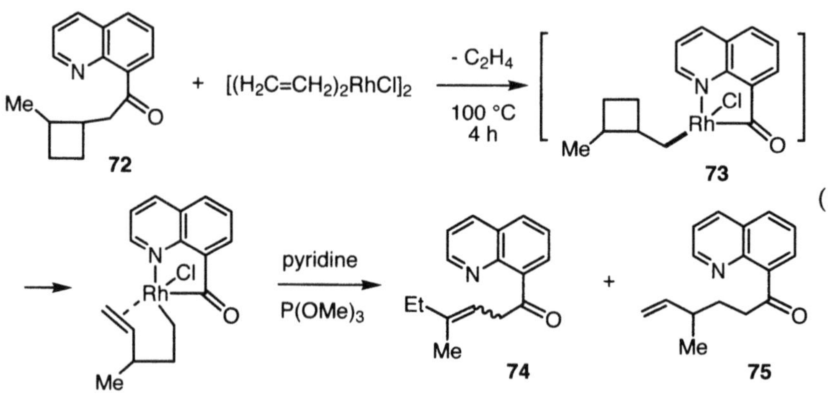

(51)

Efficient propagation pathways of ring-opening polymerization of methylenecyclobutane [87] and methylenecyclopropane [88, 89] involve β-carbon elimination at electrophilic metal centers like zirconium and samarium.

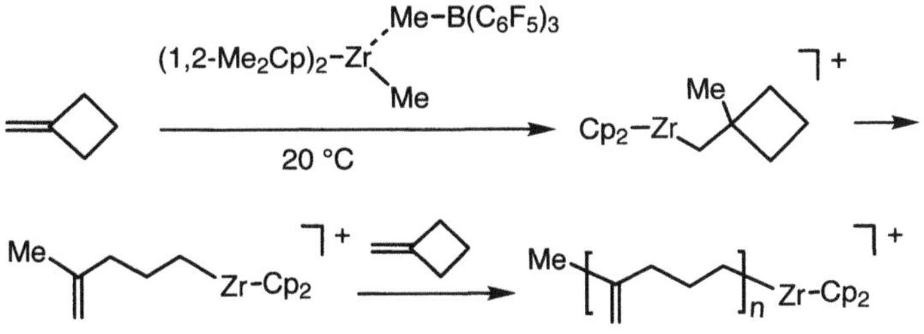

(52)

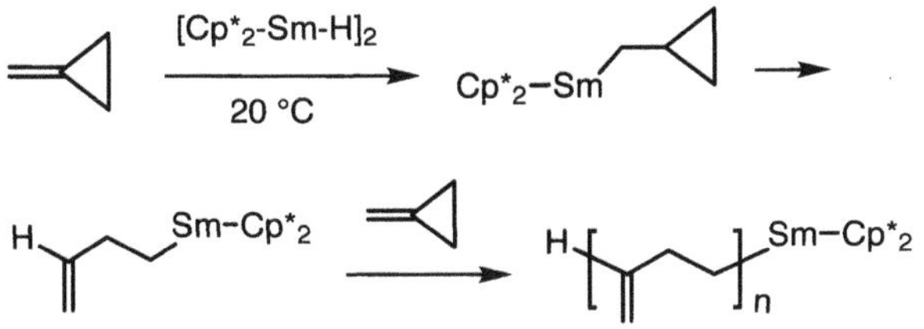

(53)

Palladium complexes derived from strained bicyclic systems like norbornene and pinene also provide examples of β-carbon elimination [90–92].

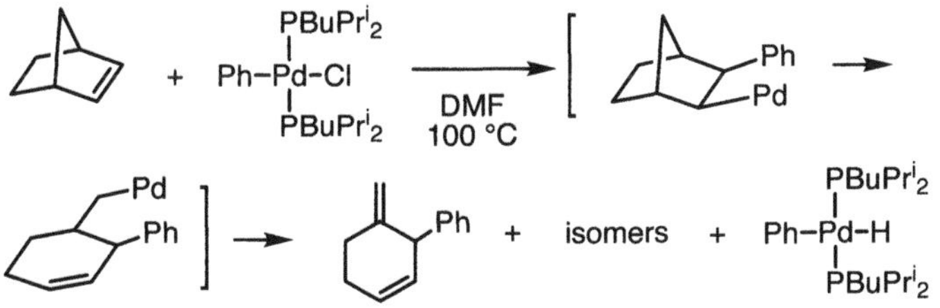

(54)

A carbon-silicon bond can be cleaved by a process of β-methyl elimination. Thermolytic rearrangement of platinum(II) complex **76** yields methyl-platinum(II) complex **78** [93]. Generation of (η^2-silene)platinum intermediate **77** by β-methyl elimination is followed by migration of the other silylmethyl ligand to the silicon terminus of the η^2-silene ligand.

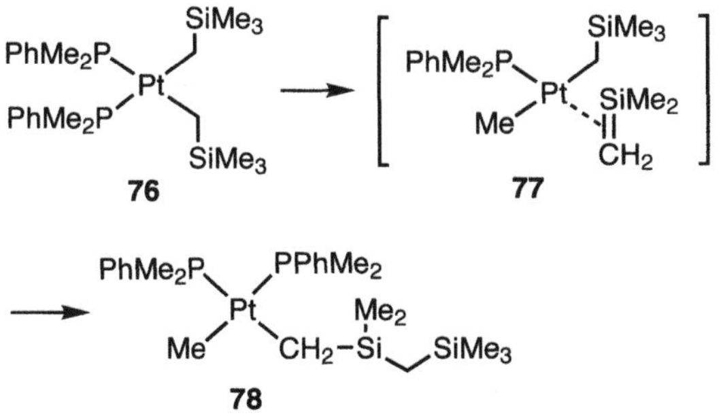

(55)

2.3
Miscellaneous Stoichiometric Reactions

The formation of a metallacyclopentane from a low-valent transition metal and an olefin is often reversible [94, 95]. In the reverse process, a C–C bond is cleaved generating two molecules of the olefin, and a two-electron reduction of the metal occurs. The following reactions provide unambiguous evidence for the reversibility of this process [96–98].

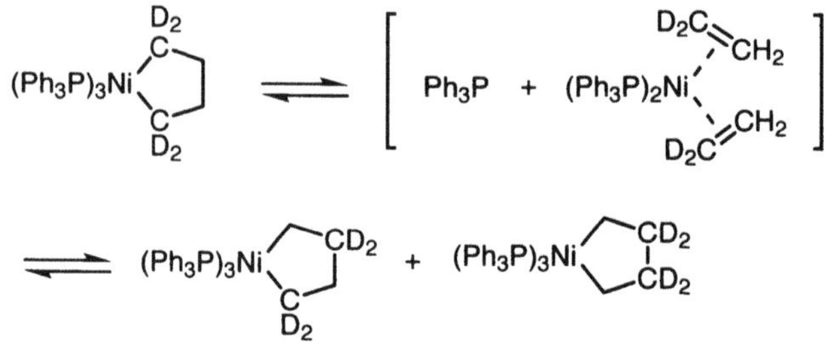

(56)

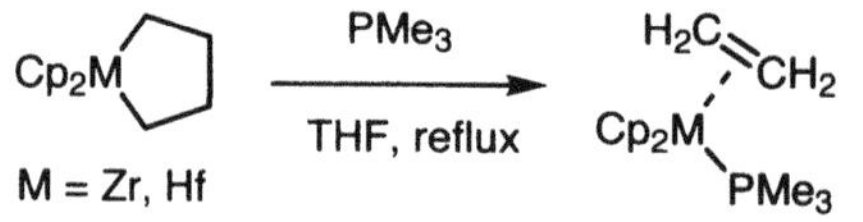

(57)

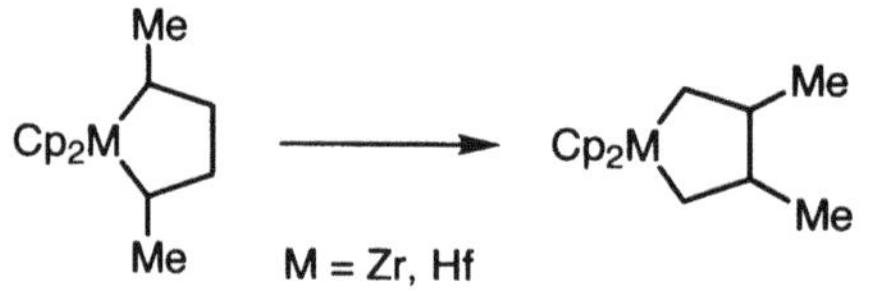

(58)

Reductive cleavage of zirconacyclopentene was utilized for the synthesis of various zirconacycles by C–C bond cleavage/formation sequences [99].

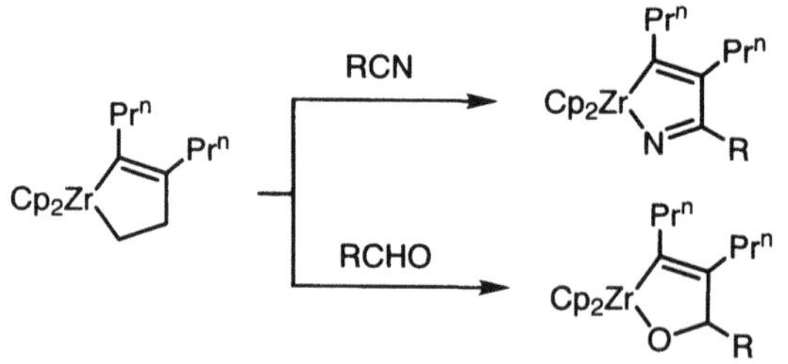

(59)

Treatment of tetrayne **79** with four equivalents of a metallocene complex leads to complex **80** resulting from cleavage of the two outer C–C single bonds [100].

$$79 + 4\,Cp_2M(\eta^2\text{-}Me_3SiC{\equiv}CSiMe_3) \longrightarrow 80 \quad (M = Ti, Zr) \tag{60}$$

3 Catalytic Reactions Involving C–C Bond Cleavage

In this section, we shall look at important applications of organometallic chemistry for C–C bond activation. The term "catalytic reaction" used herein is defined as a transformation in which organic substances are converted into different forms with the aid of a much smaller amount of a soluble transition metal complex. Examples of catalytic reactions involving C–C bond cleavage are far more rare than the corresponding stoichiometric reactions. One of the major additional difficulties a catalytic system must overcome is the thermodynamic balance of the participating organic substances. For a productive catalytic cycle, the organic products must be thermodynamically more stable than the starting materials. Although this applies to all catalytic organic transformations, it deserves special mention here. In the particular case of C–C bond cleavage, the key functionality in the starting substance is a C–C bond, which is an extremely stable bond. This must be ultimately transformed into a more stable form, without taking a metal component into consideration, since the real catalyst does not alter its form upon completion of a catalytic cycle. In addition to the large activation barrier necessary for C–C bond breaking, this thermodynamic requirement has to be fulfilled in the organic components of any catalytic system. This is mostly achieved with the assistance of a designed thermodynamic driving force. Although far from numerous, the examples presented in this section clearly demonstrate the synthetic potential of the transition metal-catalyzed C–C bond cleavage and its promising applications in organic chemistry.

Cleavage of the central C–C bond of biphenylene (Eqs. 8–12) was successfully extended to catalytic reactions. Tetraphenylene is formed quantitatively from biphenylene in the presence of a catalytic amount of $Ni(cod)(PMe_3)_2$ [101].

$$\text{biphenylene} \xrightarrow[\text{THF, }100\,^{\circ}\text{C}]{Ni(cod)(PMe_3)_2\ (10\ \text{mol\%})} \text{tetraphenylene (quant)} \tag{61}$$

Jones et al. recently found that $Pt(PEt_3)_3$ and $Pd(PEt_3)_3$ are also capable of catalyzing the formation of tetraphenylene from biphenylene (16 turnovers/day with $Pd(PEt_3)_3$ at 120°C) [102]. A detailed mechanistic study was performed with platinum. The catalytic cycle which involves sequential oxidative addition of two molecules of biphenylene onto Pt(0) and Pt(II) is proposed on the basis of kinetic analysis and identification of the intermediate Pt(II) and Pt(IV) complexes **81–83**.

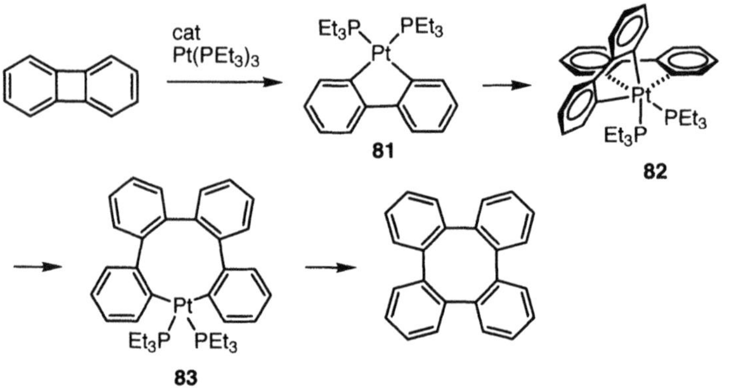

cat $Pt(PEt_3)_3$; Et_3P PEt_3 Pt **81** ; Pt PEt_3 Et_3P **82** ; Pt Et_3P PEt_3 **83**

(62)

Biphenylene is catalytically hydrogenolyzed to biphenyl (7 turnovers/7 days) in the presence of a catalytic amount of $Cp^*Rh(PMe_3)(H)_2$ [27]. Unlike the case which produces tetraphenylene (Eq. 62), the five-membered metallacycle **84**, envisioned to be involved in the catalytic cycle, fails to catalyze hydrogenolysis.

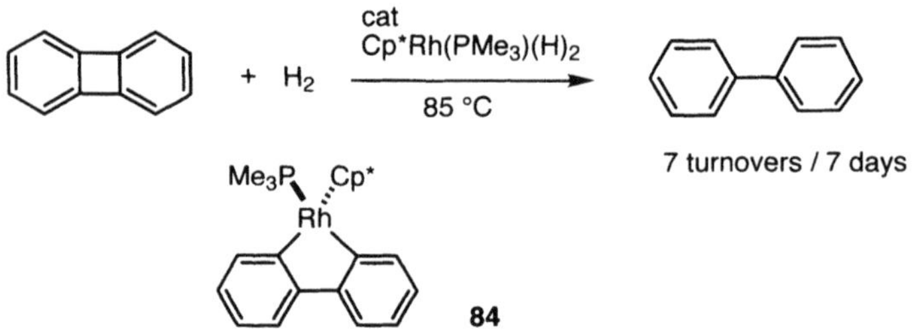

+ H_2 ; cat $Cp^*Rh(PMe_3)(H)_2$; 85 °C ; 7 turnovers / 7 days

Me_3P Cp^* Rh **84**

(63)

When biphenylene is treated with $Cp^*Rh(CO)_2$ under carbon monoxide, fluorenone is formed (ca. 1 turnover/day) [28].

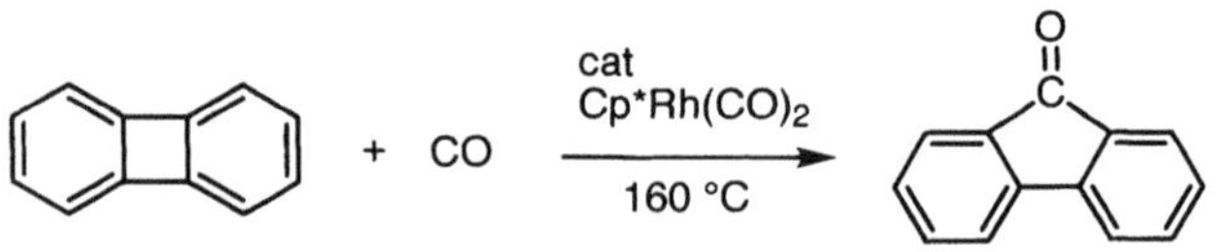

+ CO ; cat $Cp^*Rh(CO)_2$; 160 °C ; O C

(64)

The palladium-catalyzed acylation of siloxycyclopropane furnishes a 1,4-dicarbonyl compound. A C–C bond of the three-membered ring is cleaved by an electrophilic attack of a palladium(II) species [103]. An analogous electrophilic ring opening of siloxycyclopropane was induced by various so-called ligand free transition metals such as Ag^+ and $Cu2^+$ [104].

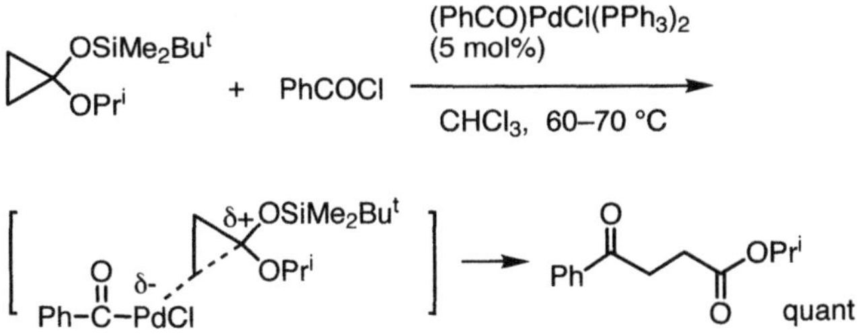

(65)

Carbonyl groups are also utilized in catalytic C–C bond cleaving reactions. Under catalytic conditions, 8-quinolyl phenyl ketone **85** reacts with ethylene to give 8-quinolyl ethyl ketone **86** and styrene in quantitative yield [105]. Styrene is formed by cleavage of the phenyl-carbonyl bond, followed by ethylene insertion into the resultant phenyl-rhodium bond, and β-hydride elimination. The accompanying formation of a rhodium-hydride complex is followed by incorporation of ethylene to furnish the ethyl ketone **86**.

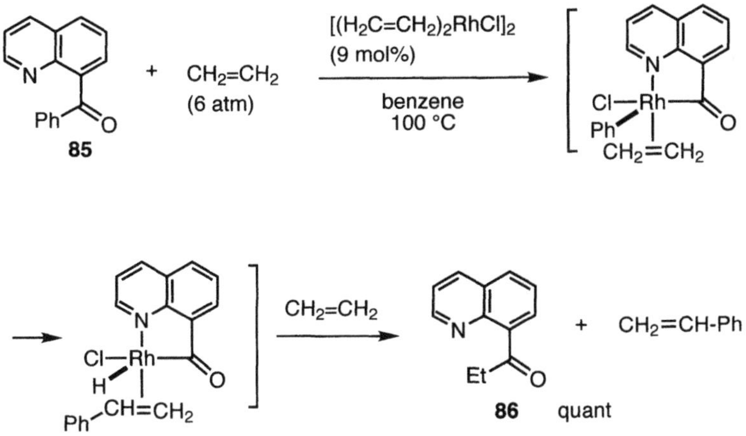

(66)

Cyclobutanones are catalytically decarbonylated by rhodium [46, 47]. Appropriate choice of the catalyst system leads to the selective formation of either a cyclopropane or an alkene.

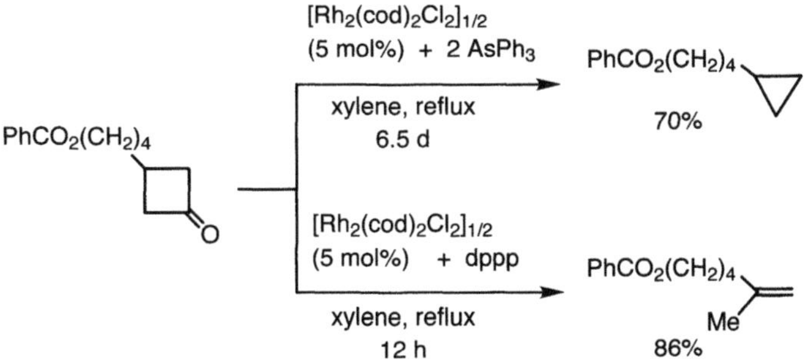

(67)

When cyclobutanone is treated under dihydrogen with a catalytic amount of a rhodium(I) complex containing a bidentate diphosphine ligand, the ring opened alcohol **89** is produced in good yield [46, 47]. The oxidative addition intermediate **87** is hydrogenated to give aldehyde **88**, which is further reduced to the alcohol **89**.

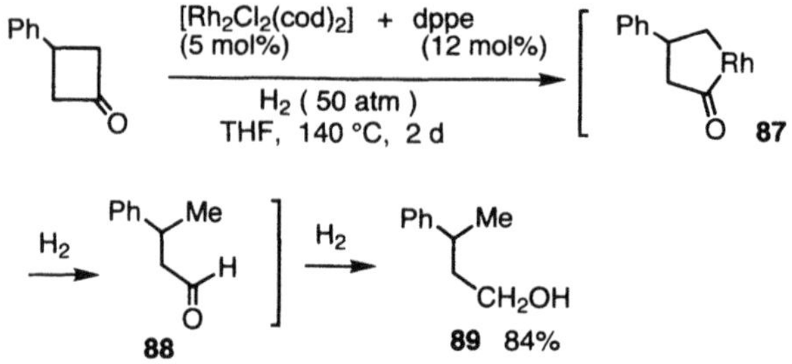

$[Rh_2Cl_2(cod)_2]$ (5 mol%) + dppe (12 mol%), H_2 (50 atm), THF, 140 °C, 2 d; **87**; H_2; **88**; H_2; **89** 84%

(68)

Substituted phenols are synthesized by the nickel(0)-catalyzed ring opening of cyclobutenones and subsequent [4+2]cycloaddition with alkynes [106].

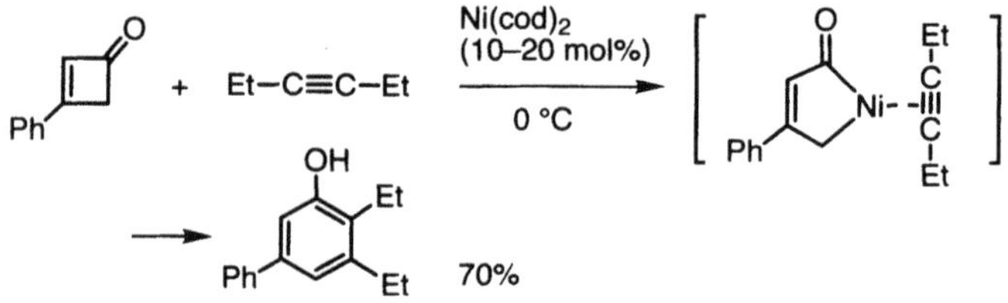

+ Et–C≡C–Et; $Ni(cod)_2$ (10–20 mol%), 0 °C; 70%

(69)

Catalytic decarbonylation of α- and β-dicarbonyl compounds is mediated by $RhCl(PPh_3)_3$ giving the corresponding mono-ketones [107].

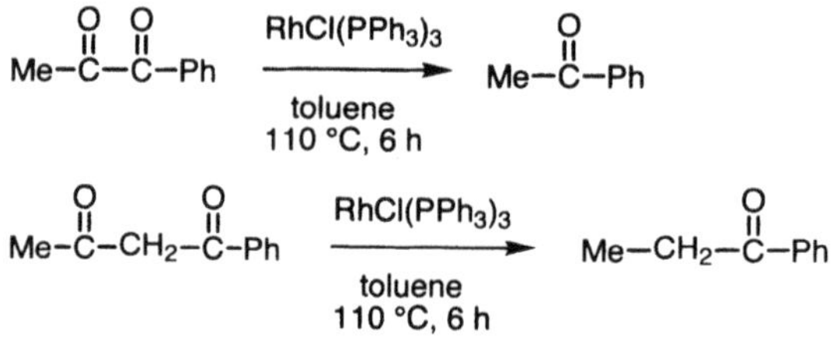

Me–C(=O)–C(=O)–Ph → ($RhCl(PPh_3)_3$, toluene, 110 °C, 6 h) → Me–C(=O)–Ph

Me–C(=O)–CH_2–C(=O)–Ph → ($RhCl(PPh_3)_3$, toluene, 110 °C, 6 h) → Me–CH_2–C(=O)–Ph

(70)

Acyl cyanides undergo oxidative addition to transition metals like rhodium and palladium, leading to catalytic decarbonylation [108, 109]. This process is also involved in the catalytic acylcyanation of terminal alkynes [110].

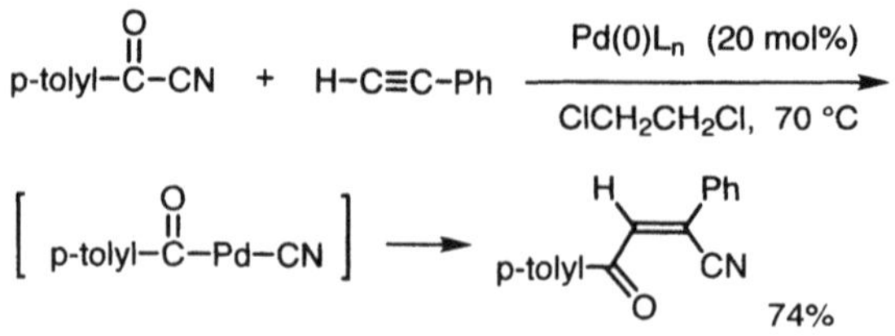

p-tolyl–C(=O)–CN + H–C≡C–Ph → ($Pd(0)L_n$ (20 mol%), $ClCH_2CH_2Cl$, 70 °C) → [p-tolyl–C(=O)–Pd–CN] → 74%

(71)

The reaction of cyclopropenone with ketene in the presence of $Ni(CO)_4$ affords cyclopentenedione as a mixture of regioisomers [111].

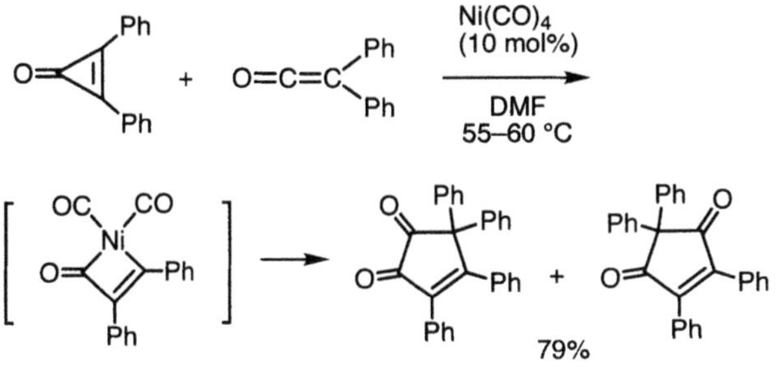

$Ni(CO)_4$ (10 mol%), DMF, 55–60 °C

79%

(72)

The cleavage of a diphosphine pincer ligand previously described (Eqs. 26, 27) was recently applied to a catalytic process. The methylene group was excised from the pincer ligand **90** under dihydrogen to afford **91** along with methane [112]. Dihydrogen can replaced by hydrosilane. Exchange of the hydrogenated ligand **91** with the starting ligand **90** is likely to be the rate-determining step.

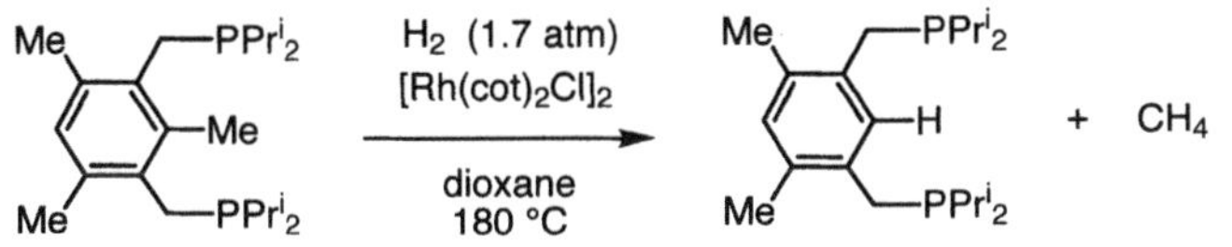

H_2 (1.7 atm), $[Rh(cot)_2Cl]_2$, dioxane, 180 °C; + CH_4

90 **91** 106 turnovers in 3 days

(73)

When norbornadiene is allowed to react under the influence of a nickel catalyst in amine, 5-tolyl-2-norbornene **93** is formed [113]. After dimerization of norbornadiene, β-carbon elimination occurs to open the norbornane ring of **92**. An analogous cleavage of a norbornane ring was observed in the palladium-catalyzed reaction of bromobenzene with norbornene [114].

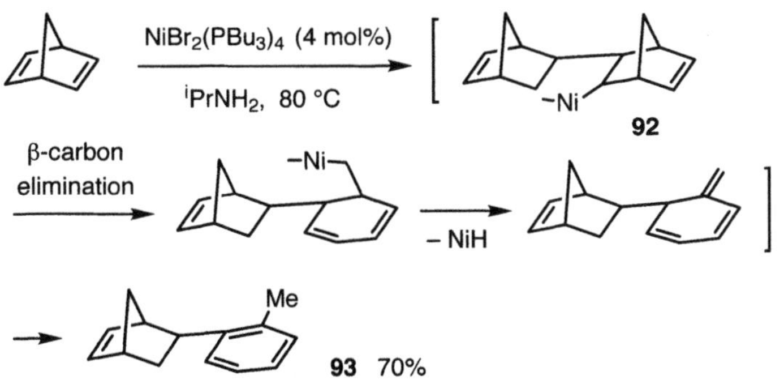

$NiBr_2(PBu_3)_4$ (4 mol%), $^iPrNH_2$, 80 °C; **92**; β-carbon elimination; – NiH

93 70%

(74)

The nickel-catalyzed allylation of norbornene affords compound **95**. This result suggests that a C–C bond cleavage process is occurring via (cyclobutylmethyl)nickel intermediate **94** [115].

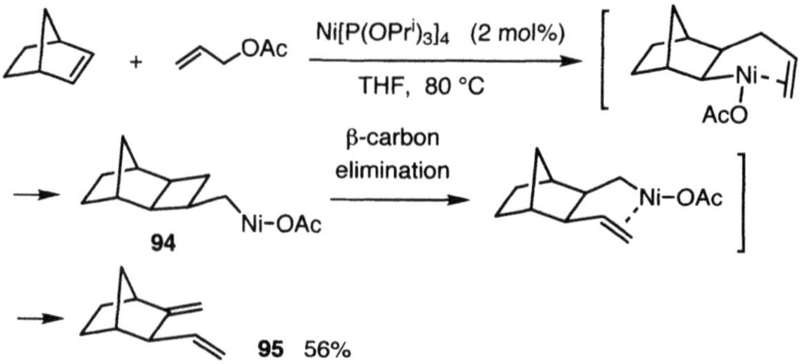

OAc
$Ni[P(OPr^i)_3]_4$ (2 mol%)
THF, 80 °C
Ni
AcO
β-carbon elimination
Ni-OAc
94
Ni-OAc
95 56%

(75)

In the presence of scandocene hydride **63**, 3-methyl-1,4-pentadiene is catalytically converted to methylenecyclopentane and its isomer via cyclobutylmethylmetal intermediate **64** [80].

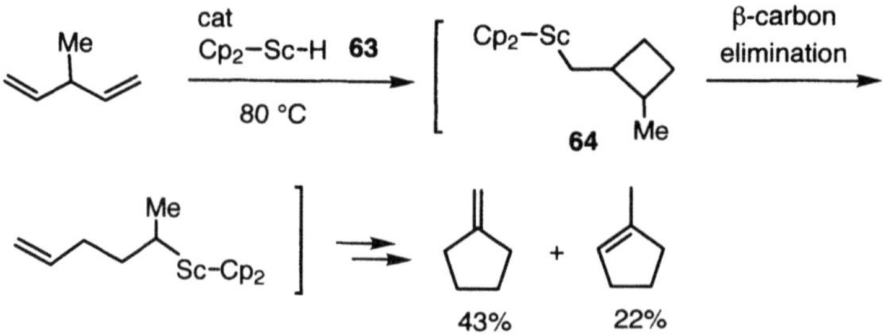

Me
cat
Cp_2-Sc-H **63**
80 °C
Cp_2-Sc
64 Me
β-carbon elimination
Me
Sc-Cp_2
43% 22%

(76)

Since oxidative addition of a C–C bond onto a transition metal results in the formation of a σ alkyl-metal complex, the two elementary steps of C–C bond cleavage, i.e., oxidative addition and β-carbon elimination could operate in sequence. Interesting examples of this sort are found in intramolecular carbocyclic ring enlargement reactions. 4-Cyclobutyl-2-cyclobutenone **96** undergoes rhodium(I)-catalyzed successive double C–C bond cleavage, giving cyclooctadienone [116].

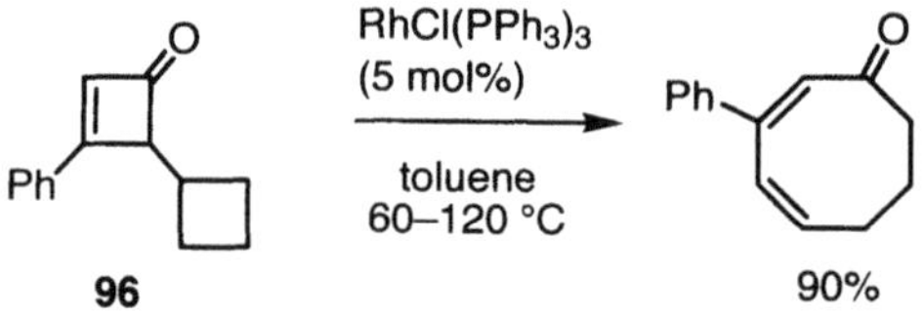

O
Ph
96
$RhCl(PPh_3)_3$ (5 mol%)
toluene 60–120 °C
Ph
O
90%

(77)

The reaction of a spiro cyclobutanone equipped with a second four-membered ring (**97**) catalyzed by $[Rh(dppp)_2]Cl$ gives rise to 2-cyclohexenone **98** [117]. Rhodium successively cleaves the two C–C bonds of **97**, the first by oxidative addition and the second by β-carbon elimination.

97 → [Rh(dppp)$_2$]Cl (5 mol%), xylene reflux, 13 h → β-carbon elimination → 98 89% (78)

The bond between the carbon atoms α and β to a C–C double bond can be broken by a transition metal with formation of a π-allyl intermediate providing the driving force. Whereas stoichiometric reactions of this sort are yet to appear, π-(allyl)metal intermediates are occasionally involved in catalytic C–C bond cleaving reactions. The nickel catalyzed skeletal rearrangement of 1,4-dienes involves the formation of an olefin coordinated π-(allyl)nickel complex (**99**) [118].

cat [NiCl$_2$(PBu$_3$)$_2$], ^{i}Bu$_2$AlCl, r t; – NiH; **99** (79)

Allylic malonate **100** completely isomerizes to the thermodynamically favored linear isomer **101** on treatment with a palladium catalyst [119]. Formation of a stabilized carbanion and π-(allyl)palladium species facilitates the C–C bond cleavage. Analogous isomerization is also catalyzed by a nickel complex [120]. These results demonstrate that the transition metal-catalyzed nucleophilic substitution of an allylic substrate with a carbon nucleophile is reversible, if the cleaved nucleophile is sufficiently stabilized.

100 C(Me)(CO$_2$Me)$_2$ → Pd(OAc)$_2$–PBu$_3$ (5 mol%), NaC(Me)(CO$_2$Me)$_2$, THF, 140 °C, 24 h → [Pd$^+$ $^-$C(Me)(CO$_2$Me)$_2$] → **101** C(Me)(CO$_2$Me)$_2$ (80)

On the other hand, a nucleophilic addition reaction of a π(-allyl)ruthenium complex to a ketone is also reversible. The deallylation of a tertiary homoallylic alcohol is catalytically mediated by $RuCl_2(PPh_3)_3$ to afford acetophenone and propene [121]. The reaction involves oxidative addition of the hydroxyl group to ruthenium and subsequent β-allyl elimination.

$RuCl_2(PPh_3)_3$ (5 mol%), CO (10 atm), allyl-OAc, THF, 180 °C → H-[Ru]-O... ; Ph(C=O)Me 91% + [Ru]-H → propene (81)

A six-membered cyclic allylic carbonate **102** undergoes a palladium-catalyzed decarboxylative C–C bond cleavage to afford dienic carbonyl compound **104** [122]. Decarboxylation of the allylic carbonate moiety provides the driving force for production of the intermediate five-membered hetero-palladacycle **103**, from which formal reductive cleavage takes place.

102 → $Pd_2(dba)_3$ (5 mol%), CH_3CN → [**103**] → **104** 82% (82)

An unusual dimerization of norbornadiene is catalyzed by Ru(cod)(cot) to afford cage compound **105** [123]. Although the precise mechanism is unclear, the reaction obviously involves multiple activations of C–C bonds.

2 norbornadiene → Ru(cod)(cot) (5 mol%), $CH_2=CHCONMe_2$, toluene, 120 °C → **105** 93 % (83)

4 Perspective

The activation of C–C bonds by transition metal complexes has been a topic of special interest in both the inorganic and the organic areas of organometallic chemistry. As a result, the chemistry has progressed significantly over the past several decades. However, we are still at a very early stage of our quest. The insertion of a transition metal into an unstrained bond between two sp3 carbon atoms in a selective fashion offers the most difficult and ultimate challenge. The authors believe that the abilities of organometallic chemists to tune the ligand set of metal complexes and to design thermodynamically downhill reaction systems will represent a true adventure in this field in the coming century.

References

1. Crabtree RH (1985) Chem Rev 85:245
2. Schröder D, Schwarz H (1995) Angew Chem Int Ed Engl 34:1973
3. Crabtree RH, Holt EM, Lavin M, Morehouse SM (1985) Inorg Chem 24:1986
4. Tipper CHF (1955) J Chem Soc:2045
5. Adams, DM, Chatt J, Guy R, Sheppard N (1961) J Chem Soc:738
6. Gillard RD, Keeton M, Mason R, Pilbrow MF, Russell DR (1971) J Organomet Chem 33:247
7. McQuillin FJ, Powell KG (1972) J Chem Soc Dalton:2123
8. Lenarda M, Ros R, Graziani M, Belluco U (1974) J Organomet Chem 65:407
9. Yarrow DJ, Ibers JA, Lenarda M, Graziani M (1974) J Organomet Chem 70:133
10. Roundhill DM, Lawson, DN, Wilkinson, G (1968) J Chem Soc (A):845
11. Powell KG, McQuillin FJ (1971) Chem Commun:931
12. McQuillin FJ, Powell KG (1972) J Chem Soc Dalton:2129
13. Cassar L, Halpern J (1970) Chem Commun:1082
14. Periana RA, Bergman RG (1986) J Am Chem Soc 108:7346
15. Noyori R, Nishimura T, Takaya H (1969) Chem Commun:89
16. Hoberg JO, Larsen RD, Jennings PW (1990) Organometallics 9:1334
17. Ikura K, Ryu I, Ogawa A, Sonoda N, Harada S, Kasai N (1991) Organometallics 10:528
18. Hemond RC, Hughes RP, Robinson DJ, Rheingold AL (1988) Organometallics 7:2239
19. Jennings PW, Johnson LL (1994) Chem Rev 94:2241
20. Bishop III KC (1976) Chem Rev 76:461
21. Ohta T, Takaya H (1991) Metal-catalyzed cycloaddition of small ring compounds. In: Trost BM, Fleming I (eds) Comprehensive organic synthesis, vol 5. Pergamon, Oxford, p 1185
22. Binger P, Büch HM (1987) Cyclopropenes and methylenecyclopropanes as multifunctional reagents in transition metal catalyzed reactions. In: de Meijere A (ed) Topics in organic chemistry, vol 135. Springer, Berlin Heidelberg New York, p 77
23. Khusnutdinov RI, Dzhemilev UM (1994) J Organomet Chem 471:1
24. Atkinson ER, Levins PL, Dickelman TE (1964) Chem Ind:934
25. Eisch JJ, Piotrowski AM, Han KI, Krüger C, Tsay YH (1985) Organometallics 4:224
26. Lu Z, Jun CH, de Gala SR, Sigalas M, Eisenstein O, Crabtree RH (1993) J Chem Soc Chem Commun:1877
27. Perthuisot C, Jones WD (1994) J Am Chem Soc 116:3647
28. Perthuisot C, Edelbach BL, Zubris DL, Jones WD (1997) Organometallics 16:2016
29. Cassar L, Eaton PE, Halpern J (1970) J Am Chem Soc 92:3515
30. Ros R, Lenarda M, Pahor NB, Calligaris M, Delisa P, Randaccio L, Graziani M (1976) J Chem Soc Dalton:1937
31. Rusina A, Vlcek AA (1965) Nature 206:295
32. Wong W, Singer SJ, Pitts WD, Watkins SF, Baddley (1972) J Chem Soc Chem Commun:672
33. Visser JP, Ramakers-Blom JE (1972) J Organomet Chem 44:C63
34. Song L, Arif AM, Stang PJ (1990) Organometallics 9:2792

35. Kolomnikov IS, Svonoda P, Vol'pin ME (1972) Izv Akad Mauk SSSR Ser Khim 12:2818. (1973) Chem Ab 78:97789k
36. Evans JA, Everitt GF, Kemmitt RDW, Russell DR (1973) J Chem Soc Chem Commun:158
37. Hamner ER, Kemmitt RDW, Smith MAR (1974) J Chem Soc Chem Commun:841
38. Liebeskind LS, Baysdon SL, South MS (1980) J Organomet Chem 202:C73
39. Liebeskind LS, Baysdon SL, South MS, Iyer S, Leeds JP (1985) Tetrahedron 41:5839
40. Huffman MA, Liebeskind LS, Pennington WT (1992) Organometallics 11:255
41. Müller E, Segnittz (1973) Liebigs Ann Chem:1583
42. Suggs JW, Jun CH (1984) J Am Chem Soc 106:3054
43. Suggs JW, Jun CH (1986) J Am Chem Soc 108:4679
44. Jun CH, Kang JB, Lim YG (1995) Tetrahedron Lett 36:277
45. Suggs JW, Wovkulich MJ, Cox SD (1985) Organometallics 4:1101
46. Murakami M, Amii H, Ito Y (1994) Nature 370:540
47. Murakami M, Amii H, Shigeto K, Ito Y (1996) J Am Chem Soc 118:8285
48. Gozin M, Weisman A, Ben-David Y, Milstein D (1993) Nature 364:699
49. Gozin M, Aizenberg M, Liou SY, Weisman A, Ben-David Y, Milstein D (1994) Nature 370:42
50. Liou SY, Gozin M, Milstein D (1995) J Chem Soc Chem Commun:1965
51. van der Boom ME, Kraatz HB, Ben-David Y, Milstein D (1996) Chem Commun:2167
52. Liou SY, Gozin M, Milstein D (1995) J Am Chem Soc 117:9774
53. Rybtchinski B, Vigalok A, Ben-David Y, Milstein D (1996) J Am Chem Soc 118:12406
54. van der Boom ME, Ben-David Y, Milstein D (1998) Chem Commun:917
55. Gandelman M, Vigalok A, Shimon LJW, Milstein D (1997) Organometallics 16:3981
56. Kang JW, Moseley K, Maitlis PM (1969) J Am Chem Soc 91:5970
57. Benfield FWS, Green MLH (1974) J Chem Soc Dalton Trans:1324
58. Eilbracht P, Dahler P (1980) Chem Ber 113:542
59. Hemond RC, Hughes RP, Locker HB (1986) Organometallics 5:2391
60. Jones WD, Maguire JA (1987) Organometallics 6:1301
61. Crabtree RH, Dion RP, Gibboni DJ, McGrath DV, Holt EM (1986) J Am Chem Soc 108:7222
62. Suzuki H, Takaya Y, Takemori T, Tanaka M (1994) J Am Chem Soc 116:10779
63. Bennett MA, Nicholls JC, Rahman AKF, Redhouse AD, Spencer JL, Willis AC (1989) J Chem Soc Chem Commun:1328
64. Nicholls JC, Spencer JL (1994) Organometallics 13:1781
65. Arce MJ, Viado AL, An YZ, Khan SI, Rubin Y (1996) J Am Chem Soc 118:3775
66. Shaltout RM, Sygula R, Sygula A, Fronczek FR, Stanley GG, Rabideau PW (1998) J Am Chem Soc 120:835
67. Watson PL, Roe DC (1982) J Am Chem Soc 104:6471
68. Hajela S, Bercaw JE (1994) Organometallics 13:1147
69. Horton AD (1996) Organometallics 15:2675
70. Etienne M, Mathieu R, Donnadieu B (1997) J Am Chem Soc 119:3218
71. Eshuis JJW, Tan YY, Teuben JH, Renkema J (1990) J Mol Cat 62:277
72. Eshuis JJW, Tan YY, Meetsma A, Teuben JH (1992) Organometallics 11:362
73. Mise T, Kageyama A, Miya S, Yamazaki H (1991) Chem Lett:1525
74. Resconi L, Piemontesi F, Franciscono G, Abis L, Fiorani T (1992) J Am Chem Soc 114:1025
75. Kesti MR, Waymouth RM (1992) J Am Chem Soc 114:3565
76. Hartwig JF, Bergman RG, Andersen RA (1991) Organometallics 10:3344
77. McNeill K, Andersen RA, Bergman RG (1997) J Am Chem Soc 119:11244
78. Tjaden EB, Stryker JM (1990) J Am Chem Soc 112:6420
79. Hosokawa T, Maitlis PM (1972) J Am Chem Soc 94:3238
80. Bunel E, Burger BJ, Bercaw JE (1988) J Am Chem Soc 110:976

81. Casey CP, Hallenbeck SL, Pollock DW, Landis CR (1995) J Am Chem Soc 117:9770
82. Larock RC, Varaprath S (1984) J Org Chem 49:3432
83. Flood TC, Bitler SP (1984) J Am Chem Soc 106:6076
84. Ermer SP, Struck GE, Bitler SP, Richards R, Bau R, Flood TC (1993) Organometallics 12:2634
85. Flood TC, Statler JA (1984) Organometallics 3:1795
86. Jun CH (1996) Organometallics 15:895
87. Yang X, Jia L, Marks TJ (1993) J Am Chem Soc 115:3392
88. Yang X, Seyam AM, Fu PF, Marks TJ (1994) Organometallics 27:4625
89. Jia L, Yang X, Yang S, Marks TJ (1996) J Am Chem Soc 118:1547
90. Larock RC, Song H, Kim S, Jacobson RA (1987) J Chem Soc Chem Commun:834
91. Marinelli AAF, Bernocchi E, Cacchi S, Ortar G (1989) J Organomet Chem 368:249
92. Portnoy M, Ben-David Y, Rousso I, Milstein D (1994) Organometallics 13:3465
93. Ankianiec BC, Christou V, Hardy DT, Thompson SK, Young GB (1994) J Am Chem Soc 116:9963
94. McDermott JX, Wilson ME, Whitesides GM (1976) J Am Chem Soc 98:6529
95. Erker G, Dolf U, Rheingold AL (1988) Organometallics 7:138
96. Grubbs RH, Miyashita A (1978) J Am Chem Soc 100:1302
97. Takahashi T, Tamura M, Saburi M, Uchida Y, Negishi E (1989) J Chem Soc Chem Commun:852
98. Takahashi T, Fujimori T, Seki T, Saburi M, Uchida Y, Rousset CJ, Negishi E (1990) J Chem Soc Chem Commun:182
99. Takahashi T, Kageyama M, Denisov V, Hara R, Negishi E (1993) Tetrahedron Lett 34:687
100. Pellny PM, Peulecke N, Burlakov VV, Tillack A, Baumann W, Spannerberg A, Kempe R, Rosenthal U (1997) Angew Chem Int Ed Engl 36:2615
101. Schwager H, Spyroudis S, Vollhardt KPC (1990) J Organomet Chem 382:191
102. Edelbach BL, Lachicotte RJ, Jones WD (1998) J Am Chem Soc 120:2843
103. Fujimura T, Aoki S, Nakamura E (1991) J Org Chem 56:2809
104. Ryu I, Ando M, Ogawa A, Murai S, Sonoda N (1983) J Am Chem Soc 105:7192
105. Suggs JW, Jun CH (1985) J Chem Soc Chem Commun:92
106. Huffman MA, Liebeskind LS (1991) J Am Chem Soc 113:2771
107. Kaneda K, Azuma H, Wayaku M, Teranishi S (1974) Chem Lett:215
108. Blum J, Oppenheimer E, Bergmann ED (1967) J Am Chem Soc 89:2338
109. Murahashi SI, Naota T, Nakajima N (1986) J Org Chem 51:898
110. Nozaki K, Sato N, Takaya H (1996) Bull Chem Soc Jpn 69:1629
111. Baba A, Ohshiro Y, Agawa T (1976) J Organomet Chem 110:121
112. Liou SY, van der Boom ME, Milstein D (1998) Chem Commun:687
113. Yoshikawa S, Aoki K, Kiji J, Furukawa J (1974) Tetrahedron 30:405
114. Catellani M, Chiusoli GP, Dradi E, Salerno G (1979) J Organomet Chem 177:C29
115. Catellani M, Chiusoli GP (1983) J Organomet Chem 247:C59
116. Huffman MA, Liebeskind LS (1993) J Am Chem Soc 115:4895
117. Murakami M, Takahashi K, Amii H, Ito Y (1997) J Am Chem Soc 119:9307
118. Golden HJ, Baker DJ, Miller RG (1974) J Am Chem Soc 96:4235
119. Nilsson YIM, Andersson PG, Bäckvall JE (1993) J Am Chem Soc 115:6609
120. Bricout H, Carpentier JF, Mortreux A (1997) Tetrahedron Lett 38:1053
121. Kondo T, Kodoi K, Nishinaga E, Okada T, Morisaki Y, Watanabe Y, Mitsudo T (1998) J Am Chem Soc 120:5587
122. Harayama H, Kuroki T, Kimura M, Tanaka S, Tamaru Y (1997) Angew Chem Int Ed Engl 36:2352
123. Mitsudo T, Zhang SW, Watanabe Y (1994) J Chem Soc Chem Commun:435

Activation of Si–Si Bonds by Transition-Metal Complexes

Michinori Suginome and Yoshihiko Ito

Department of Synthetic Chemistry and Biological Chemistry, Graduate School of Engineering, Kyoto University, Kyoto 606-8501, Japan
E-mail: suginome@sbchem.kyoto-u.ac.jp and yoshi@sbchem.kyoto-u.ac.jp

Transition-metal complexes have enabled the effective activation of Si–Si bonds, having led to the synthetic utilization of new, reactive organisilyl transition-metal species thus generated. Recent detailed investigations on this subject have shed new light on application on the basis of the activation for efficient synthesis of useful organosilicon compounds. This chapter describes the stoichiometric and catalytic generation of organosilyl transition-metal complexes through the activation of the Si–Si bonds with emphasis on the catalytic reactions of organodisilanes mediated by transition-metal complexes. The application of these catalytic reactions to organic synthesis is also mentioned briefly.

Keywords: Organopolysilane, Oxidative addition, bis-Silylation, σ-Bond metathesis, Silylene transfer

1	**Introduction**	132
2	**Stoichiometric Activation**	133
2.1	Oxidative Addition onto Low Valent Transition-Metals	133
2.1.1	Nickel	133
2.1.2	Palladium	133
2.1.3	Platinum	136
2.1.4	Other Transition Metals	136
2.2	Double Oxidative Addition Reactions	137
2.3	Silylene-Migration in Disilanyl-Transition-Metal Complexes	139
3	**Catalytic Activation**	140
3.1	Bis-Silylation	140
3.1.1	Alkyne	140
3.1.2	Diene	143
3.1.3	Alkene	145
3.1.4	Carbonyl Compounds	148
3.1.5	Isonitrile	149
3.2	Silylene Transfer	150
3.3	Si–Si σ-Bond Metathesis	151
3.3.1	Ring-Enlargement Oligomerization	151

3.3.2 Ring-Opening Polymerization 152
3.4 Silylation of Organic Halides and Allylic Esters. 153

4 **Application to Organic Synthesis**. 154

4.1 Via Allylsilanes 154
4.2 Via Peterson-Type Elimination 154
4.3 Via Oxidation of Silicon-Carbon Bonds 155

5 **Concluding Remarks**. 156

References 157

1 Introduction

Silicon-silicon bonds are nonpolarized, thermally stable σ-bonds with a dissociation energy of ca. 300 kJ/mol [1]. Thus, not only disilanes having isolated Si–Si bonds but also oligosilanes and polysilanes, which have contiguous, "conjugated" Si–Si bonds, have been successfully synthesized in stable forms and characterized [2]. The Si–Si σ-bonds are characterized by their high-energy σ- and low-energy σ*-orbitals, which have been of interest from the viewpoint of application of the oligo- and polysilanes for functionalized materials. The characteristic nature of the bonds also enables interaction of the σ-bonds with transition-metal complexes.

Recently, there has been increasing demand for the development of new synthetic methods for organosilicon compounds. As has been demonstrated by transition-metal catalyzed reactions of hydrosilanes, generation and synthetic application of reactive organosilyl complexes of transition-metals through the activation of silicon-containing σ-bonds is highly desirable for the new synthesis of organosilicon compounds [3, 4]. This chapter deals with the generation of various organosilyl-transition-metal complexes through the activation of Si–Si bonds [5–9]. Stoichiometric reactions of disilanes with transition-metal complexes are mentioned first followed by catalytic reactions involving the activation of Si–Si bonds with transition-metal complexes. Though the mechanisms of most catalytic reactions still have to be elucidated, they may conceivably be understood on the basis of the stoichiometric reactions. In the final section, synthetic applications of these catalytic reactions are briefly mentioned.

2 Stoichiometric Activation

2.1 Oxidative Addition onto Low Valent Transition-Metals

2.1.1 *Nickel*

Nickel tetracarbonyl undergoes a rapid oxidative addition of the Si–Si bond of **1**, highly strained fluorinated disilane, at room temperature to give five-membered cyclic bis(organosilyl)nickel(II) complex **2**, which then reacts with *tert*-butylacetylene to give six-membered disilacyclohexadiene derivatives **3** as a mixture of the regioisomers (Eq. 1) [10]. A similar bis-silylation reaction of alkynes with bis(organosilyl)nickel(II) complex has been reported in the reaction of bis(trichlorosilyl)(bipy)nickel(II) (bipy: 2,2′-bipyridyl), which is prepared by dialkyl(bipy)nickel(II) with trichlorosilane [11].

t-Bu, SiF_2, SiF_2 **1** —(1eq. $Ni(CO)_4$)→ t-Bu, F_2Si, $Ni(CO)_2$, SiF_2 **2** —(t-Bu—≡)→ t-Bu, F_2Si, t-Bu, SiF_2 **3** (1)

Bis(trichlorosilyl)nickel(II) complex **4**, having an η^6-arene ligand, is prepared by reaction of hexachlorodisilane with highly reactive, vaporized nickel in the presence of toluene (Eq. 2) [12]. Worthy of note is that the arene ligand is displaced by three molecules of carbon monoxide to give **5** [13].

Ni(vapor) + $SiCl_3$–$SiCl_3$ → (η^6-toluene)Ni($SiCl_3$)($SiCl_3$) **4** —(CO)→ $(OC)_3Ni(SiX_3)_2$ **5** (2)

2.1.2 *Palladium*

Oxidative addition of Si–Si bonds onto palladium(0) has long been presumed to be involved in a number of palladium-catalyzed bis-silylation reactions of unsaturated carbon compounds. The oxidative addition and its reverse reaction, i.e., reductive elimination, may be in rapid equilibrium, whose direction is influenced by the structure of disilanes and ligands on the palladium atom. In spite of early reports on the formation of bis(organosilyl)palladium(II) complexes [14, 15], a well-characterized complex was first synthesized in 1992 by reaction of hydrodisilanes with hydridepalladium complex, probably through initial activation of Si–H bond followed by silylene migration (see Sect. 2.3) [16]. Since then, some reports on the syntheses of bis(organosilyl)palladium(II) complexes through direct activation of Si–Si bonds have appeared.

Most recently, reactions of bis($Ph_{(3-n)}Me_n$ P)(styrene)palladium(0) (n=1–3) **6** with fluorinated disilanes, symmetrical and unsymmetrical ones, were examined at low temperature (Eq. 3) [17]. The reactions provide the corresponding *trans*-bis(organosilyl)Pd(II) complexes **7** quantitatively, but the reaction of Me_3SiSiF_2Ph with the Ph_2MeP-Pd complex generates an equilibrium mixture of the starting disilane and bis(organosilyl)Pd(II) complex. It is noted that the unsymmetrical bis(organosilyl)palladium complexes show higher reactivity toward stoichiometric bis-silylation of diphenylacetylene than the corresponding symmetrical ones.

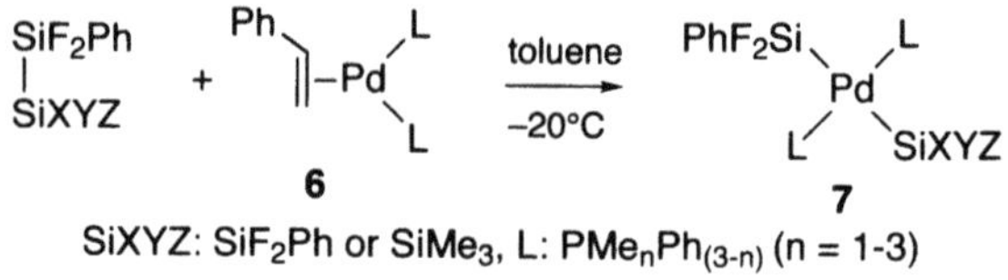

SiXYZ: SiF_2Ph or $SiMe_3$, L: $PMe_nPh_{(3-n)}$ (n = 1-3) (3)

Bis(2-phosphinoethyl)disilane **8** reacts with a dba-Pd(0) complex to give bis(organosilyl)Pd(II) complex **9** in good yield, whose *cis* orientation of the two silyl groups was confirmed by a single crystal X-ray analysis (Eq. 4) [18]. A disilane having only one 2-phosphinoethyl group failed to give the corresponding bis(organosilyl)Pd(II) complex.

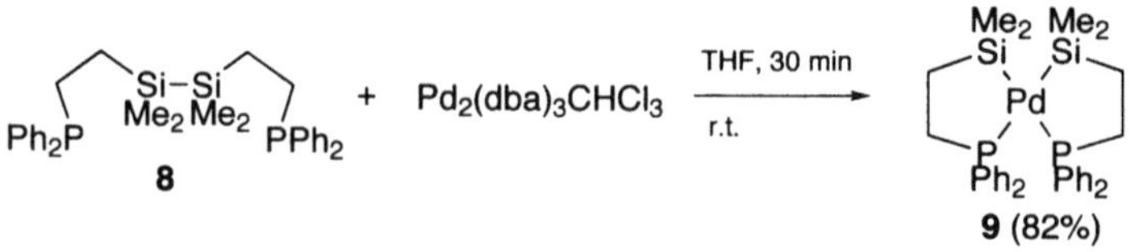

(4)

One promising approach to bis(silyl)Pd complexes may involve an oxidative addition of strained, cyclic disialanes with Pd(0) complexes [14]. Benzodisilacyclobutene derivative **10a** reacts with $Pd(PPh_3)_4$ under reflux in benzene to give five-membered ring complex **11** quantitatively, although it is not isolated (Eq. 5) [19]. The complex shows a unique reactivity toward terminal alkynes with bulky substituents; thus, reaction with 2,4,6-trimethylphenylacetylene provides alkynylsilane **13** in high yield, while reaction with phenylacetylene gives normal bis-silylation product **12**.

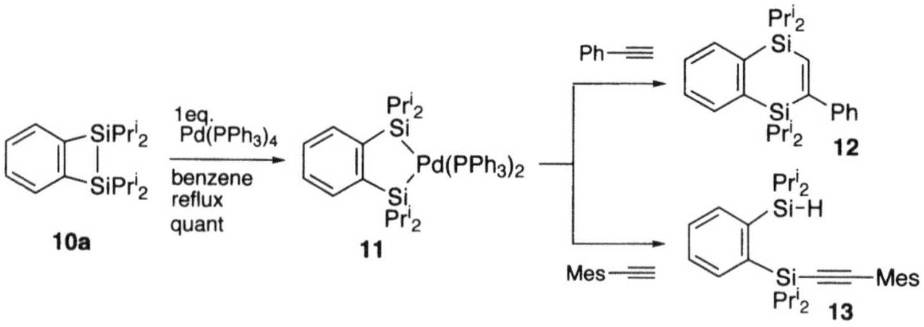

(5)

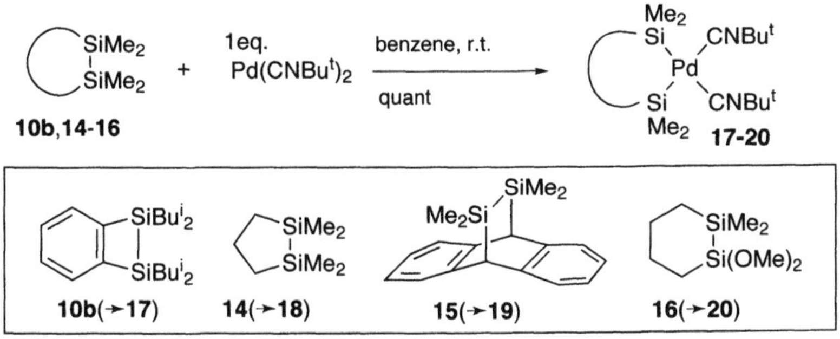

Scheme 1

Isonitrile ligands on palladium induce the activation of Si–Si bonds very efficiently. Thus, not only a four-membered disilane **10b**, five- (**14**) and six-membered disilanes (**15, 16**) furnish the corresponding cyclic bis(organosilyl)Pd(II) complexes **17–20** quantitatively in the reactions with bis(*tert*-alkyl isonitrile) palladium(0) (Scheme 1) [20].

Reaction of the isonitrile-palladium complex with a spiro trisilane **21** affords binuclear palladium complex **22** in high yield (Eq. 6) [21].

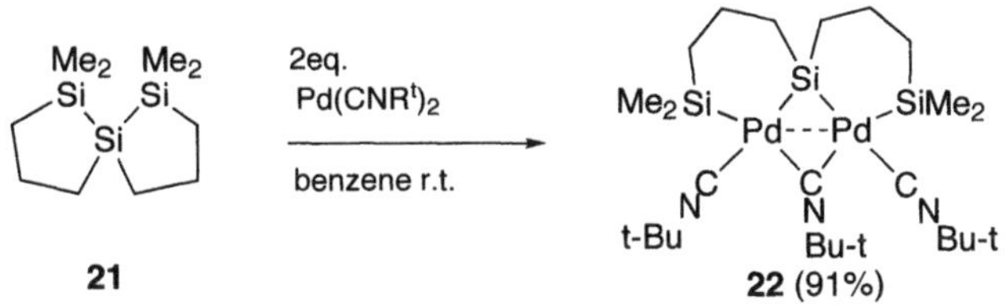

(6)

One exceptional reaction involving oxidative addition onto divalent palladium complexes has been reported. The spiro trisilane **21** is reacted with (η^3-allyl)CpPd(II) complexes at room temperature to afford Tris(organosilyl) CpPd(IV) complexes **23** in good yields (Eq. 7) [21]. The reaction may be rationalized by oxidative addition of one of the two Si–Si bonds of **21**, subsequent reductive elimination with formation of the Si-allyl bond giving mono(silyl)CpPd(II) complexes, and then oxidative addition of another Si–Si bond in the molecule onto the Pd(II).

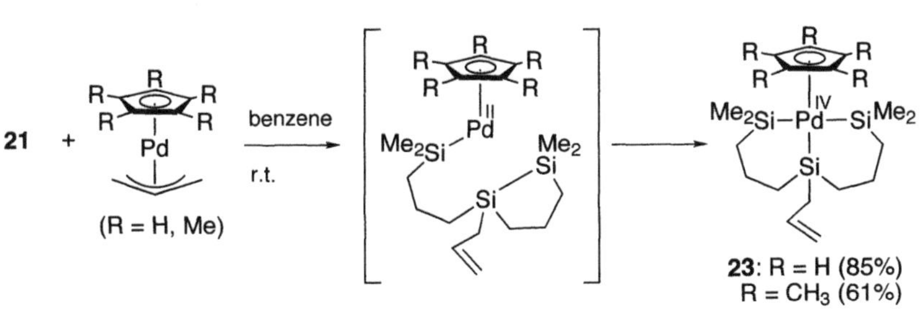

(7)

2.1.3 *Platinum*

Tris(triethylphosphine)Pt(0) reacts with various halodisilanes to give bis(organosilyl)platinum(0) complexes **24** in high yields (Eq. 8) [22]. In the case of the reaction with iodopentamethyldisilane, however, oxidative addition of the Si–I bond onto the Pt(0) complex takes place exclusively.

$$SiMe_2X{-}SiMe_2Y \text{ (ca. 2 equiv.)} + Pt(PEt_3)_3 \xrightarrow[\text{r.t., >90\% by NMR}]{\text{benzene}} (XMe_2Si)(YMe_2Si)PtL_2 \quad \mathbf{24} \tag{8}$$

[X = Y = F; X = Y = Cl; X = Cl, Y = Me; X = Br, Y = Me] **24**

Hexamethyldisilane, which is inert to the phosphine-platinum(0) complex, can react with an isonitrile-platinum(0) complex to give bis(trimethylsilyl)Pt(II) complex **25** in high yield (Eq. 9) [23].

$$SiMe_3{-}SiMe_3 \text{ (10 equiv.)} + Pt_3(CNAd)_6 \xrightarrow[\text{80°C}]{\text{benzene}} (Me_3Si)_2Pt(CNAd)_2 \quad \mathbf{25}\ (93\%) \tag{9}$$

The Si–Si double bond of stable disilene **26** is also known to react with platinum(0) complexes (Eq. 10) [24]. The produced complex **27** may be regarded as disilaplatinacyclopropane or disilene-coordinated complex. An efficient preparation of the same platinum complex is also reported by the reaction of dihydrodimesitylsilane with the platinum(0) complexes.

$$Mes_2Si{=}SiMes_2\ \mathbf{26} \xrightarrow[\text{or } (PPh_3)_2Pt(CO_2)\ /\ UV]{(PPh_3)_2Pt(CH_2{=}CH_2)} Mes_2Si{-}SiMes_2{>}Pt(PPh_3)_2 \equiv (Mes_2Si{=}SiMes_2){\rightarrow}Pt(PPh_3)_2 \quad \mathbf{27}\ (<20\%) \tag{10}$$

(Mes = 2,4,6-trimethylphenyl)

2.1.4 *Other Transition Metals*

Diiron nonacarbonyl undergoes oxidative addition of vinyl disilanes at room temperature to give oxidative addition product **28**, which is isolated by chromatography on silica gel (Eq. 11) [25]. The structure of **28** was determined by ^1H NMR spectroscopy to be (organosilyl)(η^3-1-silapropenyl)iron(II) complexes rather than simple bis(organosilyl)iron(II) complexes.

$$RMe_2Si{-}SiMe_2(CH{=}CH_2) + Fe_2(CO)_9 \xrightarrow[\text{r.t., 17 h}]{\text{benzene}} (\eta^3\text{-}Me_2Si{=}CH{-}CH_2)(RMe_2Si)Fe(CO)_3 \quad \mathbf{28}\ (60\text{-}61\%) \tag{11}$$

R = Me or CH=CH$_2$

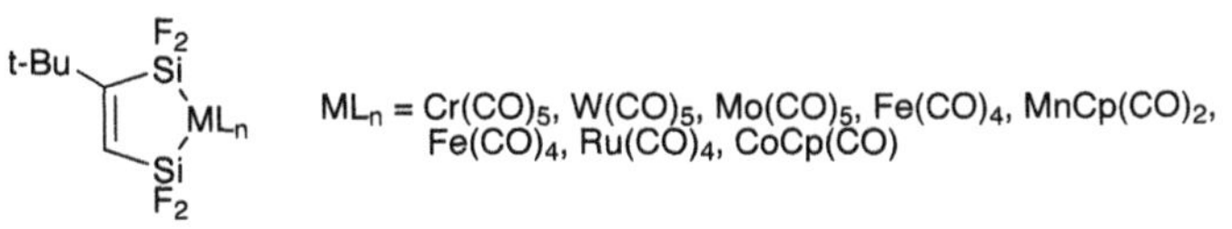

t-Bu, F_2Si, ML_n, SiF_2

ML_n = $Cr(CO)_5$, $W(CO)_5$, $Mo(CO)_5$, $Fe(CO)_4$, $MnCp(CO)_2$, $Fe(CO)_4$, $Ru(CO)_4$, $CoCp(CO)$

Fig.1

The iron carbonyl complex reacts with disilacyclobutene derivatives to give five-membered cyclic bis(silyl)Fe(II) complexes in moderate yield [26]. Similar five-membered cyclic iron complex and related bis(silyl)(carbonyl) complexes of Cr, Mo, W, Mn, Ru, and Co have been prepared by the reaction of **1** with metal carbonyls mostly under UV irradiation (Fig. 1) [27, 28]. They show interesting reactivities toward 1,3-dienes. Bis(2-phosphinoethyl)disilane **8** reacts with Ir(I) complex to give bis(organosilyl)iridium(III) complex **29** in good yield, in which two silyl groups are coordinated with *cis*-configuration (Eq. 12) [29].

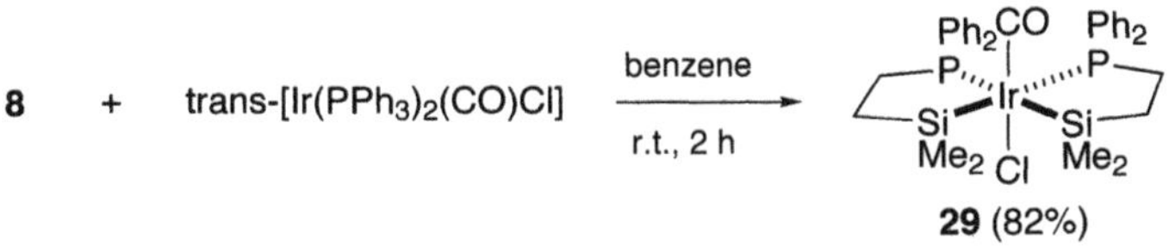

8 + trans-$[Ir(PPh_3)_2(CO)Cl]$ —(benzene, r.t., 2 h)→ **29** (82%) (12)

2.2 Double Oxidative Addition Reactions

Ruthenium carbonyl complex undergoes successive oxidative addition of Si–Si bond of **1** to give tetrakis(organosilyl)$(CO)_3$Ru(IV) complex **30** as a major, kinetic product, which further reacts with the $Ru_3(CO)_{12}$ to give bis(silyl)Ru(II) complex **31** (Eq. 13) [28].

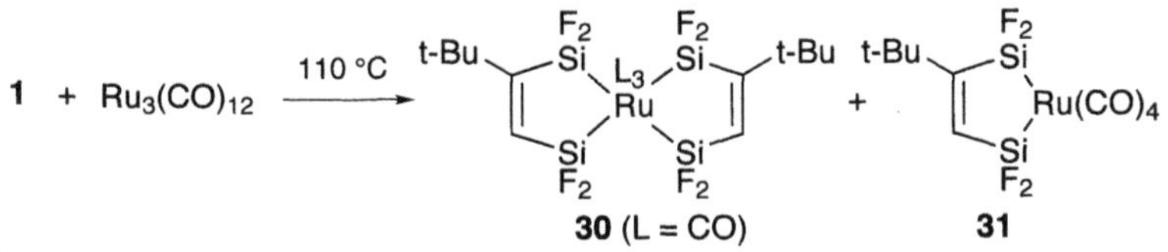

1 + $Ru_3(CO)_{12}$ —(110 °C)→ **30** (L = CO) + **31** (13)

Two Si–Si bonds in a molecule of bis(disilanyl)dithiane **32** oxidatively add to isonitrile-platinum(0) complex to give tetrakis(organosilyl)platinum(IV) complex **33**, which is isolated by column chromatography on silica gel (Eq. 14) [23].

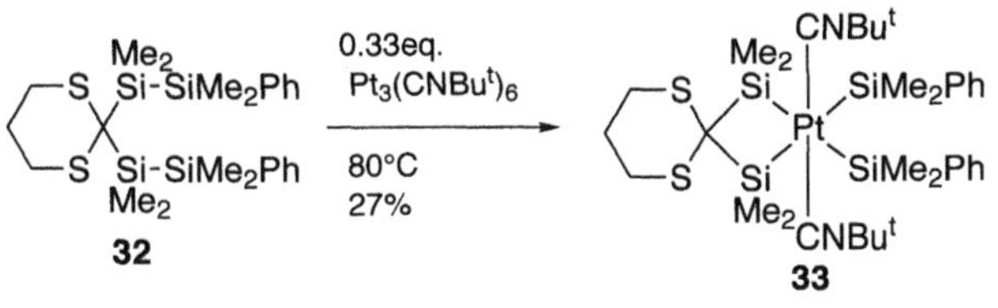

32 —(0.33eq. $Pt_3(CNBu^t)_6$, 80°C, 27%)→ **33** (14)

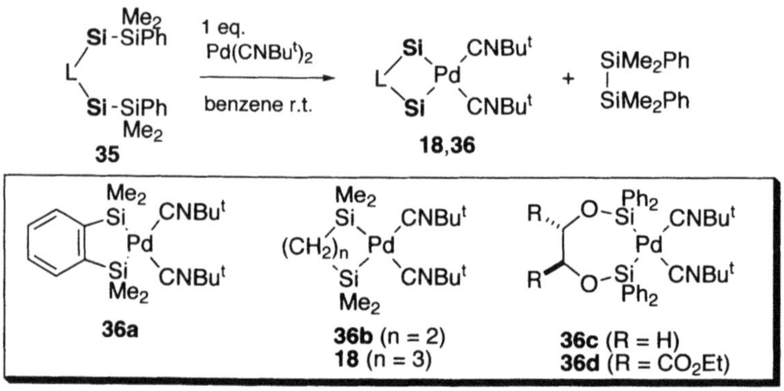

Scheme 2

Reaction of bis(disilanyl)dithiane **32** with the corresponding palladium(0)-isonitrile complex affords a four-membered cyclic bis(silyl)palladium(II) complex **34** quantitatively together with the formation of a disilane (Eq. 15) [30]. The formal intramolecular metathesis of the two Si–Si bonds of **32** may proceed through initial formation of tetrakis(silyl)Pd(IV) complex, corresponding to the platinum complex **33**. The double oxidative addition of the two Si–Si bonds may be followed by reductive elimination of the disilane with accompanying formation of four-membered bis(silyl)palladium complex **34**, due to difficulty in reductive elimination leading to formation of a three-membered cyclic disilane.

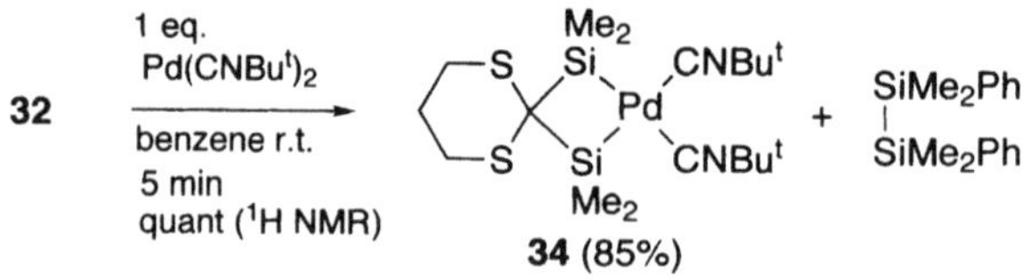

(15)

The intramolecular metathesis can be applied to the related bis(disilane)s **35** with some limitation (Scheme 2) [20]. For the bis(disilane) with a one-carbon tether linking the two disilanyl groups, the trimethylenedithio group at the carbon in **32** is essential to stabilize the four-membered complex produced. Bis(disilane)s with two and three carbon tethers undergo the intramolecular metathesis to give five- (**36a,b**) and six-membered (**18**) cyclic bis(silyl)Pd(II) complexes in reasonable reaction rate and yields. For more longer all-carbon tethers, the reaction does not proceed at all. However, bis(disilanes)s having ethylenedioxy tethers with phenyl groups on the internal silicon atoms afford the seven-membered ring **36c,d** in high yields.

2.3 Silylene-Migration in Disilanyl-Transition-Metal Complexes

Silylene extrusion or migration is often observed in the reactions involving disilanyl transition-metal complexes [6]. Typical examples are given by the following reactions of disilanyliron complexes. Alkoxydisilanyl derivative **37** affords bis(silylene) complex **38** in good yield on UV irradiation (Scheme 3) [31], whereas the corresponding derivatives having no alkoxy groups on the silicon atoms provide mono(silyl)iron complexes with a loss of the silylene [32].

The silylene migration may be involved in some stoichiometric reactions of hydrodisilanes with transition-metal complexes, forming disilanyl complexes through activation of Si–H bonds. Binuclear Ru and Os complexes **39** having terminal silyl groups and silylene bridges are isolated in the reactions of pentamethyldisilane with $Ru_3(CO)_{12}$ and $Os_3(CO)_{12}$, respectively, albeit in low yields (Eq. 16) [33]. Similar silylene migration may be involved in the reactions of hydrodisilane with other transition-metal carbonyls such as $Fe_2(CO)_9$ and $Co_2(CO)_8$ [34].

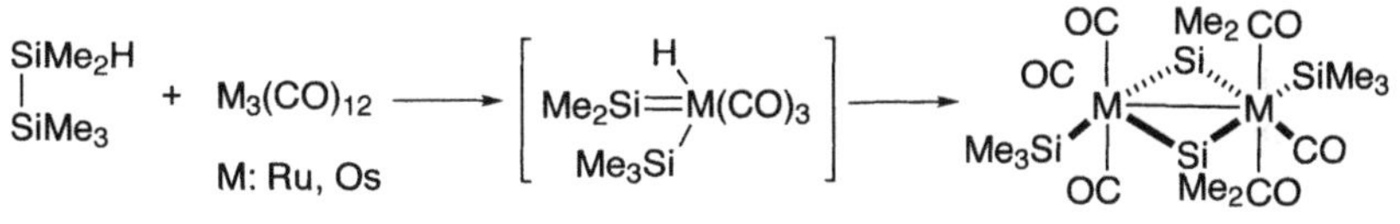

SiMe2H–SiMe3 + M3(CO)12 (M: Ru, Os) ⟶ [Me2Si=M(CO)3(H)(Me3Si)] ⟶ **39** (16)

Similarly, treatment of dihydroPt(II) complex **40** with tetrahydrodimethyldisilane provides a new access to bis(organosilyl)Pt(II) complexes **41** (Eq. 17) [35].

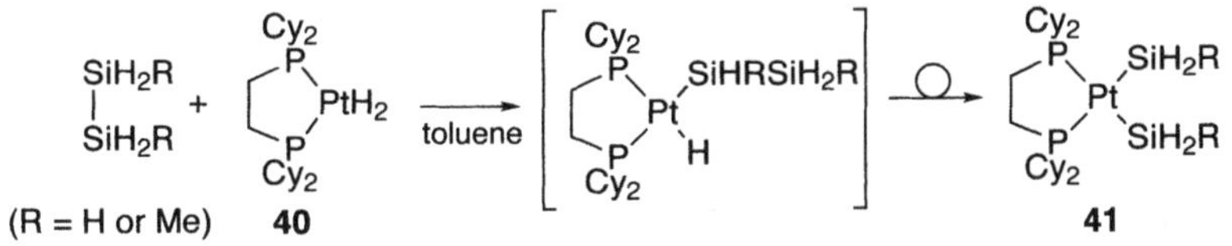

SiH2R–SiH2R (R = H or Me) + **40** (Cy2P~PCy2)PtH2 ⟶(toluene) [(Cy2P~PCy2)Pt(SiHRSiH2R)(H)] ⟶ **41** (Cy2P~PCy2)Pt(SiH2R)2 (17)

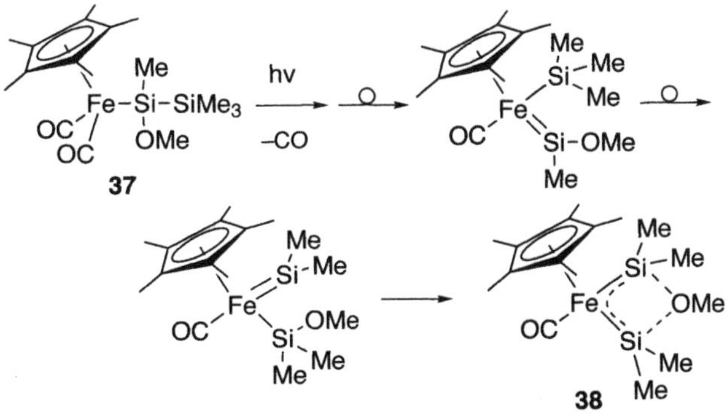

Cp*Fe(CO)2–SiMe(OMe)–SiMe3 **37** ⟶(hv, –CO) ⟶ ⟶ **38**

Scheme 3

This strategy can be applied to the synthesis of bis(silyl)Pd(II) complex **42**, which is the first successful example of X-ray analysis of bis(silyl)palladium complex (Eq. 18) [16].

SiMeRH–SiMeRH + Cy_2P–Pd(H)–PCy_2 dimer (25°C) → (HRMeSi)$_2$Pd(Cy_2P–PCy_2) (18)

(R = H or Me) **42** (R = H, Me)

3
Catalytic Activation

3.1
Bis-Silylation

A variety of catalytic bis-silylation reactions, i.e., addition of Si–Si bonds across multiple bonds, have been reported. Generally the reaction mechanism can be presented as follows: (1) formation of bis(organosilyl) transition-metal complexes through activation of Si–Si bonds, (2) insertion of unsaturated organic molecules into the silicon-transition-metal bonds, and (3) reductive elimination of the silicon-element (mostly carbon) bonds giving bis-silylation products. The final step regenerates the active low-valent transition-metal complexes. Not only appropriate choice of transition metal, but also choice of suitable ligand on the transition metal is crucially important for the success of the bis-silylation reaction. In addition, substituents on the silicon atoms of disilane are also of importance.

3.1.1
Alkyne

Addition reactions of the Si–Si bonds across carbon-carbon triple bonds have been most extensively studied since the 1970s by means of palladium catalysts. In the early reports, palladium complexes bearing tertiary phosphine ligands, mostly PPh_3, were exclusively employed as effective catalysts, enabling the alkyne bis-silylation with "activated" disilanes, i.e., disilanes with electronegative elements on the silicon atoms such as hydro [36], fluoro [37], chloro [38], and alkoxy-disilanes [39, 40] and those with cyclic structure (Scheme 4) [41–44]. The bis-silylation reactions could be successfully applied to terminal alkynes and acetylenedicarboxylates to give (*Z*)-1,2-bis(silyl)alkenes, which are otherwise difficult to synthesize.

Recent advances in the phosphine-palladium-catalyzed bis-silylation involve the use of dimethylphenylphosphine ligand, which makes bis-silylation of internal alkynes with fluorodisilanes possible (Eq. 19) [17].

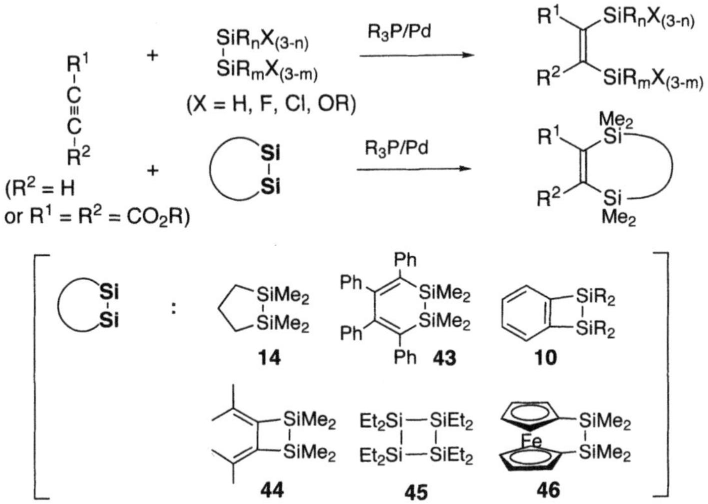

R^1–C≡C–R^2 + $SiR_nX_{(3-n)}$–$SiR_mX_{(3-m)}$ (X = H, F, Cl, OR) → (R_3P/Pd)

(R^2 = H or R^1 = R^2 = CO_2R)

14 **43** **10** **44** **45** **46**

Scheme 4

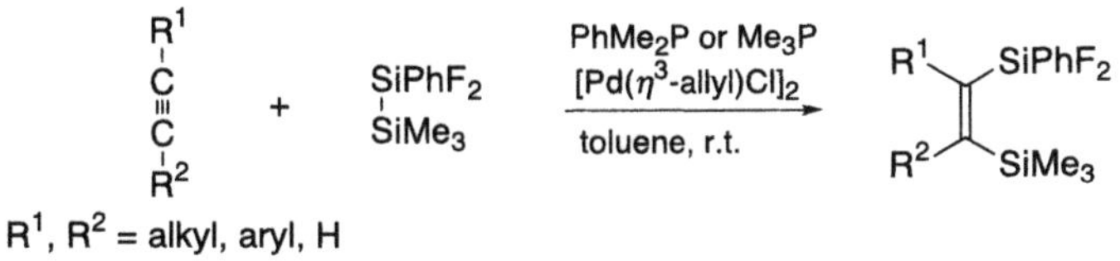

R^1–C≡C–R^2 + $SiPhF_2$–$SiMe_3$ → ($PhMe_2P$ or Me_3P, $[Pd(\eta^3\text{-allyl})Cl]_2$, toluene, r.t.)

R^1, R^2 = alkyl, aryl, H (19)

Two new classes of ligands, bicyclic phosphate **47** [45] and *tert*-alkyl isonitriles [46] on palladium, enable bis-silylation with hexaalkyldisilanes, which have been regarded to be much less reactive than the "activated" disilanes (Eq. 20). Reactions of the terminal alkynes with hexamethyldisilane in the presence of these palladium catalysts afford (*Z*)-1,2-bis(trimethylsilyl)alkenes **48** in high yields.

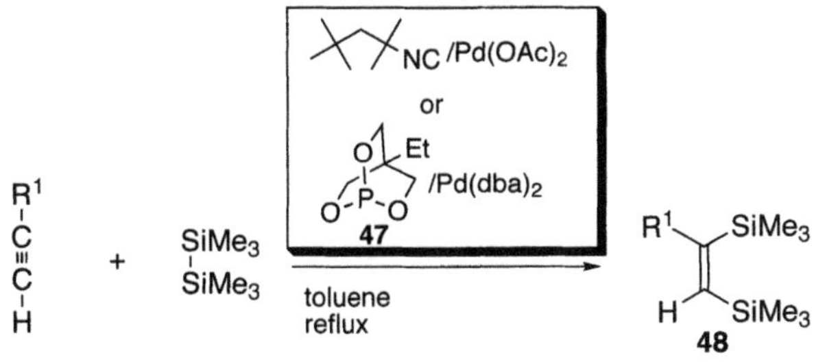

R^1–C≡C–H + $SiMe_3$–$SiMe_3$ → (NC/$Pd(OAc)_2$ or **47**/$Pd(dba)_2$, toluene reflux) **48** (20)

The latter isonitrile-palladium complexes effectively catalyze intramolecular bis-silylation of carbon-carbon triple bonds (Eq. 21) [46, 47]. Not only terminal alkynes but also internal alkynes including those having aryl, alkenyl, silyl, and ester groups **49** successfully undergo addition of the intramolecular Si–Si bonds with 5-exo cyclizations.

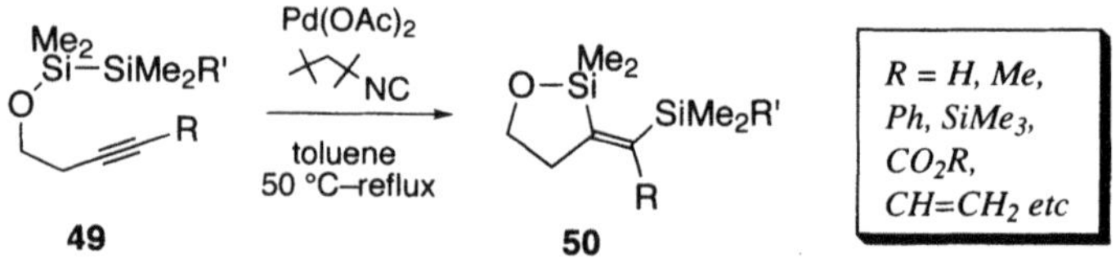

(21)

Under extremely high pressure, the isonitrile-palladium catalyst promotes intramolecular bis-silylation of bis(silyl)acetylenes **51** to give tetrakis(silyl) alkenes **52**, which are otherwise difficult to synthesize; the reaction under atmospheric pressure hardly proceeds even at 200°C (Eq. 22) [48, 49].

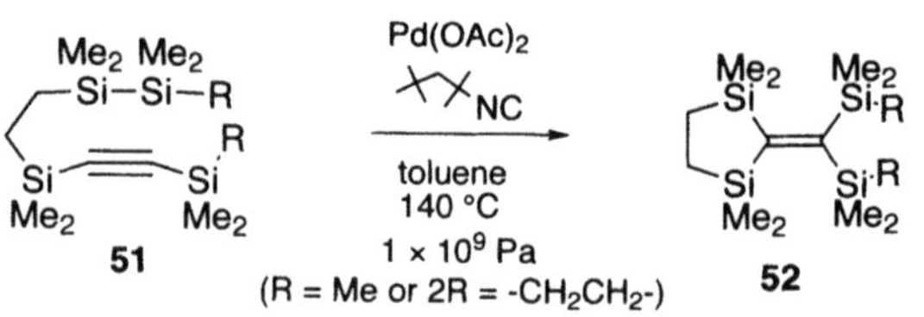

(22)

Application of the intramolecular bis-silylation to disilanyl ethers of propargylic alcohols **53** provides a new entry into oxasilethanes **54**, highly strained four-membered ring compounds including Si–O bonds (Eq. 23) [50, 51].

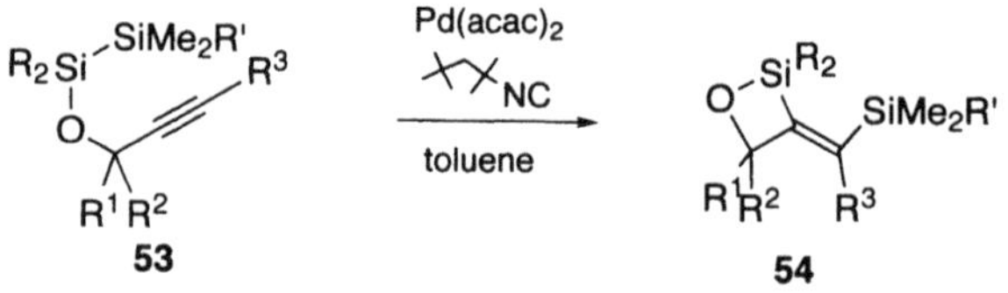

(23)

Bis(disilanyl)dithiane **32** is a good precursor for intermolecular bis-silylation of alkynes. In the presence of the isonitrile-palladium catalyst **34**, the facile intramolecular Si–Si bond metathesis produces reactive four-membered cyclic bis(organosilyl)palladium(II) intermediate **34**, which then reacts with alkynes to afford five-membered cyclic products **55** in high yields (Eq. 24) [20, 30].

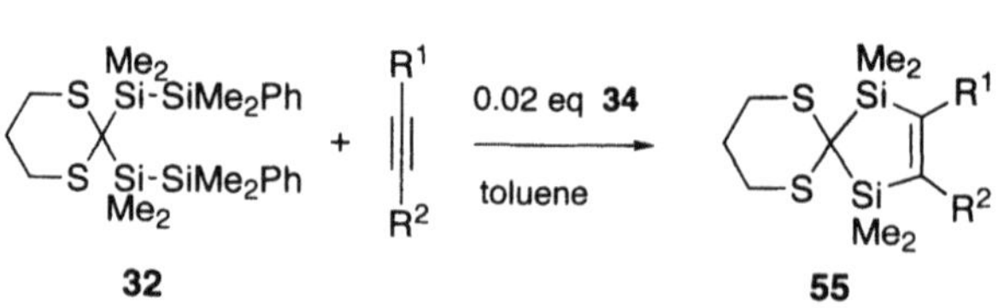

(24)

It should be mentioned that successive insertion of alkynes into Si–Si bonds is achieved in the reactions of certain cyclic disilanes with alkynes, giving cyclic dienes **56**, although these bis-silylative dimerizations were reported only as side reactions (Eq. 25) [41, 43, 44].

$$R^1-C\equiv C-R^2 + \text{(cyclic disilane) } \mathbf{14,44,46} \xrightarrow{\text{Pd cat.}} \mathbf{56} \text{ (low yields)} \quad (25)$$

Bis-silylation of alkynes is also promoted by a platinum complex; use of $Pt(dba)_2$ under CO pressure at 120°C effects bis-silylation of phenylacetylene with phenylpentamethyldisilane [52]. Nickel-catalyzed bis-silylation of diphenylacetylene with a benzodisilacyclobutene derivative has also been reported [53].

3.1.2
Diene

Reactions of 1,3-dienes with various disilanes were reported by means of group 10 transition-metal catalyst, giving synthetically useful allylic silane derivatives. Two major pathways, i.e., formation of 1:1 and 1:2 adducts, are known and controlled by the choice of disilanes and catalyst.

Reactions of 1,2-dihydrotetramethyldisilane with 1,3-butadiene and isoprene in the presence of $NiCl_2(PEt_3)_2$ afford 1,4-bis-silylation products **57** in good yields (Eq. 26) [54]. The bis-silylation of isoprene is accompanied by formation of a small amount of 1,4-hydrosilylation product **58**. The hydrosilylation becomes the major pathway in the reaction of 2,3-dimethyl-1,3-butadiene.

$$\text{diene } (R^1, R^2) + HMe_2Si-SiMe_2H \xrightarrow{NiCl_2(PEt_3)_2} \underset{\mathbf{57}}{HMe_2Si \ \ SiMe_2H} + \underset{\mathbf{58}}{H \ \ SiMe_2SiMe_2H} \quad (26)$$

$R^1 = R^2 = H$: 69% + 0%
$R^1 = H$, $R^2 = Me$: 66% + 15%
$R^1 = R^2 = Me$: 20% + 50%

Palladium-phosphine complexes catalyze the similar 1,4-bis-silylation with the "activated" disilanes. In the presence of palladium-PPh_3 complexes, fluoro- [55] and chloro-disilanes [56] provide 1,4-products **59** in good yields with high stereoselectivity giving *Z*-alkenes (Eq. 27). In the reaction of fluorinated disilane with isoprene, a minor amount of 1:2 adduct **60**, which arises from regioselective head-to-head coupling of the isoprene, is produced [55]. In the reactions of chlorinated disilanes, however, no 1:2-adduct is formed [56].

Me + $SiMe_{(3-n)}X_n$–$SiMe_{(3-n)}X_n$ (X = F, Cl) → [$(PPh_3)_4Pd$ or $(PPh_3)_2PdCl_2$] → $X_nMe_{(3-n)}Si$ / $SiMe_{(3-n)}X_n$, Me **59** (major) + $X_nMe_{(3-n)}Si$, $X_nMe_{(3-n)}Si$, Me, Me **60** (minor) (27)

A certain platinum catalyst also promotes the reaction of 1,3-dienes with aryldisilanes, giving 1,4-bis-silylation products **61** with varying E/Z ratios (Eq. 28) [52]. Peralkyldisilanes fail to give the corresponding products in moderate yield. The remarkable effect of aryl groups on the silicon may be due to interaction of the aryl group with the platinum center prior to the oxidative addition of the Si–Si bonds.

R^1, R^2 + $SiMe_2Ph$–$SiMe_2Ph$ → [$Pt(CO)_2(PPh_3)_2$, CO (10 kg/cm^{-2})] → R^1, R^2, $PhMe_2Si$ $SiMe_2Ph$ **61** (E/Z = 3.0-0.5) (28)

Nickel-catalyzed reaction of 1,3-dienes with vinyldisilane is notable [57]. In addition to an acyclic diene, 1,4-bis-silylation of a cyclic diene is also achieved to give cyclic allylsilane derivatives **62** as a mixture of *cis* and *trans* isomers (Eq. 29).

+ $SiMe_2$–$SiMe_3$ → [$Ni(PEt_3)_4$, 180°C, quant.] → Me_3Si–, Si Me_2 **62** (29)

In sharp contrast to the predominant formation of 1:1 adducts, reactions of 1,3-dienes with various disilanes in the presence of phosphine-free palladium catalyst provide bis-silylative dimerization products **63** in high yields (Eq. 30) [58, 59]. In comparison with the $Pd(OAc)_2$ and $PdCl_2(ArCN)_2$ catalysts originally found [58], the use of $Pd(dba)_2$ (dba: dibenzalacetone) in DMF was found to be more effective in enabling the reaction to proceed at room temperature [59]. Worthy of note is that head-to-head coupling products **63** having carbon-carbon double bonds with E-geometry are exclusively formed in these reactions.

R^1 + $SiMe_2R$–$SiMe_2R$ → [$Pd(OAc)_2$ or $PdCl_2(PhCN)$ neat, heat or $Pd(dba)_2$ DMF, r.t.] → RMe_2Si, RMe_2Si, R^1, R^1 **63** (30)

The bis-silylative dimerization involving regioselective head-to-head coupling is also observed in the reactions of cyclic disilanes. Five- (**14**) and six-membered (**43**) disilanes [60], cyclic tetrasilane **45** [42], ferrocenyldisilane **46** [43], and a four-membered disilane **44** [44] similarly undergo the successive insertion of 1,3-dienes in the presence of palladium catalysts. In these cases, geometry of the C=C bonds may depend upon the substrates, though geometrical assignment is not given except for one case, where *trans-trans* product predominates [43]. Application of the bis-silylative dimerization reaction to bis(diene)s **64** leads to new cyclization reaction accompanying bis-silylation at both terminal carbon atoms (Eq. 31) [61].

64 + $SiMe_2Ph$–$SiMe_2Ph$ → ($Pd(dba)_2$, r.t., toluene or dioxane) 65 (31)

64: E = CO_2Et, CN, Ts; n = 1,2

65: stereochem in the ring: >95% trans
when E = CN, n = 1: (E,Z) only
when E = CO_2Et, n = 2: (E,E) only

Bis-silylation of 1,2-dienes by means of palladium catalysts has been reported. It should be noted that 1,2-butadiene undergoes the bis-silylation exclusively at the 2,3-positions to give **66** (Eq. 32) [62].

$$\text{1,2-butadiene} + SiMe_{(3-n)}X_n\text{–}SiMe_{(3-m)}X_m \xrightarrow[\text{100-150°C}]{(PPh_3)_4Pd} X_mMe_{(3-m)}Si\text{–}C(=CH_2)\text{–}CH(Me)\text{–}SiMe_{(3-n)}X_n \quad (32)$$

(X = Cl, OMe, n > m)

66 (good yields)

3.1.3
Alkene

Bis-silylation of methyl vinyl ketone with fluorinated disilane, $FMe_2SiSiMe_2F$, proceeds in the presence of phosphine-palladium catalyst to afford (*Z*)-1,4-addition product **67** in high yield (Eq. 33) [55].

$$CH_2{=}CHC(O)Me + SiMe_2F\text{–}SiMe_2F \xrightarrow[\text{100 °C, 81\%}]{Pd(PPh_3)_4 \text{ or } PdCl_2(PPh_3)_2} FMe_2SiCH_2CH{=}C(Me)OSiMe_2F \quad (33)$$

67

More recently, bis-silylation of various α,β-unsaturated ketones with $Cl_2PhSiSiMe_3$ was reported to be promoted effectively by the phosphine-palladium catalyst in a similar manner (Eq. 34) [63]. In products **68**, the geometry of the double bonds and the regioselectivity with the TMS group attached to the oxygen atom are nearly completely controlled in the reaction. Use of BINAP ligand as an optically active ligand on palladium successfully induces asymmetric bis-silylation with up to 92% ee [64].

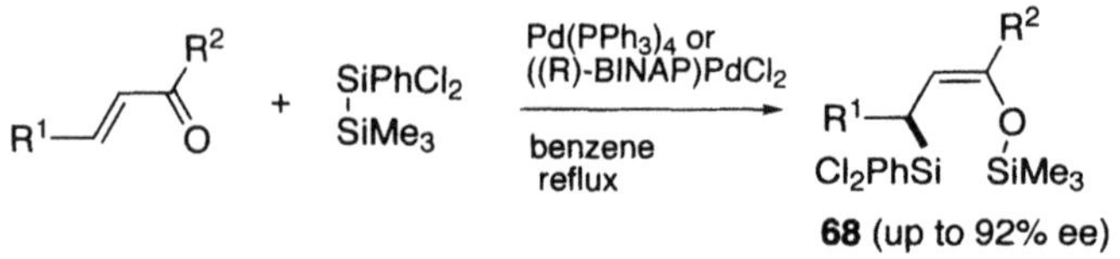

(34)

Similar 1,4-bis-silylation of α,β-unsaturated ketones with **32** is catalyzed by the isonitrile-palladium catalyst **34** to afford seven-membered ring silyl enol ether **69** in high yields (Eq. 35). Reactions of **32** with acrylic esters and acrylonitrile, however, give five-membered products **70** [20, 30].

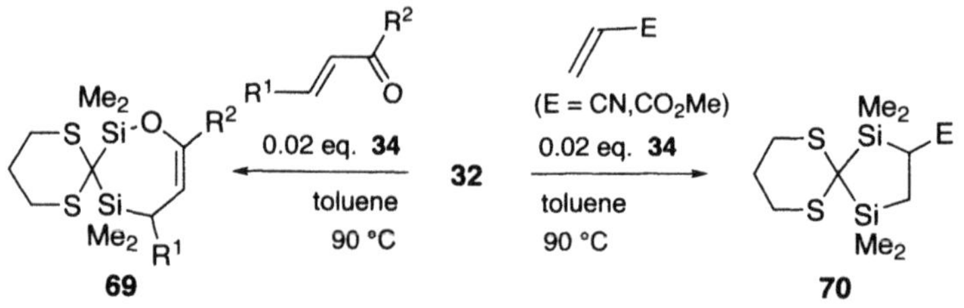

(35)

Recent detailed investigations have enabled bis-silylation of simple alkenes. Ethylene and norbornene undergo addition of activated disilanes in the presence of phosphine-platinum catalysts (Eq. 36) [65]. The stereochemical outcome of the reaction with norbornene, which selectively gave an exo-exo product **71** in 26% yield, revealed *cis*-addition of the Si–Si bond across the C=C bond. Under identical conditions, styrene and 1-hexene undergo dehydrogenative silylation to give β-silylstyrene and 1-silyl-1-hexene, respectively, as major products along with saturated silylation products [66].

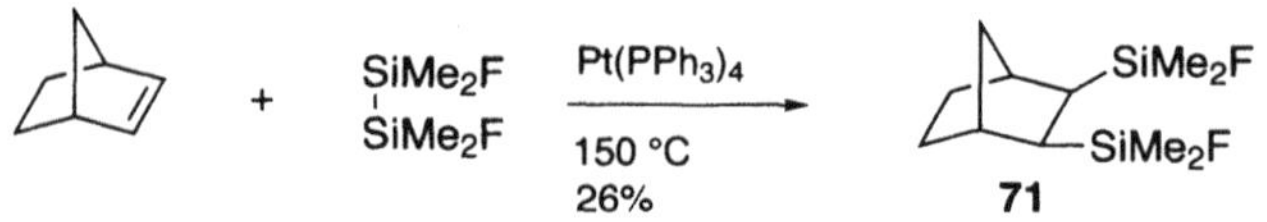

(36)

Furthermore, bis-silylation of terminal alkenes such as 1-octene and styrene with $F_2PhSiSiMe_3$ proceeds in moderate yields in the presence of palladium complex bearing two basic, sterically less demanding phosphine ligands such as $PPhMe_2$ and PMe_3 (Eq. 37) [17]. The products **72** consist of two regio isomers. The catalyst system also achieved bis-silylation of norbornene in high yield. The combination of the phosphine ligands and the unsymmetrical, fluorinated disilane is essential to attain satisfactory yields for the success of the bis-silylation.

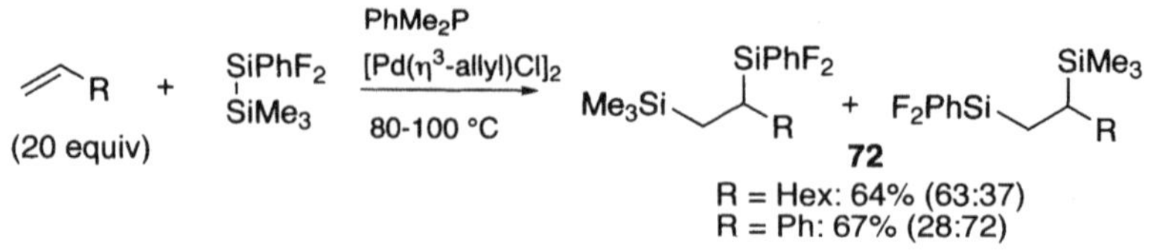

(37)

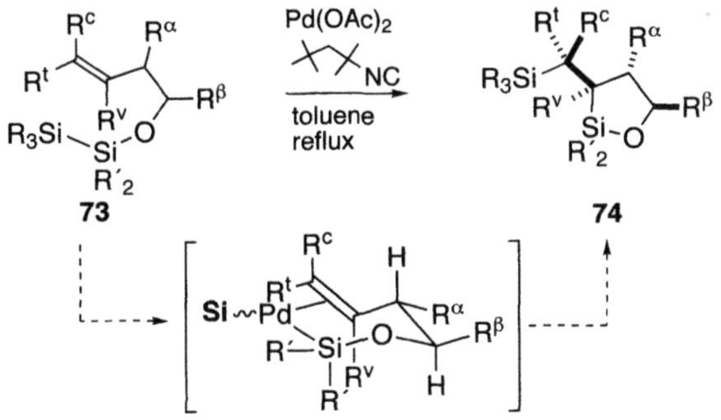

Scheme 5

In contrast to the difficulty in the intermolecular reactions, intramolecular bis-silylation of mono-substituted alkenes **73** is promoted by the isonitrile-palladium catalyst even at room temperature to give 5-exo cyclization products **74** in high yields (Scheme 5) [67–69]. Geminally as well as vicinally disubstituted double bonds also undergo the intramolecular addition of the Si–Si bonds in high yields at a higher temperature, although trisubstituted double bonds hardly react [70]. The key feature of the intramolecular bis-silylation of disilanyl ethers of homoallylic alcohols is the high diastereoselectivity observed in the reaction of those having substituents at the tethers linking the disilanyl groups to the C=C bonds; those with substituents α to the C=C give *trans* five-membered oxasilorane selectively and those with β-substituent give *cis* oxasilolanes in high diastereoselectivity.

Enantioselective intramolecular bis-silylation was accomplished by use of the optically active *tert*-alkyl isonitrile ligand **75** on palladium (Eq. 38) [71]. The bulky substituents of **76** on the silicon atom proximal to the ether oxygen are crucial to attain good enantioselectivity.

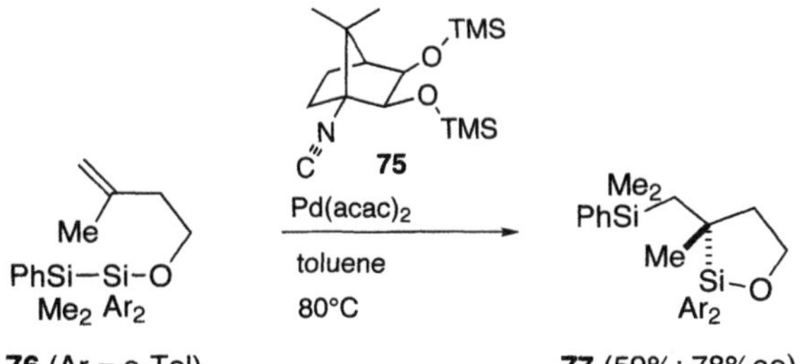

(38)

76 (Ar = o-Tol) — **75**, $Pd(acac)_2$, toluene, 80°C → **77** (59%; 78%ee)

The intramolecular bis-silylation also proceeds with disilanyl ethers of allylic alcohols **78** with vicinally disubstituted C=C bonds, providing eight-membered ring 1,5-dioxa-2,6-disilaoctane derivatives **79**, which may arise from dimerization of primarily formed four-membered 1,2-oxasilolanes **80**, in good yields (Eq. 39) [72, 73]. Worthy of note is that the bis-silylation proceeds with nearly complete diastereoselectivity to afford products **79** exclusively through the *trans* oxasilolanes **80**.

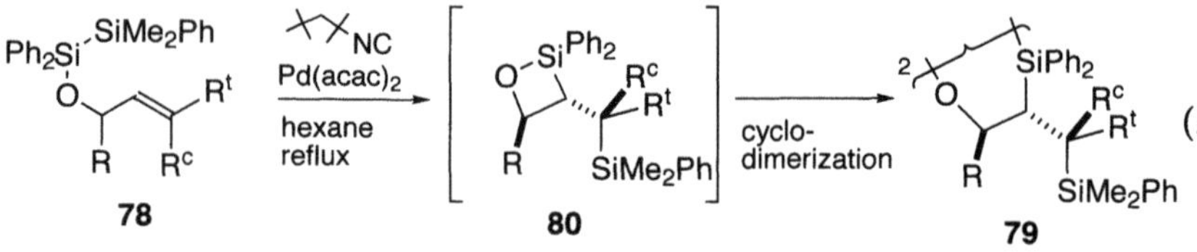

SiMe2Ph
Ph2Si
O
Rt
R
Rc
78
NC
Pd(acac)2
hexane
reflux
Si Ph2
O
Rc
Rt
R
SiMe2Ph
80
cyclo-
dimerization
2
SiPh2
O
Rc
Rt
R
SiMe2Ph
79
(39)

3.1.4
Carbonyl Compounds

Reaction of *p*-benzoquinone with fluorodisilane provides *p*-bis(siloxy)benzene **81** in 41% yield in the presence of the phosphine-palladium catalyst (Eq. 40) [55].

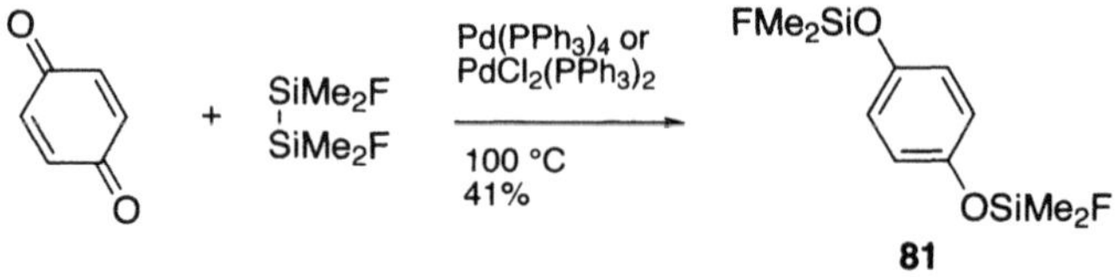

O
O
+
SiMe2F
SiMe2F
Pd(PPh3)4 or
PdCl2(PPh3)2
100 °C
41%
FMe2SiO
OSiMe2F
81
(40)

On the other hand, palladium-catalyzed reaction of *p*-benzoquinone with cyclic organosilanes having Si–Si bonds provides organosilicon polymers **82** (Eq. 41) [74]. Interestingly, dodecamethylcyclohexasilane, known as one of the least reactive oligosilanes in palladium-mediated reactions, is usable for the co-polymerization.

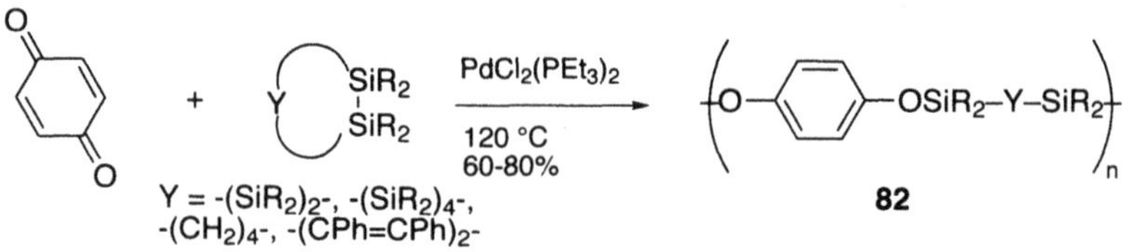

O
O
+
Y
SiR2
SiR2
Y = -(SiR2)2-, -(SiR2)4-,
-(CH2)4-, -(CPh=CPh)2-
PdCl2(PEt3)2
120 °C
60-80%
O
OSiR2–Y–SiR2
n
82
(41)

α-Diketones undergo bis-silylation with chloro- [75] as well as peralkyldisilanes [76] to give 1,2-bis(silyloxy)alkenes **83** in good yields (Eq. 42). In contrast, bis-silylative dimerization takes place with α-keto esters under identical conditions to give tartrate derivative **84** [76].

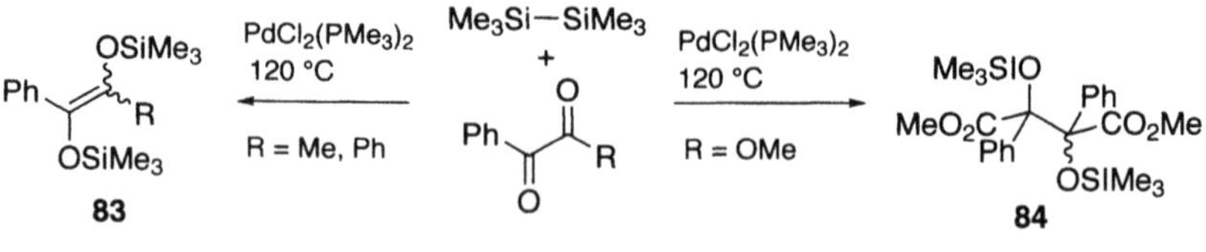

OSiMe3
Ph
R
OSiMe3
83
PdCl2(PMe3)2
120 °C
R = Me, Ph
Me3Si—SiMe3
+
O
Ph
R
O
PdCl2(PMe3)2
120 °C
R = OMe
Me3SIO
MeO2C
Ph
Ph
CO2Me
OSIMe3
84
(42)

Bis-silylation of aldehyde is also possible by use of the four-membered disilane in the presence of nickel and palladium catalyst at high temperature [77, 78].

3.1.5
Isonitrile

Insertion of isonitriles into Si–Si bonds takes place in the presence of palladium catalysts to give bis(silyl)imines **85**, which show characteristic UV absorbance around 400 nm arising from n-π^* transition (Eq. 43) [79]. Aryl and alkyl isonitriles except for tertiary alkyl isonitrile give the corresponding N-substituted imines.

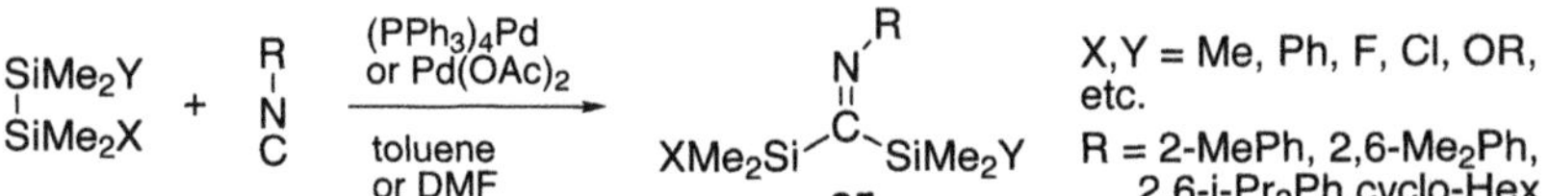

$SiMe_2Y$–$SiMe_2X$ + R–N≡C → [$(PPh_3)_4Pd$ or $Pd(OAc)_2$; toluene or DMF] → XMe_2Si–C(=N–R)–$SiMe_2Y$ **85** (43)

X,Y = Me, Ph, F, Cl, OR, etc.
R = 2-MePh, 2,6-Me_2Ph, 2,6-i-Pr_2Ph, cyclo-Hex

Linear oligosilanes up to tetradecamethylhexasilane undergo the insertion of 2,6-dimethylphenyl isonitrile into all of the Si–Si linkages to give oligoimines **86**, whereas 2,6-diisopropylphenyl isonitrile selectively inserts into the terminal Si–Si bonds to give **87** (Scheme 6) [80, 81]. Four- and five-membered cyclic oligosilanes selectively give mono-insertion products **88** (Eq. 44) [81, 82].

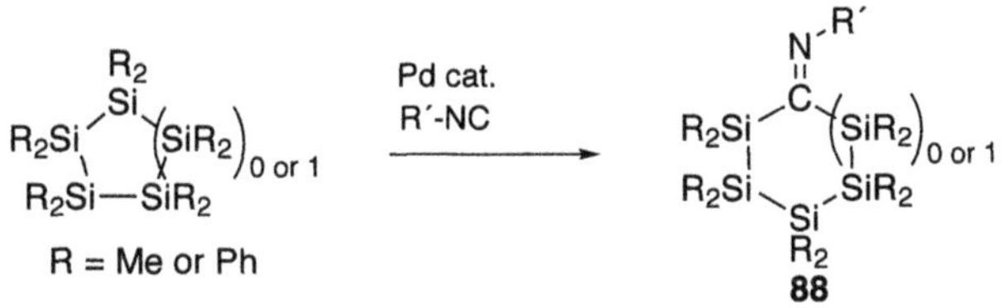

cyclo-$(SiR_2)_{4\ or\ 5}$ → [Pd cat., R′-NC] → **88** (44)

R = Me or Ph

In the insertion reactions, isonitriles may serve as effective ligands on palladium(0) for the activation of Si–Si bonds and, at the same time, they are involved as reactant. In contrast, *tert*-alkyl isonitriles, which do not insert into Si–Si bonds at all, can exclusively serve as spectator ligands for highly effective activation of Si–Si bonds [46].

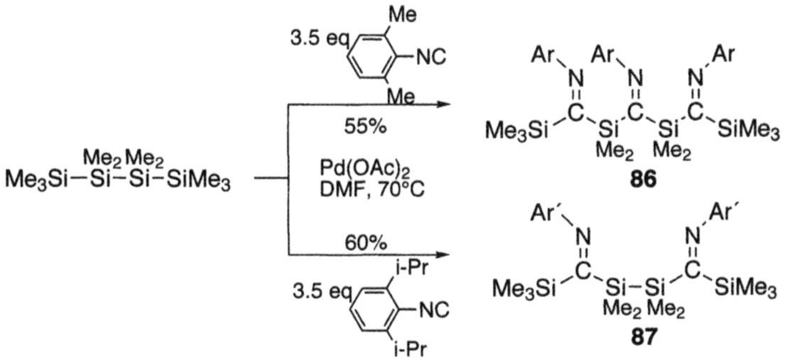

Me_3Si–$SiMe_2$–$SiMe_2$–$SiMe_3$ → [$Pd(OAc)_2$, DMF, 70°C]
- 3.5 eq 2,6-dimethylphenyl isonitrile, 55% → Me_3Si–C(=NAr)–$SiMe_2$–C(=NAr)–$SiMe_2$–C(=NAr)–$SiMe_3$ **86**
- 3.5 eq 2,6-diisopropylphenyl isonitrile, 60% → Me_3Si–C(=NAr′)–$SiMe_2$–$SiMe_2$–C(=NAr′)–$SiMe_3$ **87**

Scheme 6

3.2 Silylene Transfer

In Sect. 2.3, generation of silylene complexes of transition metals was discussed on the basis of the reactivity of disilanyl-transition-metal complexes. The formation of silylene species in the presence of a catalytic amount of transition metals is also involved in the reactions of hydrodisilanes, which may readily form disilanyl complexes through oxidative addition of the Si–H bond prior to the activation of the Si–Si bond. Platinum-catalyzed disproportionation of hydrodisilanes affords a mixture of oligosilanes **89** up to hexasilane (Eq. 45) [83]. The involvement of silylene-platinum intermediate was proven by the formation of a 1,4-disila-2,5-cyclohexadiene derivative in the reaction of the hydrodisilane in the presence of diphenylacetylene.

$$\underset{R = H, Me}{HMe_2Si{-}SiMe_2R} \xrightarrow[90\,°C]{PtCl_2(PEt_3)_2} \underset{\mathbf{89}\ (n = 1\text{-}6)}{H{-}(SiMe_2)_n{-}R} \tag{45}$$

Nickel-catalyzed reaction of 1,2-dihydrotetramethyldisilane with various internal alkynes affords silole (silacyclopentadiene) derivatives **90** in good yields (Eq. 46) [54].

$R^1{-}C{\equiv}C{-}R^2$ + $HMe_2Si{-}SiMe_2H$ $\xrightarrow[90\,°C]{NiCl_2(PEt_3)_2}$ 2,5-R^2-3,4-R^1-1-$SiMe_2$ silole **90** (46)

Recent application of the nickel-catalyzed reaction with diyne **91** provides new effective access to functionalized silole derivatives **92**, which are further utilized for the synthesis of silole-thiophene co-polymers (Eq. 47) [84].

91 (TBSO, OTBS) + $Ph_2HSi{-}SiMe_3$ $\xrightarrow[80\,°C,\ 40\%]{Ni(acac)_2,\ PEt_3,\ DIBAH}$ **92** (TBSO, OTBS, $SiPh_2$) (47)

Generation of silylene complexes from oligosilane may be involved in palladium-catalyzed skeletal rearrangement reaction of trisilanes **93** and tetrasilanes having methoxy groups on the internal silicon atoms (Eq. 48) [85]. Labeling experiments reveal that the rearrangement proceeds through silylene migration.

93 $RO{-}Si(OR)(CH_3){-}Si(OR)(CH_3){-}Si(CH_3)(CD_3)(CH_3)(CD_3){-}OR$ $\xrightarrow[80°C,\ quant]{Pd(PPh_3)_4}$ **94** $RO{-}Si(OR)(CH_3){-}Si(CH_3)(CD_3)(CH_3)(CD_3){-}Si(OR)(CH_3){-}OR$ (48)

Similar generation of palladium-silylene complexes may be involved in reaction of oligosilanes with aryl isonitriles giving four-membered rearranged products **95**, although the mechanism has not yet been clarified (Eq. 49) [81, 86].

$$Ph{-}Si(CH_3)_2{-}Si(CH_3)_2{-}Si(CH_3)_2{-}Si(CH_3)_2{-}Ph + Ar{-}N{\equiv}C \xrightarrow[110\,^{\circ}C,\ 62\%]{Pd(OAc)_2,\ ^tBu\text{-}C(Me)_2\text{-}NC} \mathbf{95} \tag{49}$$

Ar = 2,6-(i-Pr)$_2$Ph

95: $(PhMe_2Si)_2C$ in a four-membered ring with two $SiMe_2$ and N–Ar

It was reported that a substituent such as 8-dimethylaminonaphth-1-yl group, which may be able to coordinate with the generated silylene intramolecularly, facilitates the nickel and palladium-catalyzed extraction of silylene species from trisilane **96** (Eq. 50) [87].

$$\mathbf{96} \xrightarrow[\text{or } Pd(PPh_3)_4,\ 80\,^{\circ}C]{Ni(acac)_2,\ PEt_3,\ DIBAH,\ 70\,^{\circ}C} PhMe_2Si{-}SiMe_2Ph \tag{50}$$

96: $(PhMe_2Si)_2SiMe$ on C1 of naphthalene, X on C8

X = H: No Reaction
X = NMe$_2$: 76-82%

3.3
Si–Si σ-Bond Metathesis

Metathesis of Si–Si bonds is promoted by palladium catalysts presumably through bis(organosilyl)palladium(II) complexes. Though metathesis reaction of linear disilanes results only in disproportionation of the disilanes [88], use of cyclic disilane provides a useful method for the synthesis of macrocyclic or polymeric organosilicon compounds having newly formed Si–Si bonds.

3.3.1
Ring-Enlargement Oligomerization

In the presence of palladium-PPh$_3$ catalyst, a five-membered disilane **14** undergoes cyclo-dimerization to afford 10-membered cyclic **97** in moderate yield (Eq. 51) [37]. Reaction of **14** with linear disilane gives cross-metathesis product in high yield under similar conditions [37, 88].

$$\text{ring}(Si{-}SiMe_2R^1)(Si{-}SiMe_2R^2) \xleftarrow[\text{or } Pd(PPh_3)_4,\ 80\text{-}100\,^{\circ}C]{R^1Me_2Si{-}SiMe_2R^2,\ 0.01eq.\ PdCl_2(PPh_3)_2} \mathbf{14} \xrightarrow[100\,^{\circ}C,\ 39\%]{0.01eq.\ PdCl_2(PPh_3)_2} \mathbf{97} \tag{51}$$

R^1=R^2=F: 85%
R^1=Me, R^2=CCH: 33%

14: cyclic $SiMe_2{-}SiMe_2$; **97**: 10-membered ring with two $Me_2Si{-}SiMe_2$ units

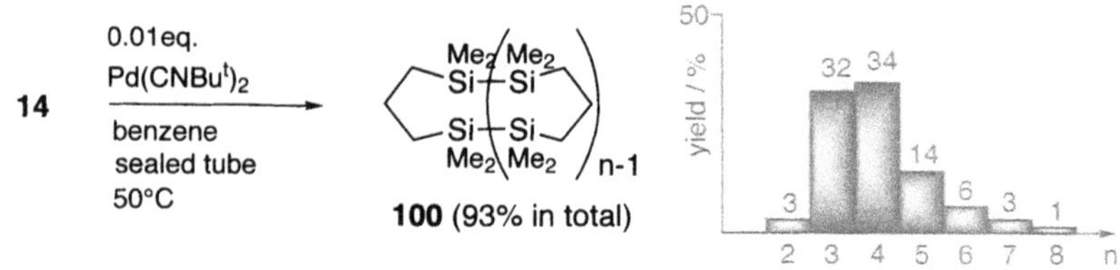

14 → 0.01eq. Pd(CNBu^t)$_2$, benzene, sealed tube, 50°C → 100 (93% in total)

yield / %: n = 2: 3; n = 3: 32; n = 4: 34; n = 5: 14; n = 6: 6; n = 7: 3; n = 8: 1

Scheme 7

Four-membered cyclic disilanes **44** [44] and **10** [44, 78, 89] also give cyclic dimers in the presence of the palladium catalyst, though silylene-migration products **99** are obtained in the reaction of a tetraethyl derivative of **10** (Eq. 52) [78, 89].

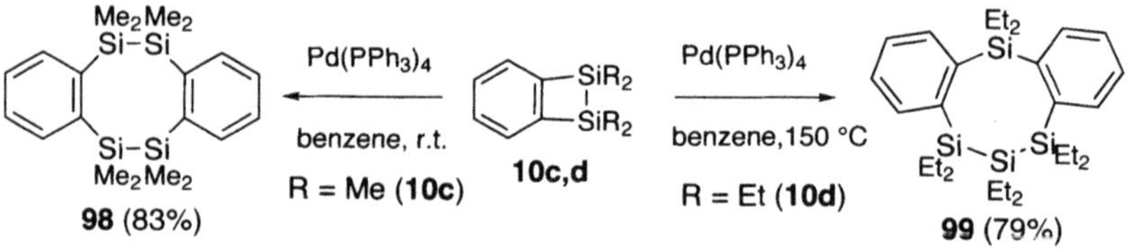

98 (83%) ← Pd(PPh$_3$)$_4$, benzene, r.t., R = Me (**10c**) — **10c,d** — Pd(PPh$_3$)$_4$, benzene, 150 °C, R = Et (**10d**) → **99** (79%) (52)

In sharp contrast to the phosphine-palladium catalyzed dimerization, cyclic disilane **14** affords macrocyclic oligomers **100** up to 40-membered octamer in the presence of bis(*tert*-alkyl isonitrile)palladium(0) complex as a catalyst (Scheme 7) [90]. Investigation of the reactions with each isolated oligomer under the oligomerization conditions reveals that the cyclo-oligomerization proceeds with stepwise ring-enlargement, in which a 1,5-disilapentanediyl moiety on the cyclic bis(silyl)palladium(II) intermediate is transferred to the Si–Si bond of the oligomers through oxidative addition-reductive elimination mechanism [91].

3.3.2 *Ring-Opening Polymerization*

Polymerization of **14** is promoted by palladium catalysts to produce macromolecules **101** having multiple Si–Si bonds in the main chains (Eq. 53). Three effective catalyst systems are known: $Pd_2(dba)_3CHCl_3/PPh_3$ in CH_3CN [92], $PdCl_2(dppb)/FMe_2SiSiMe_2F$ [93], and Cp(allyl)Pd [90, 91]. Mechanistically, these polymerizations may proceed via oxidative addition of the Si–Si bond onto mono(organosilyl)palladium(II), which may be generated through any initiation steps from the palladium precursors, and subsequent reductive elimination of new Si–Si bonds from the palladium(IV) intermediate (see Sect. 2.4) [21].

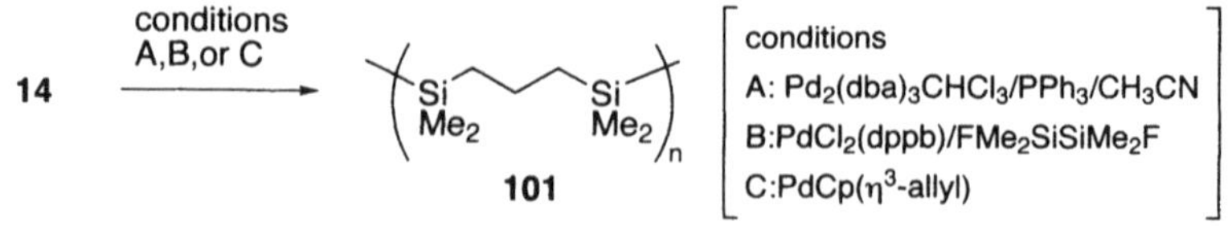

14 → conditions A, B, or C → **101** (53)

conditions
A: $Pd_2(dba)_3CHCl_3/PPh_3/CH_3CN$
B: $PdCl_2(dppb)/FMe_2SiSiMe_2F$
C: $PdCp(\eta^3\text{-allyl})$

3.4 Silylation of Organic Halides and Allylic Esters

Transition-metal catalyzed metathesis of carbon-halogen bonds with Si–Si bonds provides useful access to organosilicon compounds. Most of the reaction may involve initial oxidative addition of the carbon-halogen bond onto the transition-metal followed by activation of the Si–Si bond to give (organosilyl)(organo)palladium(II) complex, which undergoes reductive elimination of the carbon-silicon bond.

A typical example is given by the reaction of aryl halides with disilanes in the presence of palladium complexes, giving arylsilanes **102** in good yields (Eq. 54) [15, 94].

$$PhBr + SiMe_{(3-n)}Cl_n\text{–}SiMe_{(3-n)}Cl_n \xrightarrow[\text{toluene, 140 °C}]{Pd(PPh_3)_4} PhSiMe_{(3-n)}Cl_n \quad \textbf{102}\ \text{(high yields)} \tag{54}$$

Benzoylsilanes **103** are synthesized by reaction of benzoyl chlorides with hexamethyldisilane in the presence of palladium catalyst (Eq. 55) [15, 95]. The reactions are accompanied by formation of **102** as by-products via decarbonylation. In contrast, the decarbonylation producing arylsilanes **102** predominates when $ClMe_2SiSiMe_2Cl$ is used in place of hexamethyldisilane [75].

$$\underset{\textbf{102}}{PhSiMe_2Cl} \xleftarrow[\text{145 °C}]{ClMe_2Si\text{–}SiMe_2Cl,\ PdCl_2(PhCN)_2,\ PPh_3} PhC(O)Cl \xrightarrow[\text{110 °C}]{Me_3Si\text{–}SiMe_3,\ [(\eta^3\text{-allyl})PdCl]_2,\ P(OEt)_3} \underset{\textbf{103}}{PhC(O)SiMe_3} \tag{55}$$

Silylation of allylic halides and esters with disilanes is effected by use of transition-metal complex catalysts, providing a convenient method for synthesis of allylsilanes **104**, which are useful for organic synthesis (Eq. 56). Tetrakis(triphenylphosphine)palladium(0) effectively catalyzes the reactions of allyl, methallyl, and cinnamyl substrates; however, the catalyst fails to promote the reactions of allylic substrates having a primary or secondary alkyl group as R^1, which undergo β-hydride elimination to result in the formation of conjugated dienes [96, 97]. Remarkably, a new catalyst system, $Pd(dba)_2$ with LiCl in DMF, has enabled the silylation of a wide range of allylic acetate including the alkyl substituted substrates [98].

$$R^1CH{=}C(R^2)CH_2X \ \text{or}\ R^1CH(X)C(R^2){=}CH_2 + SiMe_nCl_{(3-n)}\text{–}SiMe_nCl_{(3-n)}\ (n = 0\text{–}3) \xrightarrow[\text{with or without LiCl}]{\text{Ni, Pd, or Rh catalyst}} \underset{\textbf{104}}{R^1CH{=}C(R^2)CH_2SiMe_nCl_{(3-n)}} \tag{56}$$

X = Cl, Br, OAc, OBz

4 Application to Organic Synthesis

The catalytic activation of the Si–Si bond leads to development of new methodologies for effective synthesis of organosilicon compounds. As described thus far, the reactions often enable highly regio- and stereoselective synthesis of organosilicon compounds, which are useful for stereoselective organic synthesis. In this section, utilization of organosilicon compounds obtained through the catalytic Si–Si activation to organic synthesis is briefly described.

4.1 Via Allylsilanes

Allylsilanes, which are accessible by bis-silylation of 1,3-dienes and silylation of allylic substrates, are useful allylation reagents in organic synthesis [99, 100] (*E,E*)-1,8-bis(Trimethylsilyl)octa-2,6-diene **63a,** prepared by palladium-catalyzed bis-silylative dimerization of 1,3-butadiene, was successfully applied to the synthesis of *dl*-muscone (Scheme 8) [58]. A key feature of the synthesis is regioselective reaction with acetaldehyde at the positions γ to the silyl groups.

4.2 Via Peterson-Type Elimination

Elimination of silyl groups with β-oxy groups, i.e., Peterson-type elimination, is a useful method for preparing stereodefined alkenes [101, 102]. The synthetically useful allylsilanes are effectively synthesized in geometrically and enantiomerically pure forms through the Peterson-type elimination of organosilanes prepared by palladium-catalyzed bis-silylation (Eq. 57) [72, 73]. The intramolecular bis-silylation of optically active allylic alcohols in refluxing toluene af-

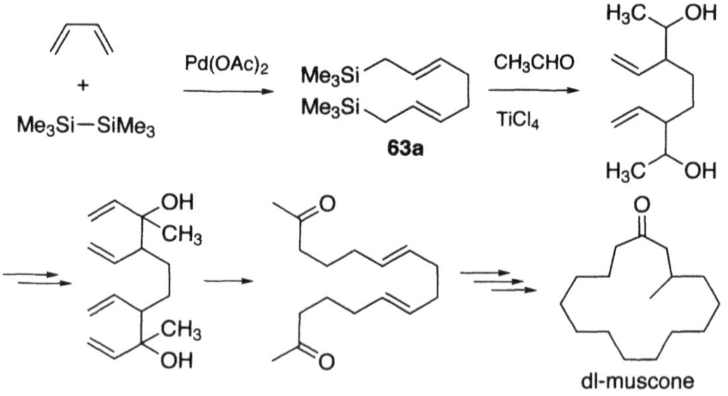

Scheme 8

fords (*E*)-allylsilanes **105** and six-membered disiladioxanes **106**, which are derived from thermal disproportionation of eight-membered **79**, in good yields. Treatment of the mixture with BuLi affords optically active allylsilanes, which completely retain the enantiomeric excesses of the starting allylic alcohols in good yields. This method has successfully been applied to the synthesis of optically active allenylsilane **107** from optically active propargylic alcohol (Eq. 58) [50].

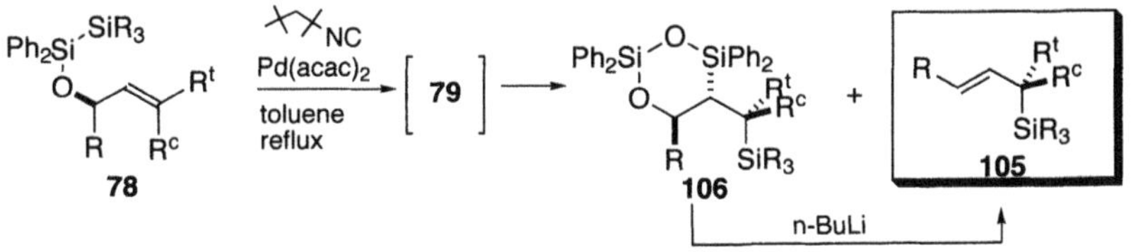

(57)

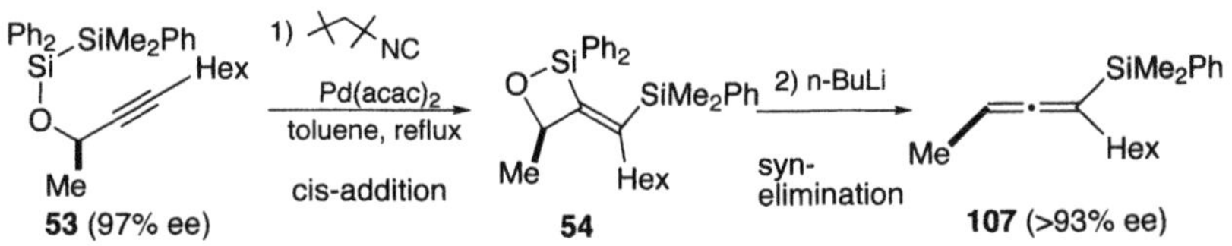

(58)

4.3
Via Oxidation of Silicon-Carbon Bonds

The silicon-carbon bonds having at least one hetero-atom substituent on the silicon atom are oxidized by H_2O_2 with a fluoride source to give the corresponding alcohols with retention of the configuration at the carbon atoms [103, 104]. The Si–C oxidation is successfully combined with bis-silylation reactions, providing new access to stereo-defined alcohols. Optically active β-hydroxy ketones **108** are synthesized by enantioselective 1,4-bis-silylation of α,β-unsaturated ketones followed by transformations including the Si–C oxidation (Scheme 9) [64].

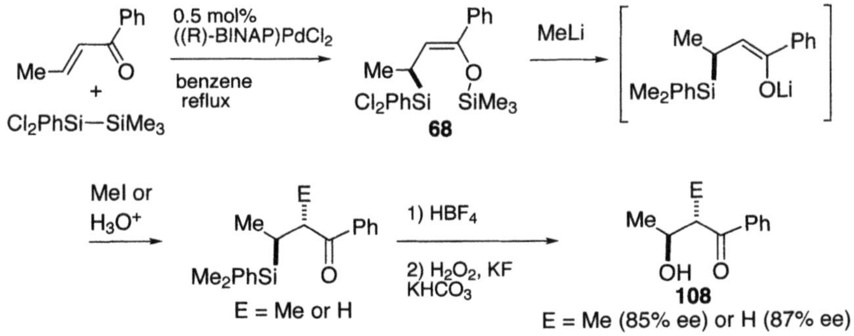

Scheme 9

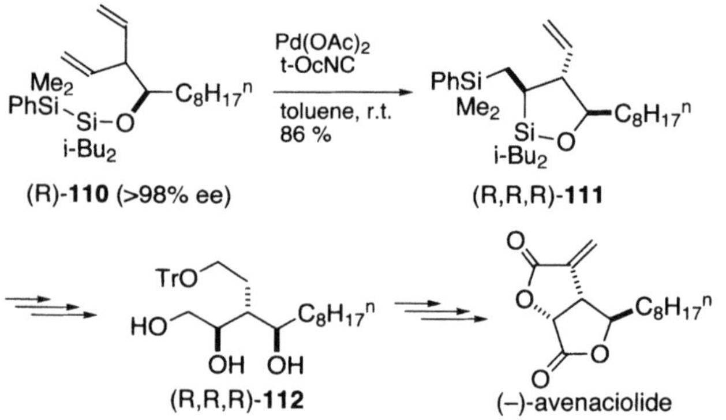

Scheme 10

Intramolecular bis-silylation of homoallylic alcohols, which proceeds with high diastereoselectivity, is applied to the stereoselective synthesis of 1,2,4-triols **109** (Eq. 59) [67–71].

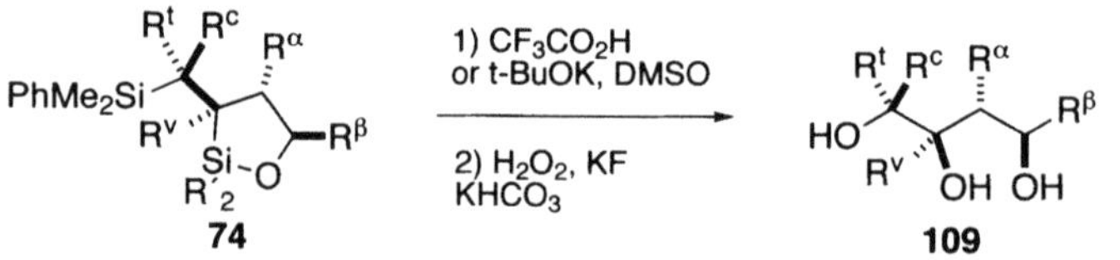

(59)

Naturally occurring (−)-avenaciolide is synthesized by intramolecular bis-silylation of optically active **110** (Scheme 10). The addition of the Si–Si bonds takes place stereoselectively at one diastereotopic face of one of the two C=C bonds, providing the anti-anti enantiomer **111** predominantly. Subsequent transformation including protection-deprotection and the Si–C oxidation affords stereodefined triol **112**, from which the target molecule is synthesized [69].

5 Concluding Remarks

A variety of catalytic reactions involving the Si–Si activation by transition-metal complexes have been developed. Studies on the stoichiometric reactions are also very important for the improvement and development of the catalytic reactions through an understanding of the reaction mechanisms. A theoretical approach has also been used to gain an understanding of the palladium- and platinum-catalyzed bis-silylation reactions [105], and may provide a useful tool for optimizing the catalytic reactions. It should be emphasized that the knowledge concerning Si–Si activation is applicable to the activation and synthetic utilization of other thermally stable element-element bonds including group 14 elements and boron, which have recently been the subject of much interest [106].

References

1. Armitage DA (1982) Organosilanes. In: Wilkinson G, Stone FGA, Abel EW (eds) Comprehensive organometallic chemistry. Pergamon Press, Oxford, p 1
2. (a) West R (1982) Organopolysilanes. In: Wilkinson G, Stone FGA, Abel EW (eds) Comprehensive organometallic chemistry. Pergamon Press, Oxford, p 365. (b) West R (1995) Organopolysilanes. In: Abel EW, Stone FGA, Wilkinson G, Davies AG (eds) Comprehensive organometallic chemistry II. Pergamon Press, Oxford, p 77
3. Curtis MD, Epstein PS (1981) Adv Organomet Chem 19:213
4. Braunstein P, Knorr M (1995) J Organomet Chem 500:21
5. Schubert U (1994) Angew Chem Int Ed Engl 33:419
6. Sharma HK, Pannell KH (1995) Chem Rev 95:1351
7. Horn KA (1995) Chem Rev 95:1317
8. Racatto CA (1995) Aldrichimica Acta 28:85
9. Suginome M, Ito Y (1998) J Chem Soc Dalton Trans: 1925
10. Liu CS, Cheng CW (1975) J Am Chem Soc 97:6746
11. Kiso Y, Tamao, K, Kumada M (1974) J Organomet Chem 76:95, 105
12. Groshens TJ, Klabunde KJ (1982) Organometallics 1:564
13. (a) Groshens TJ, Klabunde KJ (1983) J Organomet Chem 259:337. (b) Janikowski SK, Radonovich LJ, Groshens TJ, Klabunde KJ (1985) Organometallics 4:396
14. Seyferth D, Goldman EW, Escudié J (1984) J Organomet Chem 271:337
15. Eaborn C, Griffiths RW, Pidcock A (1982) J Organomet Chem 225:331
16. Pan Y, Mague JT, Fink MJ (1992) Organometallics 11:3495
17. Ozawa F, Sugawara M, Hayashi T (1994) Organometallics 13:3237
18. Murakami M, Yoshida T, Ito Y (1994) Organometallics 13:2900
19. Naka A, Okada T, Ishikawa M (1996) J Organomet Chem 521:163
20. Suginome M, Oike H, Park S-S, Ito Y (1996) Bull Chem Soc Jpn 69:289
21. Suginome M, Kato Y, Takeda N, Oike H, Ito Y (1998) Organometallics 17:495
22. Yamashita H, Kobayashi T-a, Hayashi T, Tanaka M (1990) Chem Lett:1447
23. Suginome M, Oike H, Shuff PH, Ito Y (1996) J Organomet Chem 521:405
24. Pham EK, West R (1990) Organometallics 9:1517
25. Sakurai H, Kamiyama Y, Nakadaira Y (1976) J Am Chem Soc 98:7453
26. Sakurai H, Kobayashi T, Nakadaira Y (1978) J Organomet Chem 162:C43
27. (a) Chi Y, Liu CS (1981) Inorg Chem 20:3456. (b) Lin CH, Lee CY, Liu CS (1986) J Am Chem Soc 108:1323. (c) Horng KM, Wang SL, Liu CS (1991) Organometallics 10:631
28. Huang CY, Liu CS (1989) J Organomet Chem 373:353
29. Murakami M, Yoshida T, Ito Y (1996) Chem Lett:13
30. Suginome M, Oike H, Ito Y (1994) Organometallics 13:4148
31. (a) Ueno K, Tobita H, Shimoi M, Ogino H (1988) J Am Chem Soc 110:4092. (b) Tobita H, Ueno K, Shimoi M, Ogino H (1990) J Am Chem Soc 112:3415
32. (a) Pannell K H, Cervantes J, Hernandez C, Cassias J, Vincenti S (1986) Organometallics 5:1056. (b) Tobita H, Ueno K, Ogino H (1986) Chem Lett:1777
33. Brookes A, Knox SAR, Stone FGA (1971) J Chem Soc (A):3469
34. Kerber RC, Pakkanen T (1979) Inorg Chim Acta 37:61
35. Michalczyk MJ, Recatto CA, Calabrese JC, Fink MJ (1992) J Am Chem Soc 114:7955
36. Okinoshima H, Yamamoto K, Kumada M (1975) J Organomet Chem 86:C27
37. Tamao K, Hayashi T, Kumada M (1976) J Organomet Chem 114:C19
38. Matsumoto H, Matsubara I, Kato T, Shono K, Watanabe H, Nagai Y (1980) J Organomet Chem 199:43
39. Watanabe H, Kobayashi M, Higuchi K, Nagai Y (1980) J Organomet Chem 186:51
40. Watanabe H, Kobayashi M, Saito M, Nagai Y (1981) J Organomet Chem 216:149
41. Sakurai H, Kamiyama Y, Nakadaira Y (1975) J Am Chem Soc 97:932
42. Carlson CW, West R (1983) Organometallics 2:1801

43. Finckh W, Tang B, Lough A, Manners I (1992) Organometallics 11:2904
44. (a) Kusukawa T, Kabe Y, Ando W (1993) Chem Lett:985. (b) Kusukawa T, Kabe Y, Nestler B, Ando W (1995) Organometallics 14:2556
45. (a) Yamashita H, Catellani M, Tanaka M (1991) Chem Lett:241. (b) Yamashita H, Tanaka M (1992) Chem Lett:1547
46. Ito Y, Suginome M, Murakami M (1991) J Org Chem 56:1948
47. Murakami M, Oike H, Sugawara M, Suginome M, Ito Y (1993) Tetrahedron 49:3933
48. Murakami M, Suginome M, Fujimoto K, Ito Y (1993) Angew Chem Int Ed Engl 32:1473
49. Sekiguchi A, Ichinohe M, Kabuto C, Sakurai H (1995) Organometallics 14:1092
50. Suginome M, Matsumoto A, Ito Y (1996) J Org Chem 61:4884
51. Suginome M, Takama A, Ito Y (1998) J Am Chem Soc 120:1930
52. Tsuji Y, Lago RM, Tomohiro S, Tsuneishi H (1992) Organometallics 11:2353
53. Ishikawa M, Sakamoto H, Okazaki S, Naka A (1992) J Organomet Chem 439:19
54. Okinoshima H, Yamamoto K, Kumada, M (1972) J Am Chem Soc 94:9263
55. Tamao K, Okazaki S, Kumada, M (1978) J Organomet Chem 146:87
56. Matsumoto H, Shono K, Wada A, Matsubara I, Watanabe H, Nagai Y (1980) J Organomet Chem 199:185
57. Ishikawa M, Nishimura Y, Sakamoto H, Ono T, Ohshita J (1992) Organometallics 11:483
58. Sakurai H, Eriyama Y, Kamiyama Y, Nakadaira Y (1984) J Organomet Chem 264:229
59. Obora Y, Tsuji Y, Kawamura T (1993) Organometallics 12:2853
60. Sakurai H, Kamiyama Y, Nakadaira Y (1975) Chem Lett:887
61. Obora Y, Tsuji Y, Kakehi T, Kobayashi M, Shinkai Y, Ebihara M, Kawamura T (1995) J Chem Soc Perkin Trans 1:599
62. Watanabe H, Saito M, Sutou N, Kishimoto K, Inose J, Nagai Y (1982) J Organomet Chem 225:343
63. Hayashi T, Matsumoto Y, Ito Y (1988) Tetrahedron Lett 29:4147
64. Hayashi T, Matsumoto Y, Ito Y (1988) J Am Chem Soc 110:5579
65. Hayashi T, Kobayashi T-a, Kawamoto AM, Yamashita H, Tanaka M (1990) Organometallics 9:280
66. Hayashi T, Kawamoto AM, Kobayashi T, Tanaka M (1990) J Chem Soc Chem Commun:563
67. Murakami M, Andersson PG, Suginome M, Ito Y (1991) J Am Chem Soc 113:3987
68. Murakami M, Suginome M, Fujimoto K, Nakamura H, Andersson PG, Ito Y (1993) J Am Chem Soc 115:6487
69. Suginome M, Yamamoto Y, Fujii K, Ito Y (1995) J Am Chem Soc 117:9608
70. Suginome M, Matsumoto A, Nagata K, Ito Y (1995) J Organomet Chem 499:C1
71. Suginome M, Nakamura H, Ito Y (1996) Tetrahedron Lett 38:555
72. Suginome M, Matsumoto A, Ito Y (1996) J Am Chem Soc 118:3061
73. Suginome M, Iwanami T, Matsumoto A, Ito Y (1997) Tetrahedron Asymmetry 8:859
74. Reddy NP, Yamashita H, Tanaka M (1992) J Am Chem Soc 114:6596
75. (a) Rich JD (1989) J Am Chem Soc 111:5886. (b) Rich JD (1989) Organometallics 8:2609
76. Yamashita H, Reddy NP, Tanaka M (1993) Chem Lett:315
77. (a) Ishikawa M, Sakamoto H, Okazaki S, Naka A (1992) J Organomet Chem 439:19. (b) cf. Ishikawa M, Sakamoto H, Tabuchi T (1991) Organometallics 10:3173
78. Naka A, Hayashi M, Okazaki S, Ishikawa M (1994) Organometallics 13:4994
79. Ito Y, Nishimura S, Ishikawa M (1987) Tetrahedron Lett 28:1293
80. Ito Y, Matsuura T, Murakami M (1988) J Am Chem Soc 110:3692
81. Murakami M, Suginome M, Matsuura T, Ito Y (1991) J Am Chem Soc 113:8899
82. Weidenbruch M, Kroke E, Peters K, von Schnering HG (1993) J Organomet Chem 461:35

83. (a) Yamamoto K, Okinoshima H, Kumada M (1970) J Organomet Chem 23:C7. (b) Yamamoto K, Okinoshima H, Kumada M (1971) J Organomet Chem 27:C31
84. (a) Tamao K, Yamaguchi S, Shiozaki M, Nakagawa Y, Ito Y (1992) J Am Chem Soc 114:5867. (b) Tamao K, Yamaguchi S, Ito Y, Matsuzaki Y, Yamabe T, Fukushima M, Mori S (1995) Macromolecules 28:8668
85. Tamao K, Sun G-R, Kawachi A (1995) J Am Chem Soc 117:8043
86. Ito Y, Suginome M, Murakami M, Shiro M (1989) J Chem Soc Chem Commun:1494
87. Tamao K, Tarao Y, Nakagawa Y, Nagata K, Ito Y (1993) Organometallics 12:1113
88. Sakurai H, Kamiyama Y, Nakadaira Y (1977) J Organomet Chem 131:147
89. Uchimaru Y, Tanaka M (1996) J Organomet Chem 521:335
90. Suginome M, Oike H, Ito Y (1995) J Am Chem Soc 117:1665
91. Suginome M, Oike H, Shuff PH, Ito Y (1996) Organometallics 15:2170
92. Suzuki M, Obayashi, T, Amii H, Saegusa T (1991) Polym Prepr Jpn 40:355
93. Uchimaru Y, Tanaka Y, Tanaka M (1995) Chem Lett:164
94. Matsumoto H, Nagashima S, Yoshihiro K, Nagai Y (1975) J Organomet Chem 85:C1
95. (a) Yamamoto K, Suzuki S, Tsuji J (1980) Tetrahedron Lett 21:1653. (b) Yamamoto K, Hayashi A, Suzuki S, Tsuji J (1987) Organometallics 6:974
96. Matsumoto H, Yako T, Nagashima S, Motegi T, Nagai Y (1978) J Organomet Chem 148:97
97. Urata H, Suzuki H, Moro-oka Y, Ikawa T (1984) Bull Chem Soc Jpn 57:607
98. Tsuji Y, Kajita S, Isobe S, Funato M (1993) J Org Chem 58:3607
99. Fleming I, Dunogues J, Smithers R (1989) Org React 37:57
100. Masse CE, Panek JS (1995) Chem Rev 95:1293
101. Peterson DJ (1968) J Org Chem 33:780
102. (a) Colvin EW (1985) Silicon in organic synthesis, Krieger, Malabar, Chap 12. (b) Colvin EW (1988) Silicon reagents in organic synthesis. Academic Press, London, Chap 10
103. (a) Tamao K, Kakui T, Akita M, Iwahara T, Kanatani R, Yoshida J, Kumada M (1983) Tetrahedron 39:983. (b) Tamao K (1996) Advances in silicon chemistry. JAI Press, Greenwich, London, vol 3:1
104. Fleming I, Henning R, Plaut H (1984) J Chem Soc Chem Commun:29
105. Sakaki S, Ieki M (1993) J Am Chem Soc 115:2373
106. For leading references see Suginome M, Nakamura H, Ito Y (1997) Angew Chem Int Ed Engl 36:2516

Activation of C–O Bonds: Stoichiometric and Catalytic Reactions

Yong-Shou Lin and Akio Yamamoto

Department of Applied Chemistry, Graduate School of Science and Engineering, Advanced Research Center for Science and Engineering, Waseda University, 3-4-1 Ohkubo, Shinjuku-ku, Tokyo 169-8555, Japan
**E-mail: akioyama@mn.waseda.ac.jp*

The C–O bond cleavage promoted by transition metal complexes is becoming an important process in organic synthesis. A survey of the stoichiometric reactions involving activation of the C–O bond by transition metal complexes as well as their synthetic applications is provided. After a survey of recent reports on the cleavage of allylic C–O bond that has been extensively utilized in organic synthesis the review focuses on the cleavage of single C–O bonds in esters, ethers, and anhydrides activated by transition metals. The cleavage of C–O multiple bonds is also discussed.

Keywords: C–O bond cleavage, Activation of C–O bonds, Oxidative addition, Transition metal complexes, Allylic compounds, Esters, Ethers, Anhydrides, Alcohols

1 Introduction 162

2 Activation of Allylic C–O Bonds by Transition Metal Complexes Involving Formation of η^3-Allyltransition Metal Complexes 163

2.1 Stoichiometric Reactions 165
2.2 Catalytic Reactions 166
2.2.1 Catalytic Allylation Reactions 167
2.2.2 Catalytic Reductions 171
2.2.3 Catalytic Carbonylation 172

3 Activation of Allylic C–O Bonds Without Involving η^3-Allyltransition Metal Complexes 172

4 Cleavage of the C–O Single Bond in Esters, Lactones, and Anhydrides 175

5 Cleavage of the C–O Single Bond in Ethers, Alcohols, and Acetals . 179

5.1 Cleavage of the C–O Single Bond in Acyclic Ethers 179
5.1.1 Cleavage of the sp^3-C–O Single Bond in Acyclic Ethers 179
5.1.2 Cleavage of the sp^2-C–O Single Bond in Acyclic Ethers 181
5.2 Cleavage of the C–O Single Bond in Cyclic Ethers 182
5.3 Cleavage of the C–O Single Bond in Alcohols and Acetals 184

Topics in Organometallic Chemistry, Vol. 3
Volume Editor: S. Murai

6 **Cleavage of the C–O Multiple Bonds in Acetones, Aldehydes, Esters, CO, and CO_2** 185

7 **Concluding Remarks** 189

References 189

1 Introduction

Cleavage of carbon-halogen bonds in organic halides promoted by transition metal complexes, notably by palladium complexes, has been extensively used in organic syntheses. The utility of the process arises from ease of oxidative addition of organic halides to low valent transition metal complexes to give organotransition metal halides and their subsequent facile conversions to afford a variety of organic products. For example, cross-coupling processes, Mizoroki-Heck type olefin arylation processes, as well as Heck type carbonylation processes have been widely applied in various organic syntheses with aryl and vinyl halides as the starting compounds. However, the processes involving the carbon-halogen bond cleavage have inherent problems due to the use of organic halides. Although organic halides are very convenient starting compounds, being capable of undergoing easy carbon-halogen bond cleavage, the halide employed has to be eventually removed to prepare the end products such as arylalkanes, olefins and carbonyl compounds with use of a base. Thus the total efficiency of the process is not high and discarding the salts of hydrogen halides presents environmental problems. On the other hand, if simple methods to cleave the carbon-oxygen bonds can be found to generate organotransition metal complexes that can be converted into useful organic products, there are certain advantages in affording economically and environmentally more preferable processes.

In the previous review [1] various reactions involving the cleavage of C–O bonds in organic compounds activated by transition metal complexes were treated. The present review is mainly concerned with the later development in the C–O bond activation with inclusion of essential concepts related to the C–O bond cleavage reactions promoted by transition metal complexes.

The C–O bond cleavage in organic compounds promoted by transition metal complexes can be divided into two categories: one involving the one-step cleavage of the C–O bond and the other proceeding through consecutive processes such as insertion of an unsaturated compound into metal-hydride, metal alkyl or metal alkoxide bonds to be followed by elimination involving the C–O bond cleavage. Since the two-step processes have been reviewed previously [1] and there have been few new developments, we shall be mainly concerned here with the one-step processes.

The direct cleavage of an organic compound promoted by a transition metal complex proceeds by net oxidative addition with cleavage of the A–B bond resulting in formation of M–A and M–B bonds (Eq. 1).

$$ML_n + A\text{–}B \rightleftharpoons L_nM\begin{matrix}A\\B\end{matrix} \tag{1}$$

For the oxidative addition of an organic compound to a transition metal complex to proceed, the enthalpy change is expressed by Eq. 2.

$$\Delta H_{ox.add.} = D(A\text{-}B) - \{D(L_nM\text{-}A) + D(L_nM\text{-}B)\} \tag{2}$$

Although the presently available data of bond dissociation energies of metal-carbon bonds and of metal-oxygen bonds are limited, there is accumulating evidence that these bonds are stronger than previously assumed and the values expressed by Eq. 2 are considerably negative (exothermic process) [2–5]. Thus after taking into account of the $T\Delta S$ value, which may be approximated to 10 kcal/mol around room temperature, the oxidative addition process is estimated to be thermodynamically favorable, if there is no change in the number of ligand L attached to the metal before and after the reaction. In certain cases, dissociation of the ligand L is involved in the oxidative addition process. In these cases dissociation of a ligand from the starting complex provides a kinetic barrier for undergoing the oxidative addition.

2 Activation of Allylic C–O Bonds by Transition Metal Complexes Involving Formation of η^3-Allyltransition Metal Complexes

The cleavage of allylic C–O bond in allylic organic compound on interaction with a low valent transition metal complex proceeds by net oxidative addition to give η^3-allyltransition metal complexes (Eq. 3).

$$\text{CH}_2{=}\text{CHCH}_2\text{OX} + ML_n \rightleftharpoons (\eta^3\text{-allyl})M(OX)(L) \overset{L}{\rightleftharpoons} [(\eta^3\text{-allyl})ML_2]OX^- \tag{3}$$

X = Ac, COOR', R', H, $PO(OR)_2$, etc.

Since the C–O bond dissociation energies in allylic compounds are weaker than in alkyl C–O and aryl C–O bonds and the η^3-allyl-metal bond formed is stronger than in the η^1-allyl-metal bond [2, 3], cleavage of the C–O bonds in allylic compounds by an oxidative addition process is a thermodynamically favorable process. Thus there are many examples of the oxidative addition process involving the allylic C–O bond cleavage, and the process utilizing the allylic C–O bond cleavage has become one of the most important means in organic systems [6, 7].

The other factor to enhance the ease of the allylic C–O bond cleavage is the interaction of the transition metal with the C=C bond in the allylic entity. The

S_N2' type interaction of the metal with the allylic double bond will facilitate the cleavage of the allylic C–O bond, thus making the process kinetically more favorable.

The oxidative addition process can be combined with nucleophilic attack on the η^3-allyl ligand to afford allylation products of the nucleophiles. Combination of the oxidative addition process with the nucleophilic attack provides important synthetic means to give allylation products of the nucleophiles catalytically as developed by Tsuji and Trost. In contrast to the progress of application to organic synthesis, fundamental studies on the allylic C–O bond cleavage promoted by transition metal complexes have been delayed [1, 6, 7]. The palladium-catalyzed allylation of nucleophiles with allylic acetates has been established to proceed by oxidative addition of the allylic acetates with inversion of the stereochemistry (anti-elimination) giving the η^3-allylic palladium complex to be followed by anti-attack of the allylic ligand by a soft nucleophile to give the allylation product with net retention of the stereochemistry (Scheme 1) [8].

On the other hand, the other process without involving the formation of η^3-allylic complexes may operate as an alternative route in the course of the C–O bond cleavage. One is the S_N2' type attack of a ligand bound to the metal such as hydride, alkyl or alkoxide on the terminal carbon of the allylic entity. The process is followed by elimination of the OX group (acetate or alkoxide) in a concerted manner as shown in Eq. 4. The other mode of cleavage is insertion-elimination type as shown in Eq. 5. The process proceeds by insertion of the olefinic moiety of the allylic entity into the M-Y bond, such as hydride, alkyl, or alkoxide ligand followed by β-elimination of the acetate or alkoxide moiety.

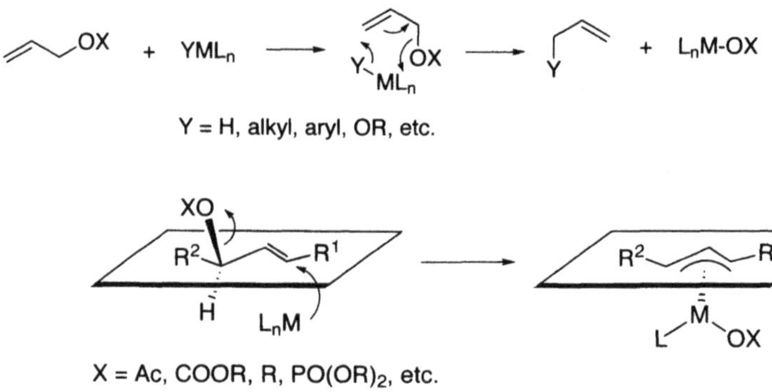

OX + YML_n → [Y, OX, ML_n] → Y + L_nM-OX (4)

Y = H, alkyl, aryl, OR, etc.

X = Ac, COOR, R, $PO(OR)_2$, etc.

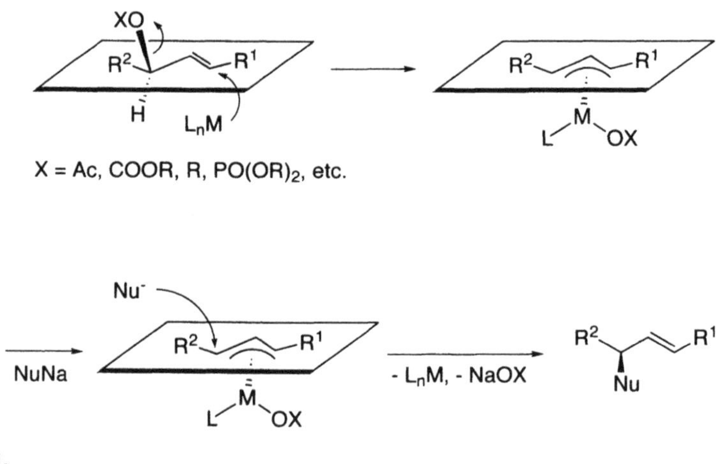

Scheme 1.

(5)

Y = H, alkyl, aryl, OR, etc.

The examples for consecutive-type reactions are quite limited. We first discuss the case through the concerted process involving the η^3-allyltransition metal complexes.

2.1 Stoichiometric Reactions

In contrast to the abundance of examples of applications utilizing various Pd(II) compounds as catalyst precursors, clear-cut fundamental studies regarding the oxidative addition of allylic compounds to a Pd(0) complex are limited.

An η^3-allylpalladium complex has been isolated in the oxidative addition of allyl acetate with $Pd(PCy_3)_2$ (Eq. 6) [9–11].

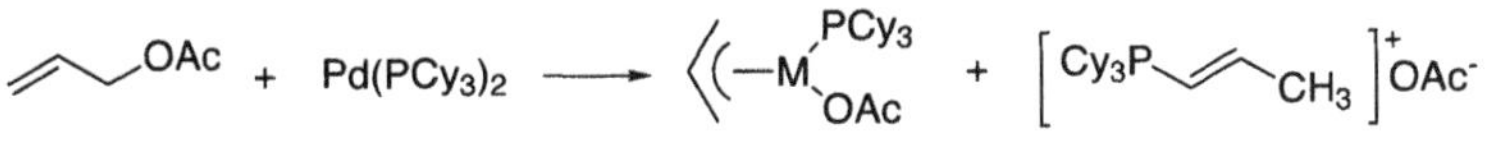

(6)

PCy_3 = tricyclohexylphosphine

Other allylic compounds having C–O bonds, such as allylic carboxylates, carbonates, phosphates, ethers, and alcohols, can also serve as a substrate in such oxidative addition reactions via the C–O bond cleavage [1].

The C–O bond in allylic formate is cleaved on its oxidative addition to Pd(0) complexes to give η^3-allylpalladium formate complexes (Eq. 7) [12].

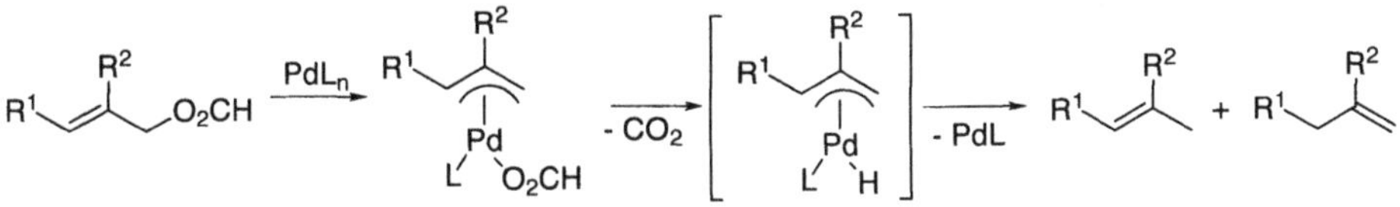

(7)

The η^3-allylpalladium formate complex is considered as a model of the intermediate in a catalytic reductive cleavage of allylic formate or allylic acetate combined with formic acid to olefins. The η^3-allylpalladium formate was revealed to be decarboxylated to release olefins upon coupling of the produced palladium hydride with the η^3-allyl ligand (Eq. 7).

As models of intermediates in palladium-catalyzed conversion of allylic carbonates into allylic ethers, various η^3-allylpalladium and -platinum carbonate complexes have been prepared by treatment of allylic carbonates with [Pd(styrene)L_2] (Eq. 8) [13].

$$\text{trans-PdEt}_2\text{L}_2 \xrightarrow[\text{- C}_2\text{H}_4,\ \text{C}_2\text{H}_6]{\text{+ styrene}} \text{Pd(styrene)L}_2 \xrightarrow{+\ \text{CH}_2\text{=C(R}^1\text{)CH}_2\text{OCO}_2\text{R}^2} [\text{R}^1\text{-}(\eta^3\text{-C}_3\text{H}_4)\text{-ML}_2]^+(\text{OCO}_2\text{R}^2)^- \quad (8)$$

R^1 = Me, H; R^2 = Me, Et; L = PMe_3, PMe_2Ph, $PMePh_2$

Although examples of catalytic conversion of allylic ethers are limited, η^3-allyl alkoxide/aryloxide complexes have been isolated by the reaction of allylic ethers with zero-valent Ni and Pd complexes [10, 14].

An η^3-allylruthenium(II) complex has been isolated by oxidative addition of allyl carboxylate to Ru(0) complex (Eq. 9) [15].

$$\text{Ru(cod)(cot)} + 3\,\text{PEt}_3 + \text{CH}_2\text{=CHCH}_2\text{OCOCF}_3 \longrightarrow \text{Ru}(\eta^3\text{-C}_3\text{H}_5)(\text{OCOCF}_3)(\text{PEt}_3)_3 \quad (9)$$

Cleavage of the C–O bond in allylic acetate by a Mo(0) complex has been achieved with stereochemical retention to afford an η^3-allylmolybdenum complex [16].

Examples of direct C–O bond cleavage of allylic alcohols are relatively rare compared to other allylic compounds. Deoxygenation of allylic alcohols by $WCl_2(PMePh_2)_4$ has been reported [17]. Several other examples of the C–O bond cleavage in allylic alcohols have been reviewed previously [1].

Compared to transition metal complexes, cleavage of the C–O bond promoted by lanthanoid complexes has been explored less. The C–O bonds of allylic ethers are cleaved on treatment with $(C_5Me_5)_2Sm(THF)_n$ to give η^3-allylsamarium complexes (Eq. 10) [18].

$$2\,\text{Cp}^*_2\text{Sm(THF)}_n + \text{R}^1\text{R}^2\text{C=C(R}^3\text{)CH(R}^4\text{)OCH}_2\text{Ph} \xrightarrow{\text{r.t.}} \text{R}^3\text{-}(\eta^3\text{-allyl; R}^1,\text{R}^2,\text{R}^4)\text{-SmCp}^*_2 + \text{Cp}^*_2\text{SmOCH}_2\text{Ph} \quad (10)$$

n = 0 or 2; R^1 = H, Me or Ph; R^2, R^3, and R^4 = H or Me

2.2
Catalytic Reactions

Utilizing the processes involving the allylic C–O bond cleavage promoted by transition metal complexes and combining them with subsequent other processes, such as nucleophilic attack, CO insertion, hydrogenolysis, etc., one can de-

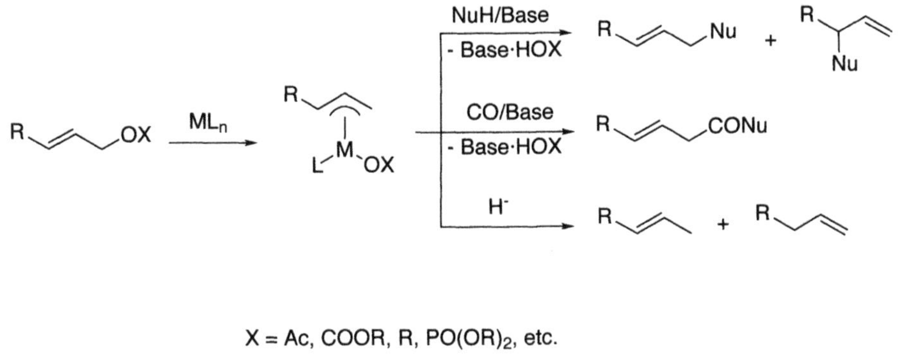

NuH/Base
- Base·HOX
R Nu + R Nu
R OX ML_n R L M OX
CO/Base
- Base·HOX
R CONu
H⁻
R + R

X = Ac, COOR, R, $PO(OR)_2$, etc.

Scheme 2.

velop various useful methods in organic syntheses (Scheme 2). The following examples illustrate concepts of catalytic processes promoted by transition metal complexes.

2.2.1
Catalytic Allylation Reactions

The catalytic processes for allylation reactions are composed of the allylic C–O bond cleavage and nucleophilic attack on the η^3-allyltransition metal complexes formed to give various organic allylation compounds (Scheme 3).

Cleavage of the C–O bond in various allylic substrates by oxidative addition to M(0) species gives η^3-allylic complexes, which undergo nucleophilic attack to produce allylic compounds catalytically. A base is needed in most cases to remove HOX and to drive the catalytic cycle. There are a lot of synthetic reactions utilizing allylic oxygen bond cleavage catalyzed by palladium complexes [6, 7, 19–21].

The experimental results that both branched and linear allylic ethers are obtained in the palladium-catalyzed decarboxylation of branched allylic carbonate indicate occurrence of direct oxidative addition involving the C–O bond cleavage followed by the nucleophilic attack of the alkoxide liberated on either the substituted or non-substituted terminus of the allylic ligand (Scheme 4) [1].

Ruthenium-catalyzed allylation of primary alcohols by allylic acetates to give α,β-unsaturated ketones has been reported (Eq. 11) [22].

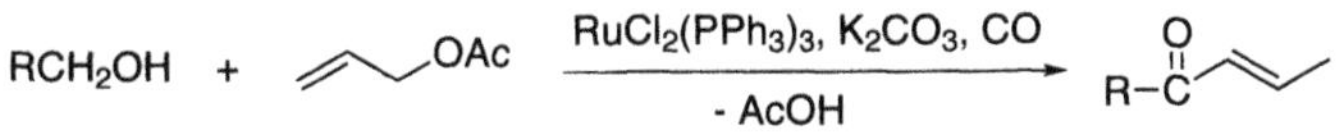

$$RCH_2OH + \text{CH}_2{=}\text{CHCH}_2\text{OAc} \xrightarrow[\text{- AcOH}]{RuCl_2(PPh_3)_3,\ K_2CO_3,\ CO} R{-}\overset{O}{\overset{\|}{C}}{-}CH{=}CHCH_3 \qquad (11)$$

As another example of ruthenium-catalyzed allylation, a coupling reaction of allylic carbonates with acrylic amide to give 3,5-dienoic acid derivatives has been achieved by using Ru(cod)(cot) as catalyst (Eq. 12) [23].

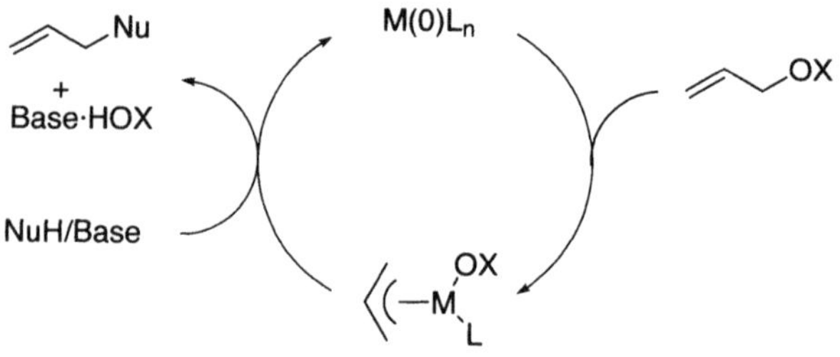

X = Ac, COOR, R, PO(OR)2, etc.

Scheme 3.

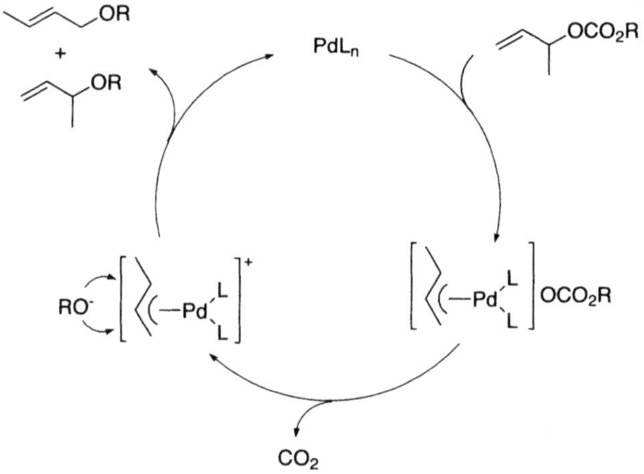

Scheme 4.

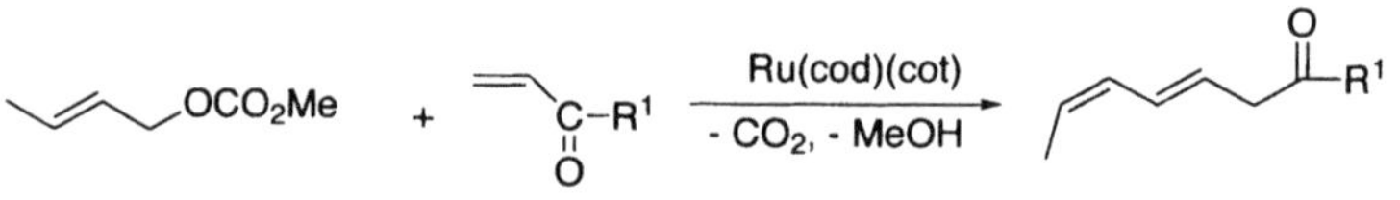

$$\text{CH}_3\text{CH=CHCH}_2\text{OCO}_2\text{Me} + \text{CH}_2\text{=CH–C(=O)–R}^1 \xrightarrow[\text{- CO}_2,\ \text{- MeOH}]{\text{Ru(cod)(cot)}} \text{product} \tag{12}$$

R^1 = OR, NR_2

In this reaction, cleavage of the C–O bond in allylic carbonates promoted by Ru(0) to form η^3-allylruthenium intermediate with liberation of CO_2 is proposed.

Recently, interesting processes for the direct activation of the C–O bonds in allylic alcohols have been realized by the promotion of CO_2 [24]. In this process the effect of CO_2 was interpreted by formation of hydrogen allyl carbonate by the reaction between allyl alcohol and CO_2. The hydrogen allyl carbonate thus produced is more susceptible to the allyl-O bond cleavage than in the parent allyl alcohol in interaction with a Pd(0) catalyst as shown in Scheme 5. Combination of

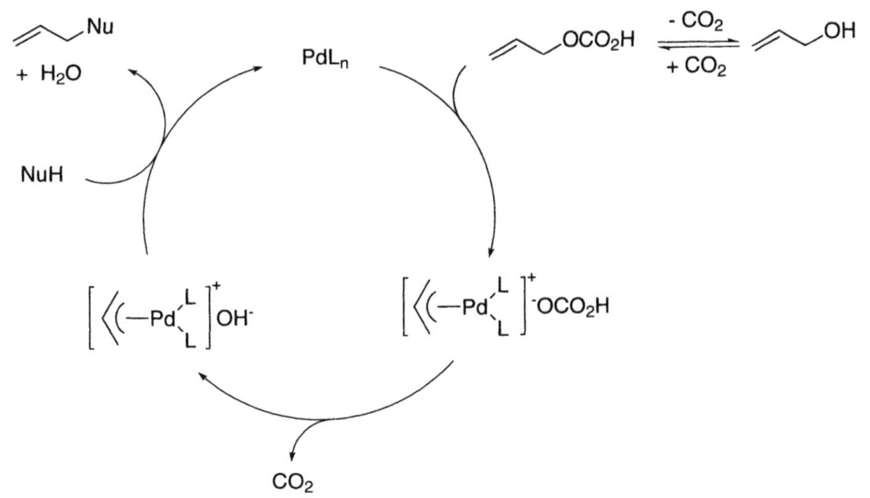

Scheme 5.

the C–O bond cleavage with nucleophilic attack or CO insertion provides new catalytic applications to promote allylation of amines and carbonylation reactions, respectively (Eqs. 13, 14).

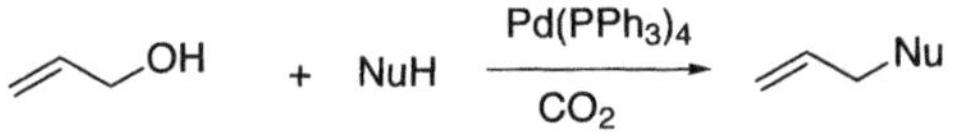

$$CH_2{=}CHCH_2OH + NuH \xrightarrow[CO_2]{Pd(PPh_3)_4} CH_2{=}CHCH_2Nu \qquad (13)$$

NuH = Et_2NH, β-keto esters, β-diketones

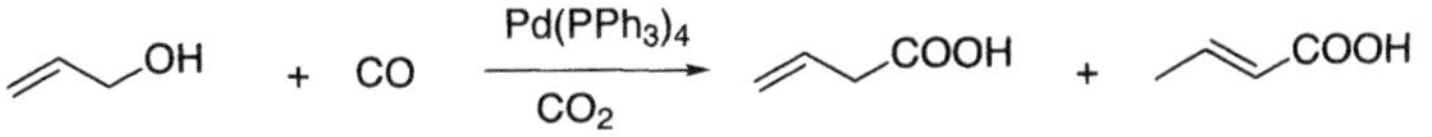

$$CH_2{=}CHCH_2OH + CO \xrightarrow[CO_2]{Pd(PPh_3)_4} CH_2{=}CHCH_2COOH + CH_3CH{=}CHCOOH \qquad (14)$$

Allylation of aromatic compounds with allylic alcohols and esters through C–O bond cleavage catalyzed by molybdenum, tungsten, and palladium complexes has been reported recently [25, 26]. In addition, molybdenum-catalyzed aromatic substitution with alcohols has been achieved [27].

Enantioselective elimination of the allylic bicyclic carbonate by using chiral phosphine-palladium catalyst leads to synthesis of chiral dienes [28].

Asymmetric allylic alkylations catalyzed by transition metal complexes have been developed significantly in recent years [29–31]. When the allylation of a nucleophile proceeds by a mechanism such as shown in Scheme 1 involving anti-elimination and anti-nucleophilic attack on the allylic ligand, control of nucleophilic attack is required to achieve the regiochemical formation of the allylated nucleophile. Generally aryl-substituted allylic esters undergo the attack on the less substituted terminus of the allylic ligand and a special method is required to direct the attack of a nucleophile on the more substituted site. Several research groups reported methods of circumventing the difficulty by employing specially designed ligands.

Hayashi and coworkers have achieved the alkylation of 1- and 3-substituted 2-propenyl acetates with high regio- and enantioselectivities by using a palladium catalyst in the presence of a chiral ligand, (*R*)-2-diphenylphosphino-2'-methoxy-1,1'-binaphthyl, (*R*)-MeO-MOP (Eq. 15) [32, 33].

$$R\text{-}CH{=}CH\text{-}CH_2OAc \xrightarrow[\text{[PdCl}(\eta^3\text{-}C_3H_5)]_2,\ \text{dppe or PPh}_3]{NaCMe(CO_2Me)_2} R\text{-}CH{=}CH\text{-}CH_2CMe(CO_2Me)_2 \quad \text{major} \tag{15a}$$

$$R\text{-}CH(OAc)\text{-}CH{=}CH_2 \xrightarrow[\text{[PdCl}(\eta^3\text{-}C_3H_5)]_2,\ (R)\text{-MeO-MOP}]{NaCMe(CO_2Me)_2} Ar\text{-}CH(CMe(CO_2Me)_2)\text{-}CH{=}CH_2 \quad \text{major} \tag{15b}$$

R = Ph, p-$MeOC_6H_4$, p-ClC_6H_4, Me (R)-MeO-MOP = [MeO, PPh_2 binaphthyl structure]

In contrast to the formation of linear achiral allylation product on usage of the catalytic system with dppe ligand, employment of the (*R*)-MeO-MOP ligand gave the branched product in a high regiochemistry and high enantioselectivity.

Another method of directing the attack of a nucleophile at a specific site on the allylic ligand is to use a special chiral chelating ligand containing both P and N donors. Helmchen's group has developed a new type of asymmetric P,N-chelete ligand in allylic substitutions catalyzed by palladium complexes with very high enantioselectivity [34, 35]. By utilizing the electronic and steric differences of the P- and N-containing special ligands having bulky substituents on the P atom, one can direct the nucleophilic attack and achieve the high regio- and enantioselectivities [36].

Molybdenum-catalyzed alkylation of aryl-substituted allylic carbonates has been directed to give the alkylation at the substituted site affording the branched isomer as the major product with a high enantioselectivity by employing a specially designed diamine type ligand (Eq. 16) [37].

$$Ar\text{-}CH{=}CH\text{-}CH_2OCO_2CH_3 \ \text{or}\ Ar\text{-}CH(OCOR')\text{-}CH{=}CH_2 + NaRC(CO_2CH_3)_2 \xrightarrow[\text{L, THF}]{(C_2H_5CN)_3Mo(CO)_3} Ar\text{-}CH(CR(CO_2CH_3)_2)\text{-}CH{=}CH_2\ (\text{major}) + Ar\text{-}CH{=}CH\text{-}CH_2\text{-}CR(CO_2CH_3)_2\ (\text{minor}) \tag{16}$$

Ar = Ph, 2-furyl, 1-naphthyl, 2-pyridyl, 2-thienyl

R = H, Me, $CH_2CH{=}CH_2$

R' = OMe, Me

L = [trans-1,2-cyclohexanediamine bis(2-pyridinecarboxamide)]

On the other hand, linear products rather than branched isomers were obtained in the allylic alkylation of 1-aryl-2-propenyl acetates and 3-phenyl-2-propenyl acetate with soft carbon nucleophiles catalyzed by the Pd/PPh_3 system when a catalytic amount of LiI was used [38].

Activation of the C–O bond in the allylic carbonates promoted by a palladium complex with a chiral ligand as shown in Eq. 17 leads to asymmetric O- and C-alkylation of phenols [39].

$(dba)_3Pd_2{\cdot}CHCl_3$; L, CH_2Cl_2, r.t.; $Eu(fod)_3$; $CHCl_3$, 50 °C (17)

R = OMe, F; R' = Me, *t*-Bu; n = 1,2 and 3 L =

The study by Trost's group has also revealed that enantioselectivity is independent of the configuration of the starting allylic ester in an asymmetric alkylation promoted by a palladium complex in support of a mechanism proceeding through formation of an η^3-allylpalladium intermediate [40].

2.2.2
Catalytic Reductions

Combination of the allylic C–O bond cleavage to form η^3-allyltransition metal complexes with nucleophilic attack by hydridic reagent gives alkenes as the reduction products of allylic compounds.

Catalytic conversion of allyl formate or other allylic compounds in the presence of formic acid to give olefins has been reviewed previously [1]. Here several new developments are discussed. Pd-catalyzed hydrogenolysis of allyloxytetrazoles to yield alkenes or alkanes in the presence of formic acid as a H-donor has been reported recently [41]. Palladium-catalyzed regioselective and stereospecific reduction of allylic formates leads to stereo-controlled formation of *cis* and *trans* ring junctions in hydrindane and decalin systems [42]. By using monodentate phosphine ligand in the presence of formic acid, palladium-catalyzed reduction of allylic esters to optically active olefins has been achieved [43]. On the other hand, reductive cleavage of allylic esters to give olefins in the presence of formic acid and triethylamine has been achieved catalytically with ruthenium complexes, and its application leads to a facile synthesis of α-hydroxy acids [44, 45].

Besides formic acid, many other reducing agents, such as $LiAlH_4$, borohydrides, hydrosilanes, and tin hydrides have been used for the hydrogenolysis of allylic compounds [7].

2.2.3
Catalytic Carbonylation

The combination of processes of the C–O bond cleavage in allylic substrates with CO insertion is a potential means of preparation of β,γ-unsaturated carboxylic acid derivatives.

The mechanism for the carbonylation of allylic substrates is considered to be composed of the following processes: (a) oxidative addition involving allylic C–O bond cleavage to form η^3-allyltransition metal complexes; (b) CO insertion into the metal-allylic bond; and (c) nucleophilic attack to liberate carboxylic acid derivatives (Scheme 6).

Allylic carbonates [46], acetates [47, 48], ethers [49, 50], alcohols [51, 52], and phosphates [53] have been employed for catalytic carbonylation by using palladium complexes. Pd-catalyzed conversion of allylic formates into carboxylic acids has been also achieved recently [54].

3
Activation of Allylic C–O Bonds Without Involving η^3-Allyltransition Metal Complexes

The other type of process of C–O bond activation that is different from the direct oxidative addition of the C–O bond to M(0) complexes to form η^3-allyltransition metal complexes is insertion-elimination type or S_N2' type as shown in Eqs. 4 and 5. Although the two processes are conceptually different, it is sometimes difficult to distinguish the two mechanisms. When the insertion-elimination process

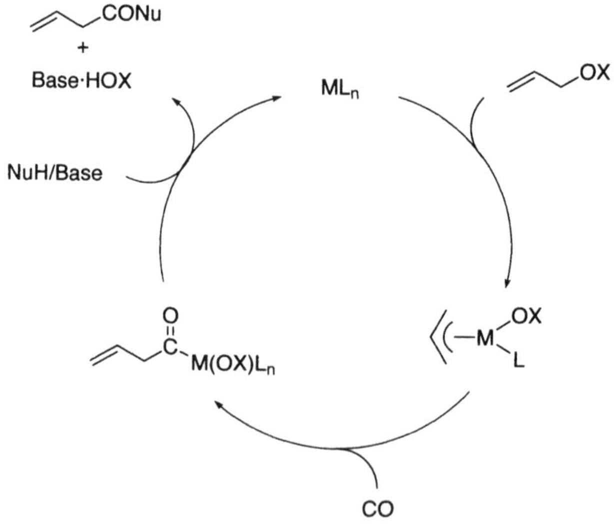

Scheme 6.

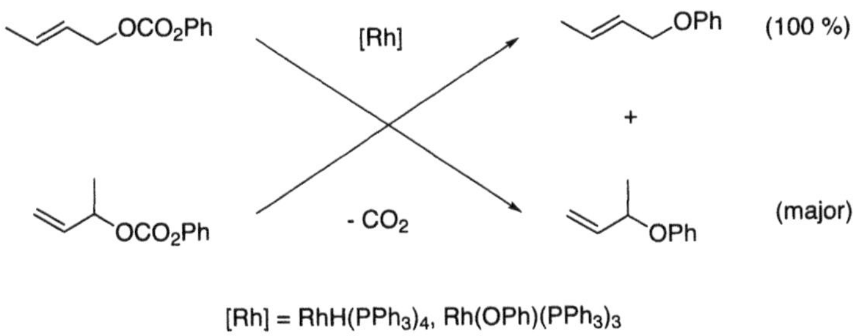

$[Rh] = RhH(PPh_3)_4, Rh(OPh)(PPh_3)_3$

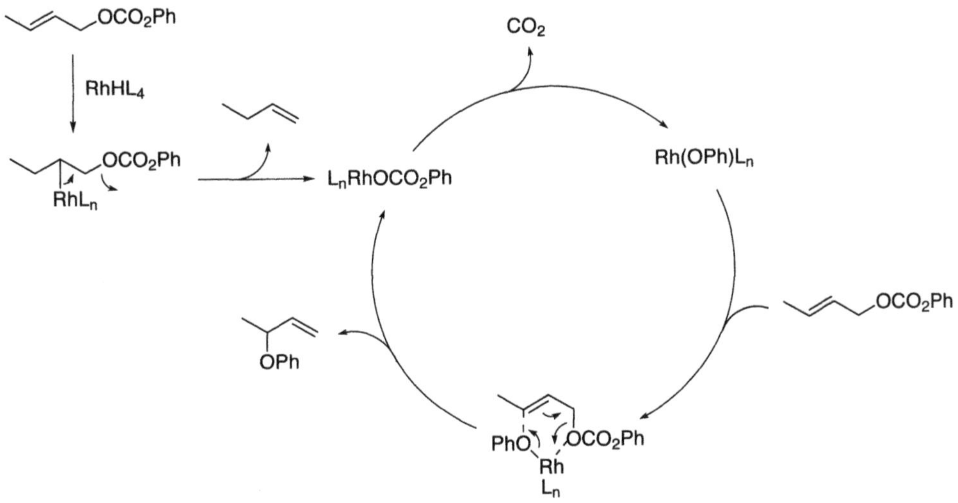

Scheme 7.

(Eq. 5) is operative and the β-elimination of OX from the intermediate alkyl complex is slow, one can observe hydrogen scrambling by using metal deuteride [55].

Operation of the insertion-elimination mechanism has been demonstrated in the reaction of rhodium hydride complex, $RhHL_4$ ($L=PPh_3$), with two isomeric allyl phenyl carbonates [56]. Unbranched 2-butenyl phenyl carbonate was found to give branched allylic phenyl ether exclusively, whereas the decarboxylation of the branched 1-methyl-2-propenyl phenyl carbonate afforded unbranched 2-butenyl phenyl ether. These results can be accounted for by assuming a precatalytic and catalytic insertion-elimination process as shown in Scheme 7.

In the precatalytic process the rhodium hydride precursor undergoes insertion into the butenyl carbonate to form an alkylrhodium complex. β-Elimination yields 1-butene and phenylcarbonatorhodium complex. Upon decarboxylation a phenoxorhodium complex is produced that undergoes the S_N2' type reaction with 2-butenyl phenyl carbonate to liberate the branched allylic ether, 1-

methyl-2-propenyl phenyl ether (Scheme 7). The conversion of $RhHL_4$ on treatment with 2-butenyl phenyl carbonate to $Rh(OPh)L_3$ with liberation of 1-butene as shown in Eq. 18 was in fact confirmed [56]. The results of exclusive formation of the branched butenyl phenyl ether from the linear butenyl phenyl carbonate and formation of the linear ether from the branched 1-methyl-2-propenyl phenyl carbonate are in contrast to the results shown in Scheme 4, where both branched and linear ethers are produced by a mechanism involving nucleophilic attack of RO^- on the substituted and unsubstituted terminal of 1-methylallyl ligand bound to palladium.

+ $RhHL_4$ → [H, Rh, OCO_2Ph, L_n] → (−1-butene)
$L = PPh_3$

$[L_nRh(O_2C\text{-}OPh)]$ $\xrightarrow{-CO_2}$ $Rh(OPh)L_3$ (18)

Cleavage of C–O bonds has also been observed in the reactions of $Pd(C_6F_5)Br(NCMe)_2$ with diallyl ether via insertion-elimination processes (Eq. 19) [57].

+ $Pd(C_6F_5)Br(NCMe)_2$ $\xrightarrow{243\ K}$ [Pd, Br, $CH_2C_6F_5$]$_2$ $\xrightarrow{293\ K}$

Br, Me, $CH_2C_6F_5$ + [Pd, Br]$_2$ + $C_6F_5CH_2CH_2CHO$

+ $C_6F_5CH_2CH{=}CH_2$ + CH_3CH_2CHO

(19)

In comparison to electron-rich late transition metal complexes that are capable of readily undergoing oxidative addition, the early transition metal complexes do not undergo ready oxidative addition. Cp_2ZrCl_2 can be treated with 2 equiv. *n*-BuLi to generate "Cp_2Zr" species [58]. This species can form a zirconacyclopropane complex on interaction with 1-phenyl-2-propenyl ether. Rearrangement of the zirconacyclopropane complex gives bis(cyclopentadienyl)-3-phenyl-2-propenylzirconium alkoxide, which on treatment with benzaldehyde followed by protonolysis provides homoallylic alcohols in high regio- and diastereoselectivities (Scheme 8).

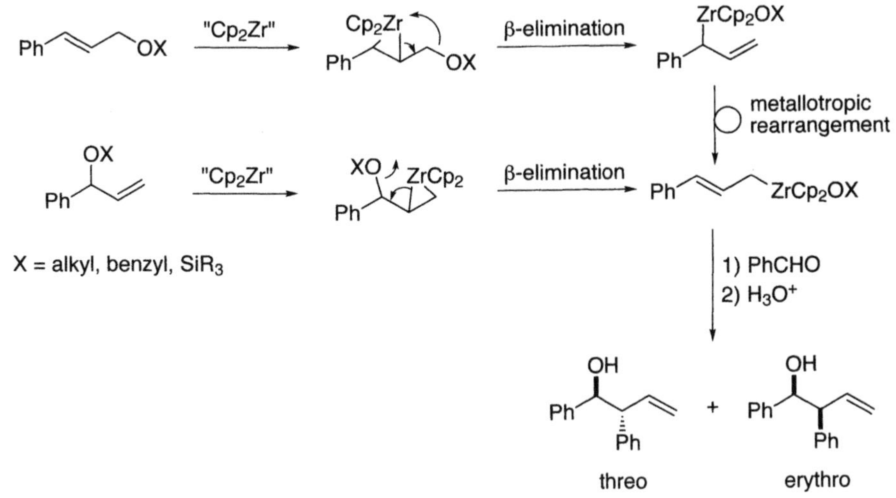

Ph⁄⁄OX —"Cp₂Zr"→ Cp₂Zr / Ph ... OX —β-elimination→ ZrCp₂OX / Ph

metallotropic rearrangement

OX / Ph —"Cp₂Zr"→ XO ZrCp₂ / Ph —β-elimination→ Ph ... ZrCp₂OX

X = alkyl, benzyl, SiR_3

1) PhCHO
2) H_3O^+

OH / Ph ... Ph threo + OH / Ph ... Ph erythro

Scheme 8.

4
Cleavage of the C–O Single Bond in Esters, Lactones, and Anhydrides

Esters, lactones, and carboxylic anhydrides having carbonyl groups interact with low valent transition metal complexes to give oxidative addition products with the C–O single bond cleavage. Phenyl acetate oxidatively adds to $Ni(cod)_2$ in the presence of 2, 2'-bipyridine (bpy) to give a methylnickel phenoxide complex involving decarbonylation of an intermediate acetylnickel species (Eq. 20) [59].

$$Ni(cod)_2 + bpy + CH_3COOPh \longrightarrow \left[(bpy)Ni\begin{matrix}COCH_3\\OPh\end{matrix} \right] \xrightarrow{-CO} (bpy)Ni\begin{matrix}CH_3\\OPh\end{matrix} \qquad (20)$$

Treatment of the methylnickel phenoxide complex with CO liberated phenyl acetate, indicating the reversibility of the C–O bond cleavage in phenyl acetate [60].

An acetylrhodium aryloxide complex was obtained as an oxidative addition product of an aryl acetate with a Rh(I) complex involving the acyl-oxygen bond cleavage (Eq. 21) [61].

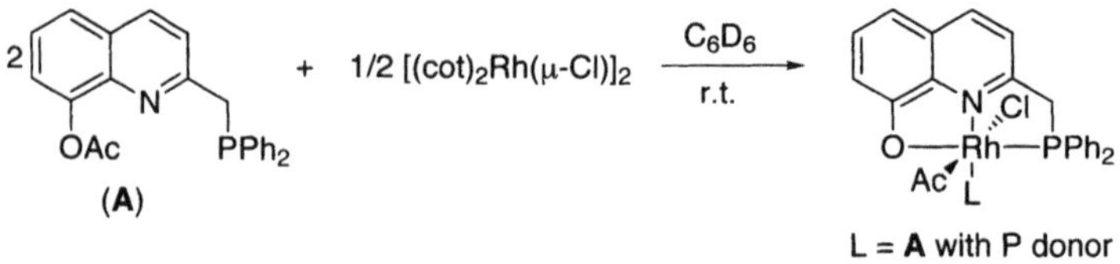

(21)

Acyl-oxygen bond in 2-hydroxyethyl methacrylate is activated by the reaction with $Ti(O\text{-}i\text{-}Pr)_4$ in toluene solution at room temperature to give a pentanuclear aggregate, $Ti_5(O\text{-}i\text{-}Pr)_9(\mu\text{-}O\text{-}i\text{-}Pr)(OC_2H_4O)_5$ [62].

Usually the C–O single bond in esters is cleaved at the acyl-O bond, whereas examples of cleavage at the other point in esters have been reported. An electron-rich iron(0) complex produced on reductive elimination of naphthalene from a hydrido(naphthyl)iron complex undergoes oxidative addition reaction with methyl benzoate to give a methyliron benzoate complex (Eq. 22) [63].

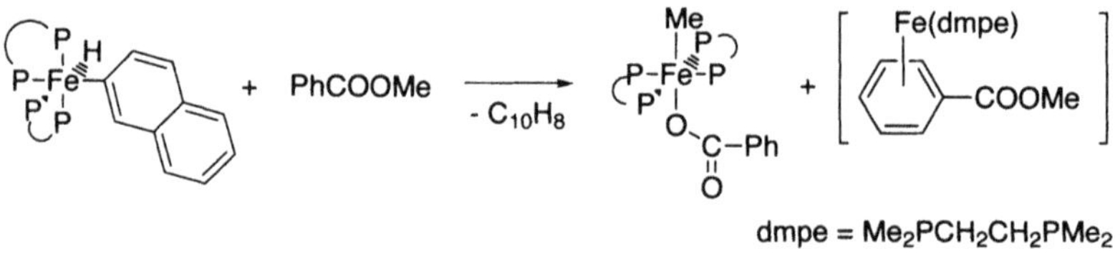

(22)

The C–O bond in vinyl esters can also be cleaved promoted by Ru(0) complex to give vinylruthenium complex (Eq. 23) [64]. Another example of the vinyl-O bond cleavage has been recently reported in the treatment of vinyl crotonate with a (perfluorophenyl)palladium complex. The reaction course has been accounted for by insertion-β-elimination processes (Eq. 24) [57].

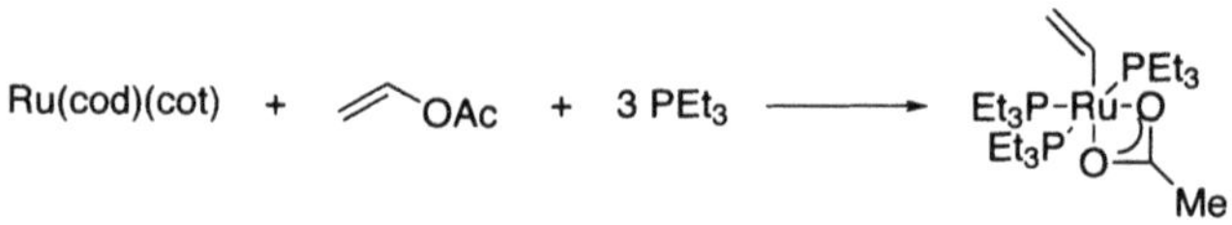

(23)

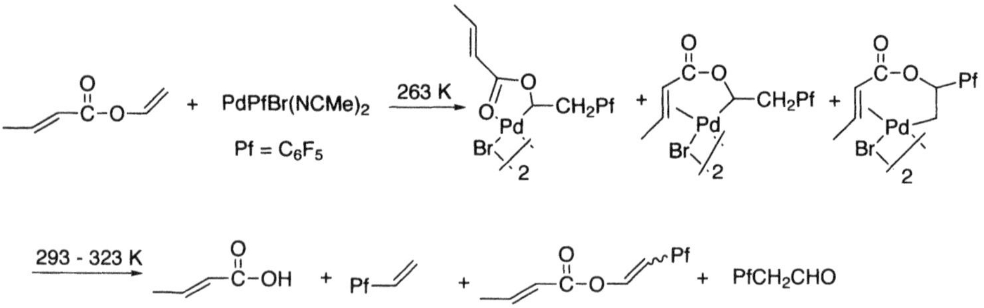

(24)

Catalytic conversion of esters through the C–O bond cleavage has been developed. Pd-catalyzed carbonucleophilic substitution of naphthylmethyl and 1-naphthylethyl esters has been achieved as shown in Eq. 25 [65]. The ease of the naphthylalkyl-O bond cleavage may be partly due to the stability of the naphthylmethyl entity that can form an η^3-allylic-palladium bond.

R
CH-OCOR' + $NaCH(CO_2CH_3)_2$ —[$Pd(dba)_2$ / dppe]→ R CH-$CH(CO_2CH_3)_2$ (25)

R = H, Me; R' = Me, CF_3, dba = dibenzylideneacetone

Palladium-catalyzed transfer hydrogenolysis of benzyl acetate to give toluene has been achieved by using ammonium formate. Hydrogen-donating abilities of various formate salts were found to depend on the counter-ion: $K^+ > NH_4^+ > Na^+ > NHEt_3^+ > Li^+ > H^+$ [66].

Combination of the processes of the C–O bond cleavage in 1-naphthylethyl esters with CO insertion catalyzed by palladium complexes in the presence of a formate salt affords a new route to 2-arylpropanoic acids [67].

Lactones are also susceptible to the C–O bond cleavage on interaction with low-valent transition metal complexes. The C–O bond in β- and γ-lactones can be cleaved to give metallalactones as exemplified in Eq. 26 [68] and Eq. 27 [69].

β-lactone + (cyclooctene)IrL_3Cl —[- cyclooctene / L = PMe_3]→ Cl, L, L, L-Ir metallalactone (26)

β-lactone + (N N)$PtMe_2$ ⟶ (N N)$PtMe_2$ metallalactone (27)

N N = 2,2'-bipyridine, 1,10-phenanthroline

Catalytic cleavage of the single bond in β-lactone promoted by Ni(0) complexes to liberate ethylene and CO_2 has been reported [70].

Metallacyclic complexes were formed when cyclic carboxylic anhydrides were used as substrates to react with zero-valent group 9 and 10 metal complexes [71–74].

Single C–O bond in acid anhydride can be also readily cleaved by oxidative addition of the anhydride to low-valent transition metal complexes to give acylcarboxylato-type complexes as reviewed previously (Eq. 28) [1].

$$L_nM + (RCO)_2O \longrightarrow RCO\text{-}ML_2\text{-}O_2CR \qquad (28)$$

M = Ir, L = PPh_3
M = Ni, L = $PtEt_3$

The first example of the oxidative addition of acyclic acid anhydrides to a Pd(0) complex has been achieved recently (Eq. 29) [75]. On the C–O bond cleavage of the anhydrides acyl(carboxylato)Pd(II) complexes can be isolated.

$$trans\text{-}PdEt_2(PMe_3)_2 \xrightarrow[\text{acetone, 50 °C; } -C_2H_4, -C_2H_6]{\text{styrene}} (Me_3P)_2Pd(\text{styrene}) \xrightarrow[\text{r.t., 2 h}]{(RCO_2)O} Pd(COR)(OCOR)(PMe_3)_2 \qquad (29)$$

R = Me, Et, i-Pr, t-Bu, Ph

Based on the fundamental studies, conversion of anhydrides to corresponding aldehydes and carboxylic acids has been found to be catalyzed by a Pd(0) complex in the presence of H_2 (Eq. 30) [75]. Prior to the report of palladium-catalyzed hydrogenation of acyclic anhydrides, cobalt carbonyl was found to convert anhydrides into aldehydes and carboxylic acids under more severe conditions [76].

$$(RCO)_2O \xrightarrow[\text{THF, 80 °C}]{H_2,\ Pd(PPh_3)_4} RCHO + RCOOH \qquad (30)$$

R = C_7H_{15}, Ph

Further studies on the reactions of anhydrides with carboxylic acids in the presence of a palladium catalyst led to the discovery of the direct hydrogenation of carboxylic acids into aldehydes. The catalytic process provides new means of synthesizing various aldehydes from carboxylic acids (Eq. 31) [77].

$$RCOOH + H_2 \xrightarrow[\text{Pd catalyst}]{(R'CO)_2O} RCHO \qquad (31)$$

Another type of application of the concept of the C–O bond cleavage of anhydrides to arylation of olefins has been reported very recently (Eq. 32) [78]. In this reaction, a C–O bond rupture was accompanied by liberation of CO.

$$(ArCO)_2O + CH_2{=}CHCOOBu \xrightarrow[\text{NaBr, 160 °C; } -CO]{PdCl_2} ArCH{=}CHCOOBu + ArCOOH \qquad (32)$$

The C–O bond activation in benzoic anhydride has also been applied to benzoylation of styrene and its derivatives catalyzed by a rhodium complex (Eq. 33) [79].

$$(ArCO)_2O + Ar'CH{=}CH_2 \xrightarrow[H_2,\ (PhO)_3P]{[RhCl(cod)_2]} ArCOCH(CH_3)Ar' + ArCOCH_2CH_2Ar' \qquad (33)$$

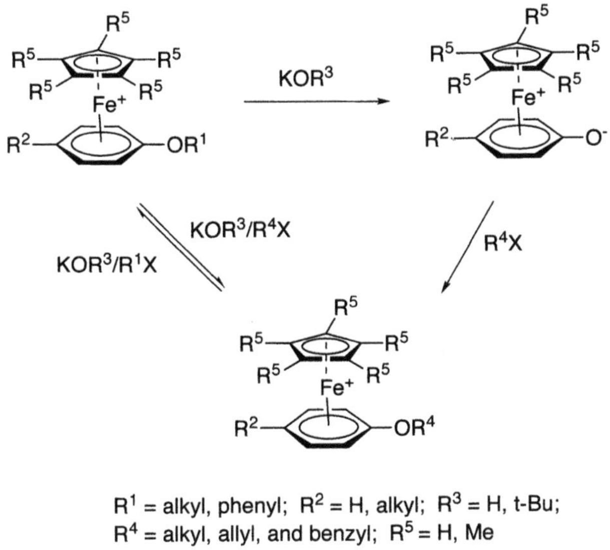

R^1 = alkyl, phenyl; R^2 = H, alkyl; R^3 = H, t-Bu;
R^4 = alkyl, allyl, and benzyl; R^5 = H, Me

Scheme 9.

5
Cleavage of the C–O Single Bond in Ethers, Alcohols, and Acetals

In comparison to the activation of C–O bond in esters, lactones, and anhydrides, reported examples of C–O bond cleavage in ethers, alcohols, and acetals are relatively rare, presumably due to the absence of activation effect by an electron-withdrawing carbonyl group. Examples of the cleavage of C–O bond in ethers will be first discussed below.

5.1
Cleavage of the C–O Single Bond in Acyclic Ethers

5.1.1
Cleavage of the sp^3–C–O Single Bond in Acyclic Ethers

Cleavage of sp^3–C–O bond in alkyl aryl ethers coordinated to a cationic cyclopentadienyliron has been reported recently [80]. The alkyl-O cleavage with KOR^3 (R^3=H or *t*-Bu) leads to an aryloxide-coordinated complex which can be converted into another ether-coordinated complex on treatment with alkyl halides, R^4X (R^4=alkyl, allyl, or benzyl), as shown in Scheme 9.

Since aryl-O bond energy is in the order of 90 kcal/mol and greater than the alkyl-O bond energy (~80 kcal/mol), the first site expected to be cleaved in aryl alkyl ethers is the alkyl-oxygen bond. A recent paper by Milstein et al. reports that the point of cleavage can be directed by using different metal complexes (Eq. 34) [81].

$$\text{C}_6\text{H}_3(\text{CH}_2\text{PtBu}_2)_2\text{-OCH}_3 \xrightarrow[\text{- [CH}_2\text{O]}_x,\ \text{- 2 C}_8\text{H}_{14}]{1/2\ [\text{RhCl(C}_8\text{H}_{14})_2]_2} \text{C}_6\text{H}_3(\text{CH}_2\text{PtBu}_2)_2\text{Rh(H)Cl}$$

$$\text{C}_6\text{H}_3(\text{CH}_2\text{PtBu}_2)_2\text{-OCH}_3 \xrightarrow[\text{- CF}_3\text{CO}_2\text{CH}_3]{\text{Pd(CF}_3\text{CO}_2)_2} \text{C}_6\text{H}_3(\text{CH}_2\text{PtBu}_2)_2\text{O-Pd-(CF}_3\text{CO}_2) \quad (34)$$

On interaction of anisol substituted with phosphine donors a nucleophilic rhodium(I) complex cleaves the aryl-O bond, whereas electrophilic Pd(II) complex activates methyl-O bond [81].

Activation of the C–O bond in ethers has been achieved by the promotion with lanthanoid complexes. The C–O bonds in Et_2O and dimethoxyethane (DME) were cleaved by ytterbium (Eq. 35) [82] or cerium and neodymium (Eqs. 36, 37) [83] complexes.

$$\text{YbI}_2 + \text{KCR}_3 \xrightarrow[2\ \text{h}]{\text{Et}_2\text{O}} [\text{Yb(CR}_3)(\mu\text{-OEt})(\text{OEt}_2)]_2 \quad (35)$$

$R = SiMe_3$

$$\text{CeCp''}_3 \xrightarrow[\text{DME}]{\text{Li}} [\text{CeCp''}_2(\mu\text{-OMe})]_2 \quad (36)$$

DME = dimethoxyethane $Cp''_2 = \eta\text{-}C_5H_3(SiMe_3)_2\text{-}1,3$ and $\eta\text{-}C_5H_3tBu_2\text{-}1,3$

$$\text{NdCp''}_3 \xrightarrow[\text{DME}]{\text{Li}} \text{NdCp''}_2(\mu\text{-OMe})_2\text{Li(DME)} \quad (37)$$

$Cp''_2 = \eta\text{-}C_5H_3(SiMe_3)_2\text{-}1,3$

The cleavage of the C–O single bond in dimethoxyethane (dme) has also been achieved by the reaction with $[La(Cp")_3]$ $[Cp"=\eta^5\text{-}C_3H_5(SiMe_3)_2\text{-}1,3]$ and K to form a complex $[\{La(Cp")_2(\mu\text{-OMe})\}_2]$ and an unidentified polynuclear lanthanum methoxide [84]. The measurement of EPR spectra confirmed the involvement of the persistent paramagnetic lanthanum(II) intermediates, $[K(dme)_x]$ $[La(Cp")_3]$ and $[La(Cp")_2(dme)_y]$, which being oxophilic and powerful reducing agents are able to cleave the C–O bond of dme at ambient temperature [83].

Organolanthanoid hydrides also promote the C–O bond cleavage in ethers (Eq. 38) [85].

$$(\text{Cp}^*_2\text{LnH})_2 + 2\ \text{ROR'} \longrightarrow \text{Cp}^*_2\text{LnOR} + \text{Cp}^*_2\text{LnOR'} + \text{R'H} + \text{RH} \quad (38)$$

Ln = Y, La, Ce; R, R' = Me, Et, nBu, tBu

5.1.2
Cleavage of the sp²–C–O Single Bond in Acyclic Ethers

On coordination to electrophilic $Cr(CO)_3$ entity, aryl ethers become susceptible to nucleophilic attack undergoing aryl-O bond cleavage (Eqs. 39, 40) [86, 87].

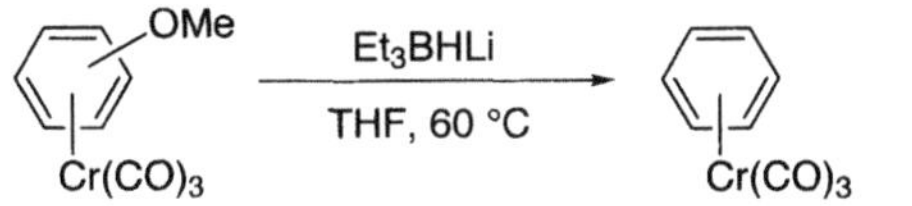

$$(\text{MeOC}_6\text{H}_5)Cr(CO)_3 \xrightarrow[\text{THF, 60 °C}]{Et_3BHLi} (C_6H_6)Cr(CO)_3 \quad (39)$$

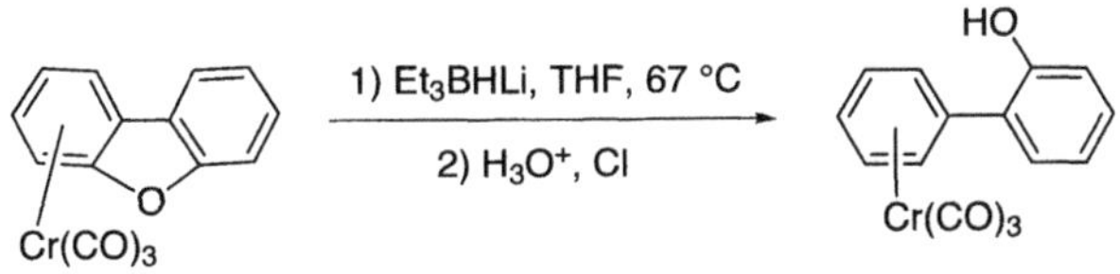

$$(\text{dibenzofuran})Cr(CO)_3 \xrightarrow[\text{2) } H_3O^+, Cl]{\text{1) } Et_3BHLi, \text{ THF, 67 °C}} (\text{2-HOC}_6\text{H}_4\text{-C}_6\text{H}_5)Cr(CO)_3 \quad (40)$$

Another example of cleavage of the ether C–O bond is dealkylation from phenyl alkyl ethers bonded to $(\eta^5\text{-}C_5H_5)Ru^+$ entity by a base such as KOH or RO^- [88].

Activation of the C–O bonds in vinyl ethers promoted by a lanthanoid hydride has been reported (Eq. 41) [85].

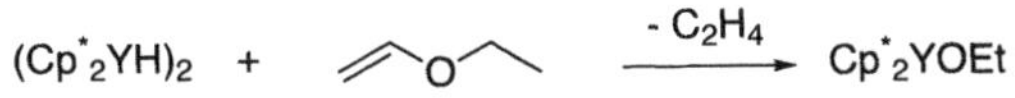

$$(Cp^*_2YH)_2 + CH_2{=}CH\text{-}O\text{-}CH_2CH_3 \xrightarrow{-\,C_2H_4} Cp^*_2YOEt \quad (41)$$

Regioselective cleavage of the sp²- or sp³-C–O bond in alkyl vinyl ethers depending on the alkyl groups has been achieved by using a samarium complex (Scheme 10) [89]. Methyl vinyl ether gives the vinylsamarium and methoxysamarium complexes by the activation of sp²-C–O bond, whereas in the case of the benzyl vinyl ether, benzyl-oxygen bond is cleaved exclusively. The cleavage processes were confirmed by the reaction of the samarium complexes formed from the C–O bond cleavage with D_2O to give corresponding deuterated organic compounds.

Heterogeneous catalytic transfer hydrogenolysis of C–O bonds in aliphatic and aromatic ethers has been reviewed [90]. As an example, catalytic transfer

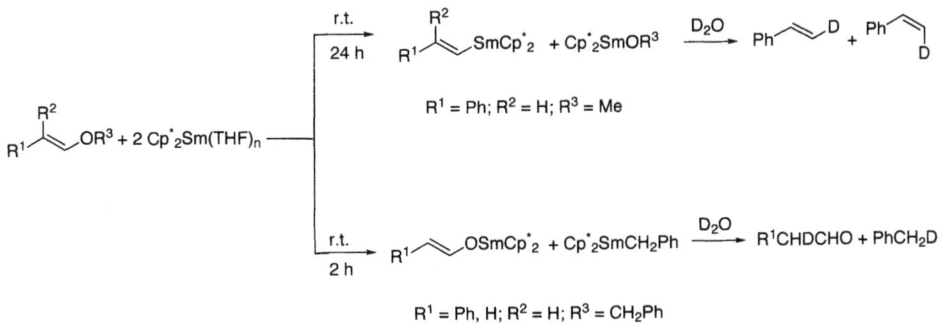

$R^1C(R^2){=}CH\text{-}OR^3 + 2\,Cp^*_2Sm(THF)_n$

(r.t., 24 h) → $R^1C(R^2){=}CH\text{-}SmCp^*_2 + Cp^*_2SmOR^3 \xrightarrow{D_2O}$ Ph–CH=CH–D + Ph–C(=CH₂)–D

$R^1 = Ph;\ R^2 = H;\ R^3 = Me$

(r.t., 2 h) → $R^1CH{=}CH\text{-}OSmCp^*_2 + Cp^*_2SmCH_2Ph \xrightarrow{D_2O} R^1CHDCHO + PhCH_2D$

$R^1 = Ph, H;\ R^2 = H;\ R^3 = CH_2Ph$

Scheme 10.

hydrogenation of C–O bonds in readily prepared heteroaromatic ethers of phenols to give arenes by using Pd/C catalyst with a hydrogen donor is shown in Eq. 42 [91].

Cl, N, S, O_2 —ArOH→ OAr, N, S, O_2 —Pd/C, Sodium Phosphinate→ O, NH, S, O_2 + ArH (42)

Activation of sp^2–C–O bond in vinyl triflates has also been reported in the electrocarbonylation with CO_2 by using $PdCl_2(PPh_3)_2$ as catalyst to afford α,β-unsaturated carboxylic acids [92]. The reaction is proposed to involve the formation of a Pd(0) complex followed by an activation by electron transfer and the formation of vinylpalladium(II) intermediate in an oxidative addition.

5.2
Cleavage of the C–O Single Bond in Cyclic Ethers

Strained cyclic ethers are more susceptible to the C–O bond cleavage. Activation of epoxides by transition metal complexes have been extensively studied [1]. Cleavage of the C–O bond in the epoxide ring was found to depend on the nature of the substituted group on the ring, causing the chemo- and regioselective isomerization of epoxides. The early study revealed selective cleavage of C–O bond at the carbon substituted with two CN groups in tricycano ethylene oxide on treatment with PtL_4 (L=PPh_3, $P(p\text{-tol})_3$, $AsPh_3$) [93].

In Pd(0)-catalyzed isomerization reaction of epoxides, alkyl-substituted epoxides afford methyl ketones (path a, Scheme 11), whereas aryl-substituted epoxides give aldehydes or ketones via cleavage of the benzylic C–O bond (path b) [94].

The C–O bonds in epoxides activated by adjacent substituents such as aryl, vinyl, silyl, or carbonyl groups can be cleaved by the promotion of Rh, Pd, Mo, Sm, and Fe complexes to give carbonyl complexes or allylic alcohols [95].

Catalytic asymmetric hydrogenolysis of epoxides catalyzed by rhodium complexes has been reported [96, 97].

By combination of the C–O bond cleavage in alkenyloxiranes and CO insertion processes, Pd-catalyzed carbonylation of alkenyloxiranes has been achieved to give unsaturated ester, β-lactone, diene, and allylic alcohol, depending on the substrate as well as the nature of the substituents (Eq. 43) [98].

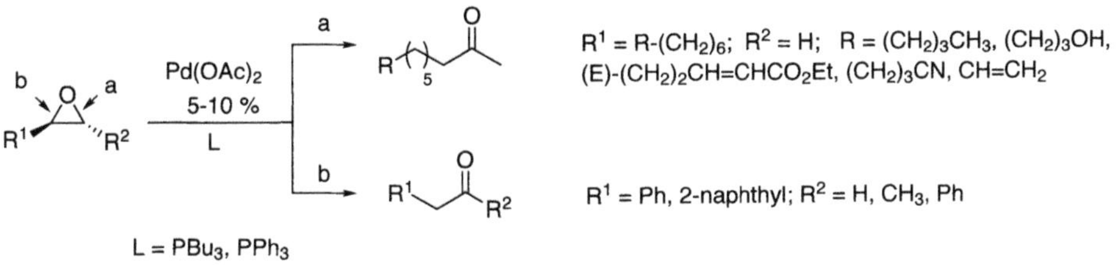

b, a; R^1, R^2 epoxide; $Pd(OAc)_2$ 5-10 %, L; a → R^1-(CH$_2$)$_5$-CH$_2$C(O)CH$_3$; b → $R^1CH_2C(O)R^2$

a: $R^1 = R\text{-}(CH_2)_6$; $R^2 = H$; $R = (CH_2)_3CH_3$, $(CH_2)_3OH$, (E)-$(CH_2)_2CH{=}CHCO_2Et$, $(CH_2)_3CN$, $CH{=}CH_2$

b: R^1 = Ph, 2-naphthyl; R^2 = H, CH_3, Ph

L = PBu_3, PPh_3

Scheme 11.

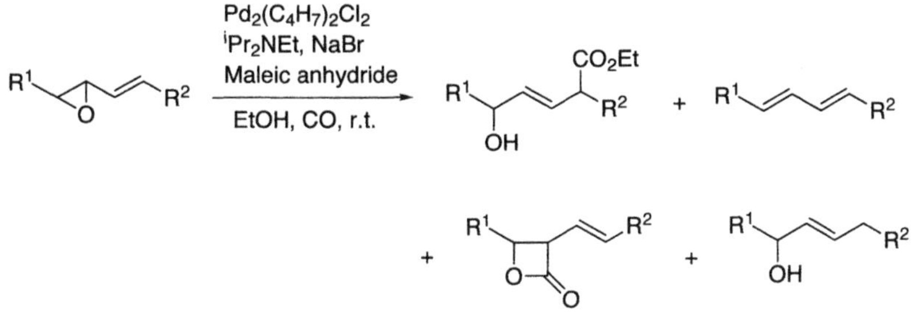

$Pd_2(C_4H_7)_2Cl_2$, $^iPr_2NEt$, NaBr, Maleic anhydride; EtOH, CO, r.t. (43)

A different type of the C–O bond cleavage involving the cleavage of both C–O bonds in the epoxides to give metal-oxo complexes has been observed in the reaction with tungsten complexes [99, 100].

The C–O bond in four- and five-membered cyclic ethers can also be cleaved on interaction with low-valent transition metal complexes [101–103]. A metalla-radical cleavage of THF has been proposed [103].

Combination of the C–O bond activation in cyclic ethers with CO insertion has been applied to catalytic reactions by transition metal complexes to corresponding lactones [104].

Epoxide isomerization through C–O bond cleavage promoted by Rh(I) and Ir(I) complexes and its application in homogeneous catalysis have been also reported [105].

Oxidative addition of an sp^2-C–O bond in 1,2-dihydrofuran to $(silox)_3Ta$ (silox=t-Bu_3SiO) takes place without involving sp^3-C–O bond cleavage (Eq. 44) [106].

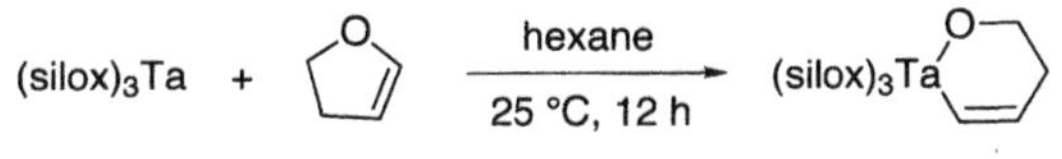

$(silox)_3Ta$ + 2,3-dihydrofuran —(hexane, 25 °C, 12 h)→ $(silox)_3Ta$ oxametallacycle (44)

The C–O bond in 3,3-dimethyloxetane was cleaved by the reaction with $(silox)_3$ Ta (Eq. 45) [106].

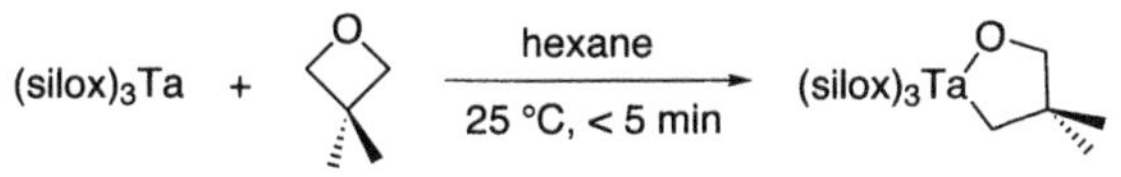

$(silox)_3Ta$ + 3,3-dimethyloxetane —(hexane, 25 °C, < 5 min)→ $(silox)_3Ta$ oxametallacycle (45)

5.3
Cleavage of the C–O Single Bond in Alcohols and Acetals

Very few examples on the cleavage of C–O bonds in alcohols except for allylic alcohols have been reported. The highly oxophilic nature of tungsten chlorides causes abstraction of oxygen from methanol or ethanol yielding oxotungsten and bis(alkoxide)tungsten complexes with liberation of alkanes (Eq. 46) [107, 108].

$$WCl_2L_4 + ROH \xrightarrow[-RH]{} W(O)Cl_2L_3 + W(OR)_2Cl_2L_2 + L_2 \qquad (46)$$

$L = PMePh_2$; R = Me, Et

Evidence for the C–O bond homolysis in the reaction of benzyl alcohol with WCl_2L_4 ($L=PMe_3$, $PMePh_2$) has been found very recently [109].

Indirect activation of the C–O bond of the aryl alcohol is further developed by conversion to corresponding triflate. The C–O bond in the aryl triflate has been achieved in the homocoupling reaction catalyzed by a palladium(0) or nickel(0) complex in the presence of an electron source (either a cathode or zinc power) (Eq. 47) [110].

$$ArOTf + Zn \xrightarrow{Pd\ or\ Ni} Ar{-}Ar + Zn(OTf)_2 \qquad (47)$$

$Ar = XC_6H_4$ (X = p-MeC, p-CN, p-CF_3, p-Cl, o-Cl, p-F, H, o-Me, p-Me, p-t-Bu), 1- and 2-naphthyl

In this reaction, most efficient catalysts in the naphthyl series have been confirmed to be $Pd(OAc)_2$+1 BINAP and $NiCl_2$(dppf). On the other hand, the single C–O bond in α-hydroxy ketones and their acetate and mesylate derivatives has been cleaved to form the corresponding ketones by the promotion of a vanadium(II) complex prepared in situ from the reaction of $VCl_3(THF)_3$ and zinc (Eq. 48) [111].

$$R^1C(O)CH(OR)R^2 \xrightarrow[CH_2Cl_2]{[V_2CL_3(THF)_6]_2[Zn_2Cl_6]} R^1C(O)CH_2R^2 \qquad (48)$$

R = H, Ac, Ms
R^1, R^2 = Ph, $CH_3(CH_2)_6$-, etc.

Activation of the C–O bond in cyclohexanol mediated by "Cp^*Ru^+" fragment, generated by protonation of $[Cp^*Ru(OMe)]_2$ with CF_3SO_3H, has been reported (Eq. 49) [112, 113].

OH

$$[Cp^*Ru(OMe)]_2 + CF_3SO_3H \longrightarrow "Cp^*Ru^{+}" \xrightarrow[\text{THF}]{+\ \text{cyclohexanol}} {}^*CpRu^{+} \qquad (49)$$

Although alkoxide ligands are often stable and fairly unreactive ancillary ligands, particularly when bonded to early transition metals, the C–O bond in an alkoxide ligand can undergo special types of reactions when a related oxo complex is accessible [114]. Two mechanisms of both homolytic and heterolytic fashions for C–O bond cleavage in alkoxide ligands have been discussed.

Oxidative addition of an acetal C–O bond to a ruthenium center, leading to the concomitant formation of carbene complexes has been reported recently (Eq. 50) [115].

OR, OR, PPh2 + $[CpRuL_3]^+(TfO)^-$ $\xrightarrow{-L}$ OR, OR, L_2, Ru, Cp, P, Ph_2, TfO^- (50)

R = Me, Et,
R + R = CH_2CH_2

$\xrightarrow[\text{- L, - ROH}]{\text{heat}}$ OR, L, Ru, Cp, P, Ph_2, TfO^-

6 Cleavage of the C–O Multiple Bonds in Acetones, Aldehydes, Esters, CO, and CO_2

The Wittig-type reaction is well known as a powerful means of converting carbonyl compounds into olefins [116]. Subsequent progress in the chemistry of transition metal-carbene complexes and metallacyclobutane complexes has introduced other new methodologies in organic synthesis. Tebbe's reagent [117], first prepared by the reaction of Cp_2TiCl_2 with $AlMe_3$, was later applied by Grubbs as an excellent agent for converting carbonyl compounds into olefins [118]. The advantage of the process lies in the easy generation of titanocene methylidene complex, "$Cp_2Ti{=}CH_2$", in situ, which reacts with organic carbonyl compounds to produce olefins. The key feature of the process is utilization of the highly oxophilic nature of the Cp_2Ti entity in the intermediate oxatitanacyclobutane intermediate. Conversion of aldehydes and ketones to olefins and esters and lactones to enol ethers can be accomplished by the process (Scheme 12).

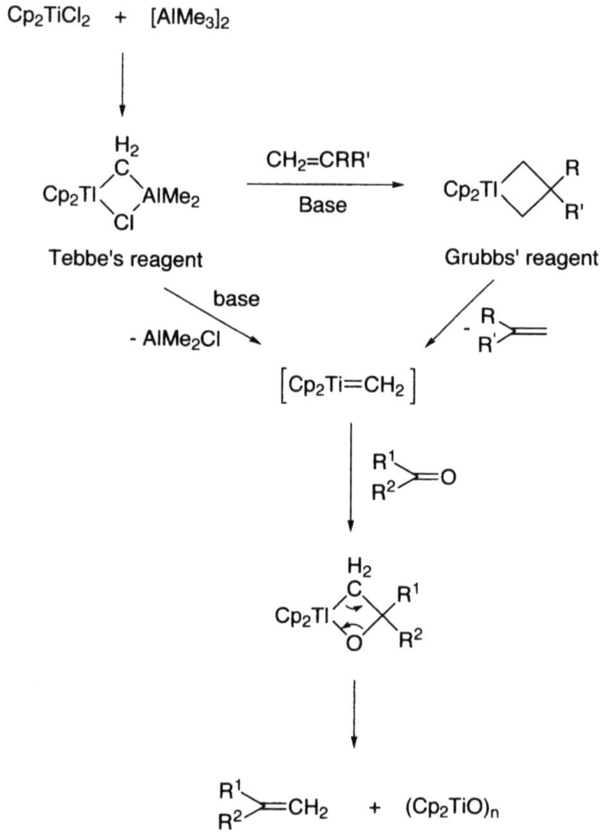

Cp_2TiCl_2 + $[AlMe_3]_2$ → Tebbe's reagent ($Cp_2Ti(\mu\text{-}CH_2)(\mu\text{-}Cl)AlMe_2$) —$CH_2$=CRR', Base→ Grubbs' reagent (titanacyclobutane); Tebbe's reagent —base, - $AlMe_2Cl$→ $[Cp_2Ti{=}CH_2]$; Grubbs' reagent — - $RR'C{=}CH_2$ → $[Cp_2Ti{=}CH_2]$ —$R^1R^2C{=}O$→ oxatitanacyclobutane → $R^1R^2C{=}CH_2$ + $(Cp_2TiO)_n$

Scheme 12.

The C=O bond in ketones can be cleaved by tungsten complexes with formation of a very strong oxo-tungsten multiple bond as has been studied extensively [119].

The C–O multiple bonds in ketones were cleaved in a C–C single-bond-forming reaction promoted by vanadium(II) complex in the presence of the allyl bromide or a catalytic amount of molecular oxygen (Eqs. 51, 52) [120–122].

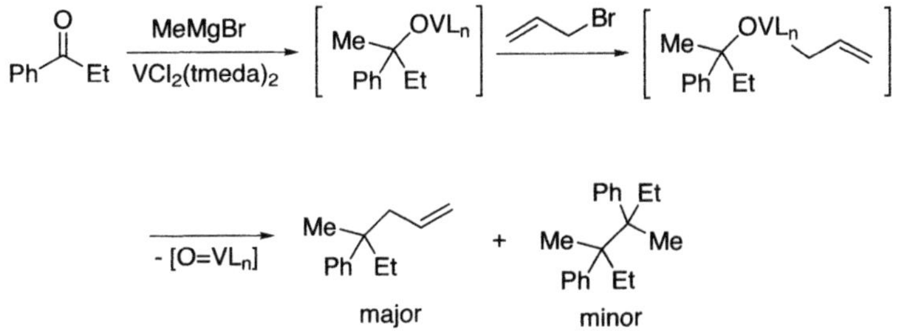

$$PhC(O)Et \xrightarrow[VCl_2(tmeda)_2]{MeMgBr} [Me(Ph)(Et)C{-}OVL_n] \xrightarrow{CH_2{=}CHCH_2Br} [Me(Ph)(Et)C{-}OVL_n(CH_2CH{=}CH_2)] \xrightarrow{-[O{=}VL_n]} \underset{major}{Me(Ph)(Et)C{-}CH_2CH{=}CH_2} + \underset{minor}{Me(Ph)(Et)C{-}C(Ph)(Et)Me} \quad (51)$$

$$PhC(O)Et + MeMgBr \xrightarrow[THF,\ 20\,°C \rightarrow reflux]{(1)\ VCl_2(tmeda)_2;\ (2)\ O_2\ (0.2\ eq.)} Me(Ph)(Et)C{-}C(Ph)(Et)Me \quad (52)$$

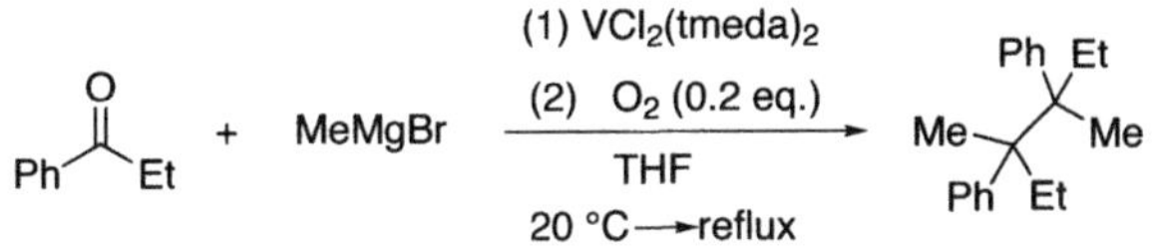

Multiple C–O bonds in RNCO and CO_2 have also been cleaved by reaction with WL_4Cl_2 ($L=PMe_3$, $PMePh_2$) (Eqs. 53, 54) [123].

$$W(PMe_3)_4Cl_2 + TolNCO \longrightarrow W(NTol)(CO)(PMe_3)_2Cl_2 + W(O)(CNTol)(PMe_3)_2Cl_2 + 2\ PMe_3 \quad (53)$$

Tol = p-tolyl

$$WL_4Cl_2 + CO_2 \xrightarrow{-2\,L} W(O)(CO)L_2Cl_2 \underset{-L}{\overset{+L}{\rightleftharpoons}} W(O)L_3Cl_2 + CO \quad (54)$$

$L = PMe_3, PMePh_2$

Another type of an interesting mode of C=O bond cleavage with cyclohexanones can be promoted by "Cp^*Ru^+" fragment generated in situ on treatment of a ruthenium methoxide with trifluoromethane sulfonic acid (Eq. 55) [112, 113].

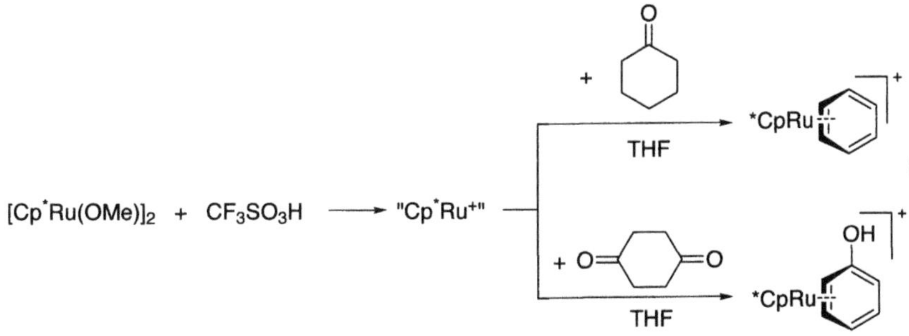

$$[Cp^*Ru(OMe)]_2 + CF_3SO_3H \longrightarrow \text{"}Cp^*Ru^{+}\text{"} \begin{cases} \xrightarrow[THF]{+\ cyclohexanone} [{}^*CpRu(\eta^6\text{-}C_6H_6)]^+ \\ \xrightarrow[THF]{+\ cyclohexane\text{-}1,4\text{-}dione} [{}^*CpRu(\eta^6\text{-}C_6H_5OH)]^+ \end{cases} \quad (55)$$

The activation of C–O bond in carbon monoxide has been considered as a key step in Fischer-Tropsch synthesis in the reactions of synthesis gas [124].

The C–O bond-breaking in carbon monoxide assisted by $(silox)_3Ta$ (silox=*t*-Bu_3SiO) has been reported (Eq. 56) [125, 126].

$$(silox)_3Ta + 1/2\ CO \xrightarrow[25\ °C]{benzene} 1/2\ (solox)_3Ta{=}O + 1/4\ [(silox)_3Ta]_2(\mu\text{-}C_2) \quad (56)$$

The C–O bond can be more easily activated when the CO molecule interacts with more than two metal atoms. Recently, the dissociative adsorption of carbon monoxide by polynuclear metal complexes, such as $[(silox)_2TaH_2]_2$ (Eq. 57) [126, 127] and $[(silox)_2WCl]_2$ (Eq. 58) [126–129], and tetratungsten alkoxides [129] has been achieved. Hydrogenation of CO to give hydrocarbons promoted by metal clusters has been reviewed [130].

$$[(tBu_3SiO)_2TaH_2]_2 \xrightarrow[-78\ °C,\ Et_2O,\ 6\ h]{CO} (tBu_3SiO)_2(H)Ta(\mu\text{-}CH_2)(\mu\text{-}O)Ta(H)(OSitBu_3)_2 \quad (57)$$

$$(tBu_3SiO)_2ClW{\equiv}WCl(OSitBu_3)_2 \xrightarrow[hexane\ or\ benzene]{CO} (tBu_3SiO)_2Cl(CO)W{=}WCl(CO)(OSitBu_3)_2 \xrightarrow[-\ CO]{toluene,\ 120\ °C} (tBu_3SiO)_2(O){=}W{=}C{=}WCl_2(OSitBu_3)_2 \quad (58)$$

Complete C–O bond cleavage in carbon monoxide in the zirconium- and titanium-assisted homologation of a pyrrole to a pyridine ring within the porphyrinogen skeleton has been achieved [131–134].

Thermal decomposition of CO_2 coordinated to Pd(0) complex caused the cleavage of one C=O bond of CO_2 to give CO and $O{=}PMePh_2$ (Eq. 59) [135].

$$(Ph_2MeP)_2Pd(\eta^2\text{-}CO_2) \xrightarrow[reflux]{CD_2Cl_2} CO + CO_2 + O{=}PMePh_2 + Pd_n(PMePh_2)_m \quad (59)$$

Electrophilic attack on the coordinated CO_2 in iron complex resulting in the C–O bond cleavage has been reported (Eq. 60) [136, 137].

$$Fe(CO_2)(depe)_2 + 2\ MeX \longrightarrow Me_2O + [FeX(CO)(depe)_2]X \quad (60)$$

depe = 1,2-bis(diethylphosphino)ethane; X = I, OTf

The activation of C–O multiple bonds in CO_2 can be seen in the course of hydrogenolysis of carbon dioxide by promotion of the transition metal complexes [138]. More examples of the C–O bond cleavage in CO_2 activated by metal complexes have been reviewed [139].

7 Concluding Remarks

Following the development of synthetic applications involving allylic C–O bond cleavage catalyzed by palladium complexes, we now see emerging applications utilizing transition metal-promoted cleavage of the C–O bonds. Further findings of elementary processes involving the C–O bond cleavage promoted by various transition metals coupled with developments of novel reactions of organotransition metal complexes will undoubtedly unravel possibilities of unexpected applications in organic syntheses.

References

1. Yamamoto A (1992) Adv Organomet Chem 34:111
2. Martinho Simões JA (1992) Energetics of organometallic species, NATO ASI Series, Kluwer Academic Publishers, Dordrecht, Netherlands
3. Marks TJ (1990) Bonding energetics in organometallic compounds, ACS Symposium. Series 428, Washington DC
4. Mondal JU, Blake DM (1982) Coord Chem Rev 47:205
5. Yamamoto A (1986) Organotransition metal chemistry: fundamental concepts and applications, Wiley-Interscience, New York, p 42
6. (a) Trost BM, Verhoeven TR (1982) Organopalladium compounds in organic synthesis and in catalysis. In: Wilkinson G, Stone FGA, Abel EA (eds) Comprehensive organometallic chemistry, vol 8. Pergamon, Oxford, p 799. (b) Trost BM (1980) Acc Chem Res 13:385
7. Tsuji J (1995) Palladium reagents and catalysts: innovation in organic synthesis. Wiley, New York
8. Hayashi T, Hagihara T, Konishi M, Kumada M (1983) J Am Chem Soc 105:7767
9. Yamamoto T, Saito O, Yamamoto A (1981) J Am Chem Soc 103:5600
10. Yamamoto T, Akimoto M, Saito O, Yamamoto A (1986) Organometallics 5:1559
11. Yamamoto T, Akimoto M, Yamamoto A (1983) Chem Lett 1725
12. Oshima M, Shimizu I, Yamamoto A, Ozawa F (1991) Organometallics 10:1221
13. Ozawa F, Son T, Ebina S, Osakada K, Yamamoto A (1992) Organometallics 11:171
14. Yamamoto T, Ishizu J, Yamamoto A (1981) J Am Chem Soc 103:6863
15. Komiya S, Kabasawa T, Yamashita K, Hirano M, Fukuoka A (1994) J Organomet Chem 471:C6
16. Faller JW, Linebarrier D (1988) Organometallics 7:1670
17. Jang S, Atagi LM, Mayer JM (1990) J Am Chem Soc 112:6413
18. Takaki K, Kusudo T, Uebori S, Makioka Y, Taniguchi Y, Fujiwara Y (1995) Tetrahedron Lett 36:1505
19. Tsuji J (1986) Tetrahedron 42:4361
20. Frost CG, Howarth J, Williams JMJ (1992) Tetrahedron. Asymmetry 3:1089
21. Tsuji J, Minami I (1987) Acc Chem Res 20:140
22. Kondo T, Mukai T, Watanabe Y (1991) J Org Chem 56:487
23. Mitsudo T-A, Zhang S-W, Kondo T, Watanabe Y (1992) Tetrahedron Lett 33:341
24. Sakamoto M, Shimizu I, Yamamoto A (1996) Bull Chem Soc Jpn 69:1065
25. Tada Y, Satake A, Shimizu I, Yamamoto A (1996) Chem Lett 1021
26. Shimizu I, Sakamoto T, Kawaragi S, Maruyama Y, Yamamoto A (1997) Chem Lett 137
27. Shimizu I, Khien KM, Nagatomo M, Nakajima T, Yamamoto A (1997) Chem Lett 851

28. Shimizu I, Matsumoto Y, Shoji K, Ono T, Satake A, Yamamoto A (1996) Tetrahedron Lett 37:7115
29. Trost BM, Van Vranken DL (1996) Chem Rev 96:395
30. Noyori R (1994) In: Asymmetric catalysis in organic synthesis. John Wiley &Sons, New York, Chap 2
31. Hayashi T (1993) In: Ojima I (ed) Catalytic asymmetric synthesis. VCH Publishers, New York, p 325
32. Hayashi T, Kawatsura M, Uozumi Y (1997) Chem Commun 561
33. Hayashi T, Kawatsura M, Uozumi Y (1998) J Am Chem Soc 120:1681
34. Sprinz J, Helmchen G (1993) Tetrahedron Lett 34:1769
35. Helmchen G, Kudis S, Sennhenn P, Steinhagen H (1997) Pure Appl Chem 69:513
36. Prétôt R, Pfaltz A (1998) Angew Chem Int Ed 37:323
37. Trost BM, Hachiya I (1998) J Am Chem Soc 120:1104
38. Kawatsura M, Uozumi Y, Hayashi T (1998) Chem Commun 217
39. Trost BM, Toste FD (1998) J Am Chem Soc 120:815
40. Trost BM, Bunt RC (1996) J Am Chem Soc 118:235
41. Cristiano MLS, Johnstone RAW, Price PJ (1996) J Chem Soc Perkin Trans 1453
42. Mandai T, Matsumoto T, Kawada M, Tsuji J (1992) J Org Chem 57:1326
43. Hayashi T, Iwamura H, Naito M, Matsumoto Y, Uozumi Y (1994) J Am Chem Soc 116:775
44. Maruyama Y, Sezaki T, Tekawa M, Sakamoto T, Shimizu I, Yamamoto A (1994) J Organomet Chem 473:257
45. Shimizu I, Tekawa M, Maruyama Y, Yamamoto A (1992) Chem Lett 1365
46. Tsuji J, Sato K, Okumoto H (1984) J Org Chem 49:1341
47. Murahashi SI, Imada Y, Taniguchi Y, Higashiura SY (1988) Tetrahedron Lett 29:4945
48. Matsuzaka H, Hiroe Y, Iwasaki M, Ishii Y, Koyasu Y, Hidai M (1988) J Org Chem 53:3832
49. Neibecker D, Poirier J, Tkatchenko I (1989) J Org Chem 54:2459
50. Bonnet MC, Coombes J, Manzano B, Neibecker D, Tkatchenko I (1989) J Mol Catal 52:263
51. Itoh K, Hamaguchi N, Miura M, Nomura M (1992) J Mol Catal 75:117
52. Naigre R, Alper H (1996) J Mol Catal A: Chemical 111:11
53. Murahashi S-I, Imada Y, Taniguchi Y, Higashiura S (1993) J Org Chem 58:1538
54. Yamamoto A (1995) Bull Chem Soc Jpn 68:433
55. Komiya S, Yamamoto A (1975) J Organomet Chem 87:333
56. Hayashi Y, Komiya S, Yamamoto T, Yamamoto A (1984) Chem Lett 977
57. Albéniz AC, Espinet P, Lin Y-S (1997) Organometallics 16:5964
58. Ito H, Taguchi T, Hanzawa Y (1992) Tetrahedron Lett 33:1295
59. Yamamoto T, Ishizu J, Kohara T, Komiya S, Yamamoto A (1980) J Am Chem Soc 102:3758
60. Komiya S, Akai Y, Tanaka K, Yamamoto T, Yamamoto A (1985) Organometallics 4:1130
61. Grotjahn DB, Joubran C (1995) Organometallics 14:5171
62. Pajot N, Papiernik R, Hubert-Pfalzgraf LG, Vaissermann J, Parraud S (1995) J Chem Soc Chem Commun 1817
63. Tolman CA, Ittel SD, English AD, Jesson JP (1979) J Am Chem Soc 101:1742
64. Komiya S, Suzuki J, Miki K, Kasai N (1987) Chem Lett 1287
65. Legros J-Y, Fiaud J-C (1992) Tetrahedron Lett 33:2509
66. Rajagopal S, Spatola AF (1997) Applied Catalysis A 152: 69, and references cited therein
67. Baird JM, Kern JR, Lee GR, Morgans Jr DJ, Sparacino ML (1991) J Org Chem 56:1928
68. Zota AA, Frolow F, Milstein D (1990) Organometallics 9:1300
69. Aye K-T, Colpitts D, Ferguson G, Puddephatt RJ (1988) Organometallics 7:1454
70. Yamamoto T, Ishizu J, Yamamoto A (1982) Bull Chem Soc Jpn 55:623
71. Yamamoto T, Sano K, Yamamoto A (1987) J Am Chem Soc 109:1092
72. Sano K, Yamamoto T, Yamamoto A (1983) Chem Lett 115

73. Uhlig VE, Fahske G, Nestler B (1980) Z Anorg Allg Chem 465:151
74. Sano K, Yamamoto T, Yamamoto A (1984) Bull Chem Soc Jpn 57:2741
75. Nagayama K, Kawataka F, Sakamoto M, Shimizu I, Yamamoto A (1995) Chem Lett 367
76. Wakamatsu H, Furukawa J, Yamakami N (1971) Bull Chem Soc Jpn 44:288
77. Nagayama K, Shimizu I, Yamamoto A (1998) Chem Lett 1143
78. Stephan MS, Teunissen AJJM, Verzijl GKM, de Vries JG (1998) Angew Chem Int Ed 37:662
79. Kokubo K, Miura M, Nomura M (1995) Organometallics 14:4521
80. Moulines F, Djakovitch L, Delville-Desbois M-H, Robert F, Gouzerh P, Astruc D (1995) J Chem Soc Chem Commun 463
81. van der Boom ME, Liou S-Y, Ben-David Y, Vigalok A, Milstein D (1997) Angew Chem Int Ed Engl 36:625
82. Hitchcock PB, Holmes SA, Lappert MF, Tian S (1994) J Chem Soc Chem Commun 2691
83. Gun'ko YK, Hitchcock PB, Lappert MF (1995) J Organomet Chem 499:213
84. Cassani MC, Lappert MF, Laschi F (1997) Chem Commun 1563
85. Deelman B-J, Booij M, Meetsma A, Teuben JH, Kooijman H, Spek AL (1995) Organometallics 14:2306
86. Rose-Munch F, Djukic JP, Rose E (1990) Tetrahedron Lett 31:2589
87. Djukic JP, Rose-Munch F, Rose E (1993) J Am Chem Soc 115:6434
88. Kimura M, Morita M, Mitani H, Okamoto H, Satake K, Morosawa S (1992) Bull Chem Soc Jpn 65:2557
89. Takaki K, Maruo M, Kamata T, Makioka Y, Fujiwara Y (1996) J Org Chem 61:8332
90. Johnstone RAW, Wilby AH, Entwistle ID (1985) Chem Rev 85:129
91. Brigas AF, Johnstone RAW (1990) Tetrahedron Lett 31:5789
92. Jutand A, Negri S (1997) Synlett 6:719
93. Lenarda M, Pahor NB, Calligaris M, Graziani M, Randaccio L (1978) J Chem Soc, Dalton Trans 279
94. Kulasegaram S, Kulawiec RJ (1994) J Org Chem 59:7195
95. Kulasegaram S, Kulawiec RJ (1997) J Org Chem 62: 6547. For the isomerization of epoxides promoted by various transition metal complexes see the references cited therein
96. Bakos J, Orosz Á, Cserépi S, Tóth I, Sinou D (1997) J Mol Ctal A 116: 85
97. Chan ASC, Coleman JP (1991) J Chem Soc Chem Commun 535
98. Shimizu I, Maruyama T, Makuta T, Yamamoto A (1993) Tetrahedron Lett 34:2135
99. Bryan JC, Geib SJ, Rheingold AL, Mayer JM (1987) J Am Chem Soc 109:2826
100. Atagi LM, Over DE, McAlister DR, Mayer JM (1991) J Am Chem Soc 113:870
101. Wang MD, Calet S, Alper H (1989) J Org Chem 54:20
102. Tatsumi T, Tominaga H, Hidai M, Uchida Y (1977) Chem Lett 37
103. Covert KJ, Mayol A-R, Wolczanski PT (1997) Inorg Chim Acta 263:263
104. Khumtaveeporn K, Alper H (1995) Acc Chem Res 28:414
105. Milstein D (1984) Acc Chem Res 17:221
106. Bonanno JB, Henry TP, Neithamer DR, Wolczanski PT, Lobkovsky EB (1996) J Am Chem Soc 118:5132
107. Jang S, Atagi LM, Mayer JM (1990) J Am Chem Soc 112:6413
108. Chiu KW, Lyons D, Wilkinson G, Thornton-Pett M, Hursthouse MB (1983) Polyhedron 2:803
109. Crevier TJ, Mayer JM (1997) J Am Chem Soc 119:8485
110. Jutand A, Mosleh A (1997) J Org Chem 62:261
111. Inokuchi T, Kawafuchi H, Torii S (1992) Chem Lett 1895
112. Rondon D, Chaudret B, He X-D, Labroue D (1991) J Am Chem Soc 113:5671
113. Chaudret B (1995) Bull Soc Chim Fr 132:268
114. Mayer JM (1995) Polyhedron 14:3273
115. Grotjahn DB, Lo HC (1996) Organometallics 15:2860

116. Smith DJH (1979) Phosphorus compounds. In:Barton D, Ollis WD (eds) Comprehensive organic chemistry, vol 2. Pergamon Press, Oxford, p 1119
117. Tebbe, FN, Parshall GW, Reddy GS (1978) J Am Chem Soc 100:3611
118. Brown-Wensley KA, Buchwald SL, Cannizzo L, Clawson L, Ho S, Meinhardt D, Stille JR, Straus D, Grubbs RH (1983) Pure Appl Chem 55:1733, and references cited therein
119. Nugent WA, Mayer JM (1988) Metal-ligand multiple bonds. Wiley Interscience, New York
120. Kataoka Y, Akiyama H, Makihira I, Tani K (1997) J Org Chem 62:8109
121. Kataoka Y, Akiyama H, Makihira I, Tani K (1996) J Org Chem 61:6094
122. Kataoka Y, Makihira I, Akiyama H, Tani K (1995) Tetrahedron Lett 36:6495
123. Hall KA, Mayer JM (1992) J Am Chem Soc 114:10402
124. Ref. 5, p 339
125. Neithamer DR, LaPointe RE, Wheeler RA, Richeson DS, Van Duyne GD, Wolczanski PT (1989) J Am Chem Soc 111:9056, and references cited therein
126. Wolczanski PT (1995) Polyhedron 14:3335
127. Miller RL, Toreki R, LaPointe RE, Wolczanski PT, Van Duyne GD, Roe DC (1993) J Am Chem Soc 115:5570
128. Miller RL, Wolczanski PT (1993) J Am Chem Soc 115:10422
129. Chisholm MH, Hammond CE, Johnston VJ, Streib WE, Huffman JC (1992) J Am Chem Soc 114:7056, and references cited therein
130. Gates BC (1993) Angew Chem Int Ed Engl 32:228
131. Jacoby D, Isoz S, Floriani C, Chiesi-Villa A, Rizzoli C (1995) J Am Chem Soc 117:2793
132. Jacoby D, Floriani C, Chiesi-Villa A, Rizzoli C (1993) J Am Chem Soc 115:7025
133. Jacoby D, Isoz S, Floriani C, Chiesi-Villa A, Rizzoli C (1995) J Am Chem Soc 117:2805
134. Crescenzi R, Solari E, Floriani C, Chiesi-Villa A, Rizzoli C (1996) Organometallics 15:5456
135. Sakamoto M, Shimizu I, Yamamoto A (1994) Organometallics 13:407
136. Komiya S, Akita M, Kasuga N, Hirano M, Fukuoka A (1994) J Chem Soc Chem Commun 1115
137. Hirano M, Akita M, Tani K, Kumagai K, Kasuga NC, Fukuoka A, Komiya S (1997) Organometallics 16:4206
138. Jessop PG, Ikariya T, Noyori R (1995) Chem Rev 95:259
139. Behr A (1988) Carbon dioxide activation by metal complexes, VCH, Weinheim, Germany

Activation of Otherwise Unreactive C–Cl Bonds

Vladimir V. Grushin[a]* and Howard Alper[b]

[a]Du Pont de Nemours and Company Inc., Central Research and Development, Experimental Station, Wilmington, DE 19880-0328, USA
E-mail: vlad.grushin-1@usa.dupont.com

[b]Department of Chemistry, University of Ottawa, 10 Marie Curie, Ottawa, Ontario K1N 6N5, Canada
E-mail: halper@oreo.chem.uottawa.ca

During the past decade, considerable progress has been made in the area of transition metal-catalyzed cleavage and functionalization of the inert C–Cl bond in nonactivated chloroaromatic compounds. This new and important field of chemistry is reviewed in the present chapter, which describes both mechanistic and synthetic aspects of C–Cl activation. Oxidative addition reactions of chloroarenes to complexes of catalytic metals are discussed, along with their applications in a wide variety of reductive dechlorination, nucleophilic displacement, olefin arylation, coupling, and carbonylation reactions.

Keywords: C–Cl activation, Ar–Cl oxidative addition, Chloroarenes, Homogeneous catalysis with metal complexes, Reductive dechlorination, Aromatic nucleophilic substitution, Heck reaction, Homocoupling, Cross-coupling, Carbonylation

1 Introduction 194

2 Activation and Cleavage of Inert C–Cl Bonds with Transition Metal Complexes 195

2.1 Nickel Complexes 195
2.2 Palladium Complexes 198
2.3 Cobalt Complexes 200
2.4 Rhodium Complexes 201

3 Catalytic Transformations of Nonactivated Chloroarenes 203

3.1 Reductive Dechlorination of Chloroarenes 204
3.2 The Heck Arylation of Olefins 206
3.3 Carbonylation of Chloroarenes 210
3.4 Homocoupling and Cross-Coupling Reactions 214
3.5 Nucleophilic Substitution 216

4 Recent Progress, Conclusions, and Perspectives 218

References 219

1 Introduction

In 1994 we published a review on catalytic activation of the C–Cl bond in chloroarenes [1]. The importance of this topic stems from the availability and low cost of chlorinated aromatic compounds, which could be used as precursors for a wide variety of valuable products. Chloroarenes are significantly less expensive than their iodo, bromo, and fluoro analogues and thus would be ideal electrophilic arylating agents. At the same time, Ar–Cl bonds are considerably stronger and hence more difficult to activate than Ar–Br and Ar–I bonds. In particular, the experimental D_{Ph-X} values are equal to 527, 402, 339, and 272 $kJmol^{-1}$ for X=F, Cl, Br, and I, respectively. It is not surprising, therefore, that unlike bromo and especially iodoarenes chloroaromatic compounds usually remain inert under $S_{RN}1$ [2] and Ullmann-type [3] reaction conditions. Reactivity of the carbon-chlorine bond in chlorobenzene can be enhanced by a variety of means, such as the introduction of a strong electron-withdrawing group (e.g., NO_2, CN) into the benzene ring [4], deprotonation at one of the ortho-positions with a strong base [5], conversion to much more reactive chloronium ions, $[Ph\text{-}Cl\text{-}R]^+$ [6], and π-coordination of the benzene ring to an electron-deficient metal fragment, e.g., $Cr(CO)_3$ [7, 8]. All of these techniques increase the electron deficiency of the substrate, making the C–Cl bond more reactive. As far as chlorinated N-heterocycles are concerned, the same effect can be achieved by their N-quaternization or N-oxidation [9]. The noncatalytic ways of C–Cl activation listed above are discussed in more detail in our previous review [1].

This chapter focuses on *transition metal-catalyzed* reactions of most unreactive chloroarenes occurring with the C–Cl bond cleavage. Stoichiometric reactions will be discussed only if they are closely related to the catalytic transformations, e.g., informative from the perspective of mechanisms of catalysis or at least can provide guidance to the catalytic chemist. Although important for both industrial and fundamental research, C–Cl activation of alkyl chlorides, polychlorinated methanes, and chlorofluorocarbons (CFCs) will not be covered. Various reactions of CFCs, proceeding with the C–Cl bond cleavage, have been recently reviewed [10, 11]. Strong carbon-chlorine bonds in alkyl and benzyl chlorides [12], CH_2Cl_2 [12, 13], $CHCl_3$ [12], and CCl_4 [14] are reactive toward nucleophiles and bases under mild conditions in the absence of a transition metal catalyst. This is also true for activated nitro and cyano aryl chlorides, which readily undergo S_NAr-type transformations via the Meisenheimer intermediate [4, 5, 12]. This chapter deals with mechanistic and synthetic aspects of catalytic cleavage and functionalization of otherwise unreactive C–Cl bonds which are found in so-called *nonactivated* chloroarenes. We suggest that the term "nonactivated chloroarene" be defined as any aryl chloride whose C–Cl bond exhibits similar or lower reactivity than that of chlorobenzene. According to this definition, tolyl chlorides, *p*-chloroanisole, *p*-chloroaniline, etc., are certainly nonactivated chloroarenes. Both isomers of chloronaphthalene, and chlorobenzenes bearing weakly electron-accepting groups on the ring (F, Cl) may be regarded as slightly activated chloroarenes. As mentioned above, strong electron-acceptors activate chloroarene substrates, making them sufficiently reactive toward nucleophiles, so that no metal complex is needed to cleave the C–Cl bond. For this reason, numerous metal-catalyzed reactions of such *activated* chloroarenes will not be fully covered in this chapter but rather touched on occasionally when needed. In the presence of transition metal complexes, vinylic chlorides are normally more reactive than nonactivated chloroarenes [15, 16], probably due to the ability of RCH=CHCl to form π-allylic complexes. A convincing illustration of this point is the fact that various vinylic chlorides readily cross-couple with 1-alkynes in the presence of Cu and Pd catalysts (the Sonogashira reaction), whereas only strongly activated chloroaromatic compounds undergo this transformation [17].

Over the last 5 years some dozens of publications have appeared in the literature, reporting new reactions of chloroarenes, catalyzed by transition metal complexes. This most recent material will be compiled with the already reviewed [1] information, in an attempt to provide comprehensive coverage to the reader. It is worth noting that the tables of data presented in our previous review [1] can serve as an informative complement to this chapter. More emphasis will be put on conceptual, mechanistic aspects of C–Cl activation, as well as experimental observations which may eventually become "points of growth" in the future, determining and directing further research in the area.

2
Activation and Cleavage of Inert C–Cl Bonds with Transition Metal Complexes

Various complexes of transition metals can activate and cleave unreactive C–Cl bonds via nucleophilic [1, 18–20], electrophilic [21, 22], and radical [23, 24] paths, under mild conditions. For a number of reasons [1], not all of these reactions can be utilized in a catalytic manner. In this chapter, we will discuss only those C–Cl bond cleavage reactions which can consequently lead to *catalytic* transformations of weakly activated or nonactivated chloroarenes.

2.1
Nickel Complexes

Nickel catalysts are most widely used for various reactions of chloroarenes. Zero-valent Ni complexes, both preformed and/or generated in situ, oxidatively add the C–Cl bond of chloroarenes under very mild conditions (Eq. 1). Although complexes like [$(cod)_2Ni$] [25] and [$Ni(CO)_4$] [26] are certainly capable of cleaving C–Cl bonds in some ArCl, the highest reactivity is normally exhibited by tertiary phosphine complexes of Ni(0). For example, the reactions of chlorobenzene with [$(Et_3P)_3Ni$] [27] and [$(Cy_3P)_2Ni$] (Cy=cyclohexyl) [28] occur rapidly at room temperature to give [$(Et_3P)_2Ni(Ph)Cl$] and [$(Cy_3P)_2Ni(Ph)Cl$], respectively.

$$[(R_3P)_nNi] + ArCl \longrightarrow trans\text{-}[(R_3P)_2Ni(Ar)Cl] + (n\text{-}2)R_3P \quad (1)$$

n = 2, 3, 4

It has been found [29] that the reaction between 1,2,4-trichlorobenzene and $[(Ph_3P)_2Ni(C_2H_4)]$ results in the formation of three isomeric complexes, $[(Ph_3P)_2Ni(Ar)Cl]$, where Ar=2,5-$C_6H_3Cl_2$ (87%), 3,4-$C_6H_3Cl_2$ (7%), and 2,4-$C_6H_3Cl_2$ (6%). This selectivity pattern is similar to that observed for organic S_NAr reactions of the same substrate, suggesting that alike mechanisms are operative in the oxidative addition and aromatic nucleophilic substitution [29, 30]. This conclusion is supported by the fact that $[(Et_3P)_3Ni]$ is more reactive toward chloroarenes than its less basic triphenylphosphine congener $[(Ph_3P)_3Ni]$ [31–33].

A meticulous kinetic study of the reaction between $[(Ph_3P)_3Ni]$ and various p-XC_6H_4Cl revealed a number of mechanistic features [32]. For strong electron-withdrawing X (σ>+0.23), the reaction was very sensitive to electronic effects of X, the ρ value determined being 8.8. Remarkably, this high value dropped down to virtually 0 for any X with σ<+0.23. In other words, the nickel(0) complex appeared to be equally reactive to p-XC_6H_4Cl, regardless of whether substituent X was a weak electron acceptor, neutral, or an electron donor of any strength! For instance, no difference in rate constants was noticed for the reactions of $[(Ph_3P)_3Ni]$ with p-XC_6H_4Cl when X was Cl (σ=+0.23) and PhO (σ=–0.32). Furthermore, when X was a strong electron acceptor, the bromides p-XC_6H_4Br reacted with the Ni(0) 2 orders of magnitude faster than their chloro analogues. However, this difference was almost negligible for less electron-withdrawing X. Clearly, two different mechanisms governed the reactions of the substrates bearing strongly electron-accepting and all other p-substituents. It was reasonably proposed that an S_NAr-type mechanism governed the reactions of the most electron-deficient chloroarenes (ρ=8.8) (Eq. 2), whereas in all other cases ($\rho\approx$0) the oxidative addition occurred via an unsymmetrical three-center transition state (Eq. 3).

$O_2N{=}C_6H_4(Cl)Ni(PR_3)_n \longrightarrow O_2N\text{–}C_6H_4\text{–}Ni^+(PR_3)_2 + Cl^- \longrightarrow O_2N\text{–}C_6H_4\text{–}Ni(Cl)(PR_3)_2$ (2)

$C_6H_5\cdots Cl\cdots Ni(PR_3)_2 \longrightarrow C_6H_5\text{–}Ni(Cl)(PR_3)_2$ (3)

R = Ph

Because the oxidative addition was first order in the substrate, slowing down considerably in the presence of excess PPh_3, it was concluded that both $[(Ph_3P)_3Ni]$ and $[(Ph_3P)_2Ni]$ participated, the latter being considerably more reactive. The 14e dicoordinate complex emerged from phosphine dissociation from the $[(Ph_3P)_3Ni]$ employed for the study [32].

In general, oxidative addition reactions of low-valent transition metal complexes can be governed by three different mechanisms, namely nucleophilic displacement, template (concerted three center), and radical [34]. Oxidative addition reactions between chloroarenes and Ni(0) complexes are usually clean and selective, implying no participation of radical intermediates. Only in a very few cases, however, has the formation of paramagnetic Ni(I) impurities been detected in such reactions [33, 35]. It is worth noting at this point that Ni(I) complexes of the type $[L_3NiX]$ (X=I, Br) are commonly produced in substantial quantities when the corresponding Ni(0) compounds are reacted with iodo and bromoarenes [33]. In their classical work, Tsou and Kochi [33] investigated, by kinetic, electrochemical, and ESR methods, the reaction between $[(Et_3P)_4Ni]$ and various haloarenes. The reaction gave rise to two Ni-containing products, the diamagnetic organonickel(II) complex, $[(Et_3P)_2Ni(Ar)X]$, and paramagnetic nickel(I) halide (Eq. 4).

$$[(Et_3P)_4Ni] + Ar\text{-}X \longrightarrow [(Et_3P)_2Ni(Ar)X] + [(Et_3P)_3NiX] \quad (4)$$

Firm evidence was obtained for a *common* reaction intermediate in all cases, despite the fact that the Ni(I) to Ni(II) ratio varied in a broad range, depending on the nature of Ar, X, and the solvent used. The mechanism (Scheme 1) involves the rate limiting step of single electron transfer (SET) from the electron-rich Ni(0) to ArX, followed by the fast formation and decomposition of the tight radical ion pair $[L_3Ni^{\cdot +}\ ArX^{\cdot -}]$.

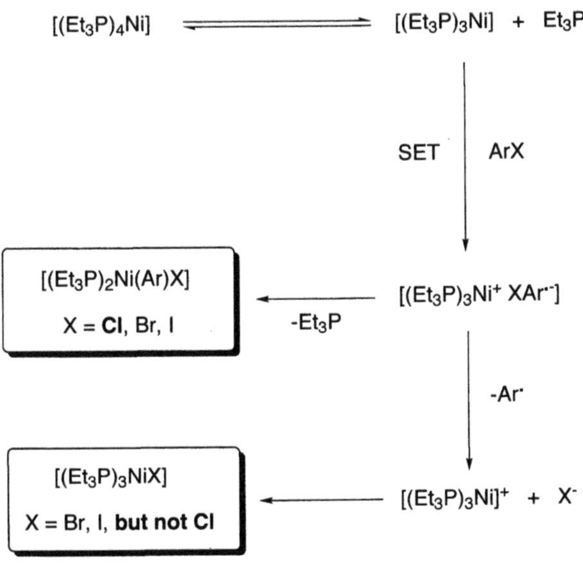

Scheme 1.

The relatively weak Ar–Br and especially Ar–I bonds would readily dissociate, giving rise to the Ni(I) paramagnetic complex and free aryl radical. This decomposition path is normally disfavored for aryl chlorides with considerably stronger Ar–Cl bonds. As a result, no Ni(I) species formed in the reactions of all chloroarenes studied, the only exception being $[p\text{-}Me_3NC_6H_4Cl]^+$. The ρ value of 5.4 obtained by Tsou and Kochi [33] is close to that (8.8; see above) previously reported by Foà and Cassar [32], suggesting that SET (Scheme 1) may play a certain role in some of the reactions of triphenylphosphine nickel(0) complexes with chloroarenes. It is still unclear if every reaction between any chloroarene and Ni(0) always involves the SET step. However, the excellent selectivity of the σ-aryl Ni(II) complex formation from ArCl and highly reactive Ni(0) makes chloroarenes especially attractive substrates for various arylation reactions catalyzed by Ni complexes.

2.2 Palladium Complexes

Palladium is certainly one of the most versatile catalytic metals. Soluble complexes of Pd are excellent catalysts for a number of catalytic reactions of haloarenes, which have been extensively reviewed in recent years [1, 8d, 15, 17, 36–42]. In most cases, aryl iodides are used for such reactions, due to the ease of the oxidative addition of ArI to triphenylphosphine Pd(0) complexes at room temperature [43–53]. Aryl bromides are only slightly less reactive, whereas nonactivated chloroarenes exhibit very poor reactivity toward $[(Ph_3P)_nPd]$. In particular, the reaction between PhCl and $[(Ph_3P)_4Pd]$ or $[(Ph_3P)_2Pd(dba)]$ (dba= dibenzylideneacetone) without a solvent requires hours at 140°C to go to completion [44, 54]. As a result, the expected organometallic complex, $[(Ph_3P)_2Pd(Ph)Cl]$, is formed in good yield [54]. A serious problem arises as soon as other nonactivated $p\text{-}XC_6H_4Cl$ (X=Me, MeO, etc.) are reacted with triphenylphosphine Pd(0) complexes. These high temperature (140°C) reactions give mixtures of complexes $[(Ph_3P)_2Pd(p\text{-}XC_6H_4)Cl]$ and $[(Ph_3P)(p\text{-}XC_6H_4PPh_2)Pd(Ph)Cl]$ [54] due to the facile exchange between the σ-aryl and phenyls on the PPh_3 ligands. The exchange readily occurs at 60°C [54, 55], i.e., below the temperature required for the oxidative addition of the Ar–Cl bond. It is noteworthy that when X in $[(Ph_3P)_2Pd(p\text{-}XC_6H_4)Cl]$ is an electron-withdrawing group, such as NO_2, CN, CHO, no aryl-aryl exchange takes place [54]. Importantly, unlike the Ni(0) compounds (see above) [56], zero-valent Pd complexes are not prone to one-electron oxidation [53, 57], resulting in the formation of paramagnetic Pd(I) species.

Tertiary phosphine palladium (0) complexes, $[L_nPd]$, are more reactive toward chloroarenes [35, 58–63] if L is a basic, bulky trialkylphosphine, such as Cy_3P [59–61], $i\text{-}Pr_3P$ [59], and $i\text{-}Pr_2BuP$ [62]. For instance, $[(Cy_3P)_2Pd(dba)]$ reacts with PhCl at 60°C to give $[(Cy_3P)_2Pd(Ph)Cl]$ in good yield [59] (Eq. 5). Other chloroarenes also oxidatively add to this Pd(0) complex, the order of reactivity being $p\text{-}NO_2C_6H_4Cl > p\text{-}EtOOCC_6H_4Cl >> C_6H_5Cl > p\text{-}MeO_6H_4Cl$ [59].

$$[(Cy_3P)_2Pd(dba)] + PhCl \longrightarrow [(Cy_3P)_2Pd(Ph)Cl] \quad (5)$$

The phase-transfer catalyzed [61] reaction between $[(Cy_3P)_2PdCl_2]$ and alkali results in the generation of a highly reactive metal complex, presumably $[(Cy_3P)Pd]$, which readily activates the C–Cl bond of chlorobenzene at 100°C [60]. The oxidative addition, followed by Cl/OH ligand exchange, produces the binuclear organopalladium hydroxo complex which has been isolated in 80% yield (Eq. 6) [61].

$$[(Cy_3P)_2PdCl_2] \xrightarrow[-2KCl,\ -Cy_3PO,\ -H_2O]{2KOH} '[(Cy_3P)Pd]' \xrightarrow{PhCl} 1/2\ [(Cy_3P)(Ph)Pd(\mu\text{-}Cl)_2Pd(Ph)(PCy_3)] \xrightarrow[-2KCl]{2KOH} 1/2\ [(Cy_3P)(Ph)Pd(\mu\text{-}OH)_2Pd(Ph)(PCy_3)] \quad (6)$$

Both high basicity and optimal cone angle of the phosphine ligand on Pd appear to be crucial for C–Cl activation. For instance, Pd(0) complexes of less basic Cy_2PhP, m-Tol_3P, and o-Tol_3P with similar or larger cone angles fail to react with chlorobenzene [59, 64]. Originally, no catalytic C–Cl bond activation was observed when Cy_3P was replaced by bulkier, basic phosphines, such as t-Bu_3P and t-Bu_2PhP [59]. However, the most recent results obtained by Koie's group and others (Sect. 3.5 and 4) indicate that t-Bu_3P complexes of Pd(0) can easily cleave the C–Cl bond of nonactivated chloroarenes.

Electron-rich *bidentate* phosphines containing i-Pr or Cy groups form complexes with zero-valent palladium, which activate the C–Cl bond in nonactivated chloroarenes under mild conditions [35, 62, 63]. The reactivity toward PhCl has been shown [62] to decrease in the order $[(dippp)_2Pd]>[(i\text{-}Pr_2BuP)_3Pd]>>[(dippe)_2Pd]>>[(dppp)_2Pd]$, revealing a dramatic influence of the chelate effect on the reaction rate (dippp=1,3-bis(diisopropylphosphino)propane; dippe=1,2-bis(diisopropylphosphino)ethane; dppp=1,3-bis(diphenylphosphino)propane). Because of the high stability of the chelate rings in $[(dippe)_2Pd]$, the lack of coordinative unsaturation on the metal results in the slower oxidative addition. Taking into consideration the noticeably weaker chelate effects of electron-rich dippp and dippb (dippb=1,4-bis(diisopropylphosphino)butane) ligands, the enhanced reactivity of their Pd(0) complexes toward oxidative addition is not surprising. It has been demonstrated by Portnoy and Milstein [35] that $[(dippp)_2Pd]$ readily reacts with PhCl in dioxane at 90°C to give a mixture of the *cis* and *trans* organopalladium complexes (Eq. 7). Having formed independently from the Pd(0) complex and PhCl, the two isomers then exist in equilibrium with one another. The oxidative addition reaction slows down considerably if performed in solvents of low polarity and/or in the presence of extra dippp. As suggested by these observations and the results of the inversion transfer NMR experiment, the reaction between $[(dippp)_2Pd]$ and PhCl proceeds via a 14-e Pd(0) intermediate, $[(dippp)Pd]$, arising upon loss of one of the two dippp ligands. The kinetic data obtained were indicative of an S_NAr-type charged transition state involved in the oxidative addition, with the Pd center partially coordinating with the Cl atom of chlorobenzene [35].

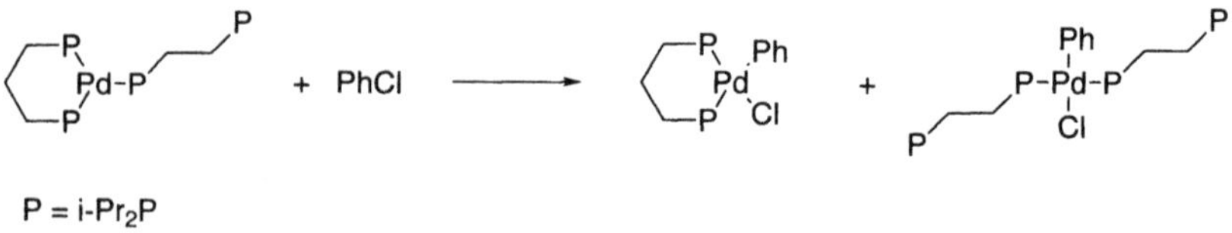

(7)

P = i-Pr_2P

A highly reactive Pd(0) complex has been generated by the UV-induced elimination of CO_2 from [(dcpe)Pd(C_2O_4)] (dcpe=1,2-bis(dicyclohexylphosphino)ethane) in MeCN [63]. Under such conditions, the resulting mononuclear species, [(dcpe)Pd], dimerizes rapidly to [(μ-dcpe)$_2$Pd$_2$]. The latter has been isolated in 85% yield and characterized by single crystal X-ray diffraction, revealing the presence of a Pd-Pd contact [2.7611(5) Å]. If treated with chlorobenzene, the binuclear complex easily undergoes oxidative addition of the C–Cl bond, giving rise to [(dcpe)Pd(Ph)Cl] (Eq. 8).

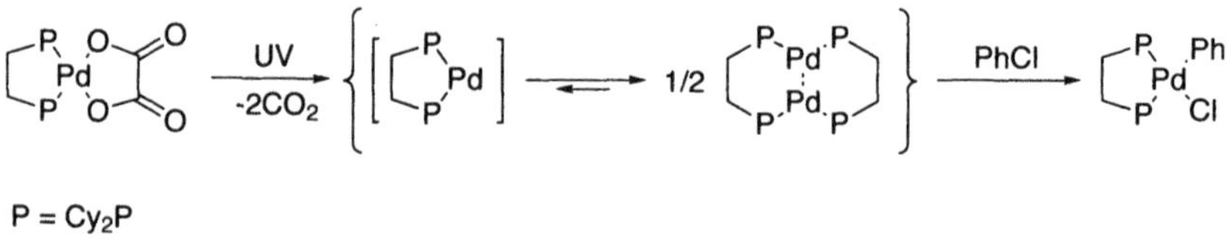

(8)

P = Cy_2P

A recent report [65] describes the photoinduced reaction between [(dppm)$_3$Pd$_3$(CO)]$^{2+}$ and various chlorinated organic compounds, chlorobenzene included. The reaction with PhCl gives [(dppm)PdCl$_2$], biphenyl, PPh$_3$, and a variety of other products which have not been identified. It has been proposed that one of the routes leading to C–Cl activation is the intermediate formation of two reactive complexes, "[(dppm)$_2$Pd$_2$(CO)]$^{2+}$" (stabilized by solvent molecules) and 14-e [(dppm)Pd] [65]. It is noteworthy, however, that because dppm is not as bulky and basic as dippp and dcpe (see above) one might question the ability of [(dppm)Pd] to oxidatively add PhCl. Moreover, no formation of [(dppm)Pd(Ph)Cl] was observed [65], though one would anticipate this complex to emerge, should the oxidative addition of PhCl to [(dppm)Pd] occur.

2.3
Cobalt Complexes

Cobalt complexes have been used to catalyze the carbonylation of chloroarenes to the corresponding carboxylic acids and their esters (Sect. 3.3). Some complexes of cobalt in the oxidation state –1 activate the Ar–Cl bond via an $S_{RN}1$-type mechanism [2] involving single electron transfer from the metal to chloroarene, followed by elimination of Cl^-. The simplest Co(–I) carbonyl species, [Co(CO)$_4$]$^-$, is not electron-rich enough to react with haloarenes. However, its reactivity has been shown to enhance tremendously in the presence of Caubère's "complex bases," mixtures of NaH and NaOAlk [23, 66, 67]. For instance, the stoichiometric carbonylation of chlorobenzene has been performed with the

$NaH/NaOCH_2CMe_3/Co(OAc)_2$ system, in which $[Co(CO)_4]^-$ is generated in situ (Eq. 9) [68]. Although a plausible $S_{RN}1$-type reaction path has been proposed to account for the unusual reactivity of the cobalt system [23], the intimate mechanism of this carbonylation process remains unknown.

$$PhCl \xrightarrow[2.\ H_3O^+]{1.\ NaH/NaOCH_2CMe_3/Co(OAc)_2/CO} PhCOOH \quad (9)$$

Light can often be used to promote $S_{RN}1$ reactions [2]. Indeed, the photochemically induced, cobalt-catalyzed carbonylation of haloarenes, PhCl included, readily occurs under phase-transfer conditions. This interesting methodology was first developed by Brunet, Sidot, and Caubère [23, 69] and subsequently used for the carbonylation of various chloroarenes in the presence of catalytic amounts of cobalt compounds (Sect. 3.3).

There is another way to increase the reducing ability of the metal in $[Co(CO)_4]^-$, making it reactive toward some chloroarenes [70]. Alkylation of $[Co(CO)_4]^-$ with MeI or Me_2SO_4 results in the formation of $[MeCo(CO)_4]$, which readily adds alkoxide anions to produce $[MeCo(CO)_3(COOR)]^-$. This anionic complex is electron-rich enough to cleave the C–Cl bond in slightly activated chloroarenes, presumably via single electron transfer (Eq. 10). Various complexes of the type $[(ZCH_2)Co(CO)_3COOR]^-$ (Z=COOR', F, CN, and H) have been used as catalysts for the single and double carbonylation reactions of chloronaphthalenes, 2-chlorothiophene, and 2-chlorofuran (Sect. 3.3).

$$[(CO)_4CoMe] + RO^- \longrightarrow [(CO)_3Co(Me)(COOR)]^- \xrightarrow{ArCl} [(CO)_3Co(Ar)(Me)(COOR)] + Cl^- \quad (10)$$

2.4 Rhodium Complexes

Two different strategies have been developed for C–Cl activation with rhodium compounds: (a) oxidative addition of C–Cl bonds to electron-rich Rh(I) complexes and (b) π-coordination of coordinatively unsaturated, electron-deficient Rh(III) species with the benzene ring of ArCl, followed by aromatic nucleophilic substitution of chlorine in the thus activated aromatic system.

There have been several articles reporting oxidative addition of various C–Cl bonds to Rh complexes [71–78]. Only a few Rh species, however, are capable of activating the C–Cl bond of nonactivated chloroarenes in a catalytic manner [73, 77, 78]. Rhodium complexes containing bulky basic phosphines, such as Cy_3P and $i\text{-}Pr_3P$, are excellent catalysts for the biphasic hydrogenolysis of the C–Cl bond of chloroarenes under mild conditions (Sect. 3.1) [77, 78]. The mechanism of this catalytic C–Cl activation seems to be rather complex (Scheme 2).

Added catalyst for the reduction of ArCl to ArH (Scheme 2) is a mixture of co-crystallized $[L_2Rh(H)Cl_2]$ and paramagnetic $[L_2RhCl_2]$ ($L=Cy_3P$ or $i\text{-}Pr_3P$), which is rapidly and quantitatively converted to $[L_2Rh(H)_2Cl]$ in the presence of

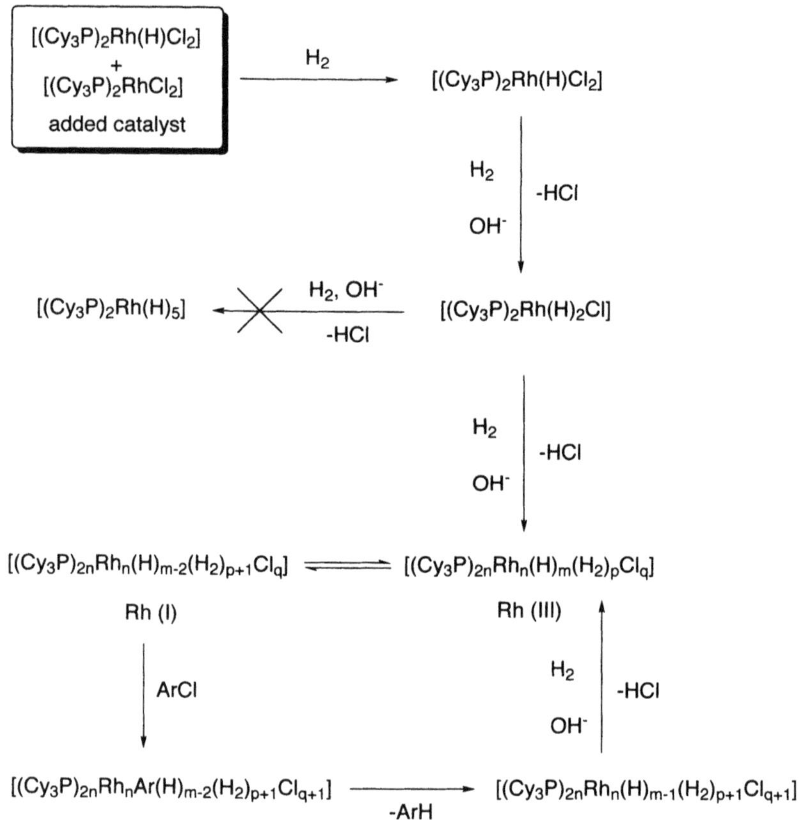

Scheme 2.

H_2 and alkali [78–80]. Unlike its Ir analogues [78, 79, 81, 82], the Rh dihydride does not produce the pentahydride upon prolonged exposure to H_2/OH^- but rather gives a mixture of highly reactive, electron-rich Rh species which easily activate aromatic C–Cl and even C–H [83a] and C–F [83b] bonds via oxidative addition. The structure of these species remains unknown, though evidence has been obtained [83, 84] for the presence of nonclassical hydrids in the catalytic system.

In the "Introduction," we mentioned the enhanced reactivity of chloroarenes activated by π-coordination with the $Cr(CO)_3$ fragment [7, 8]. The interaction between π-electron density on the benzene ring of PhCl in $[(PhCl)Cr(CO)_3]$ with vacant d orbitals on the metal has the same effect on the reactivity of the C–Cl bond toward nucleophiles as the introduction of strong electron-withdrawing groups (e.g., NO_2) into the benzene ring of chlorobenzene. The dicationic fragment, $[(C_5EtMe_4)Rh]^{2+}$, can be used instead of $Cr(CO)_3$ to activate chloroarenes in the same manner (Eq. 11) [85, 86].

$$[(C_5EtMe_4)Rh(\eta^6\text{-Me}_3C_6H_2Cl)]^{2+} \xrightarrow[-HCl]{MeOH} [(C_5EtMe_4)Rh(\eta^6\text{-Me}_3C_6H_2OMe)]^{2+} \quad (11)$$

Although conceptually similar from the perspective of Ar–Cl activation, the Cr and Rh systems are certainly distinct, as far as catalysis is concerned. In the case of chloroarene chromium tricarbonyl complexes C–Cl activation cannot be rendered catalytic in Cr. Once the chlorine in $[(ArCl)Cr(CO)_3]$ is replaced by a nucleophile Nu, the π-arene ligand in the resulting pseudo-octahedral 18e complex, $[(ArNu)Cr(CO)_3]$, is normally too inert to be replaced by another ArCl molecule. In contrast, the $[(C_5EtMe_4)Rh]^{2+}$ moiety can transfer from one π-aromatic ring to another, thus opening up the possibility for catalysis (Sect. 3.5).

3
Catalytic Transformations of Nonactivated Chloroarenes

In this section, we will describe and discuss various reactions of chloroarenes, catalyzed by transition metal complexes. Finding a complex which can cleave aromatic C–Cl bonds does not mean that a catalyst for their functionalization has been developed. As will be shown below, in many instances no catalysis takes place at all, despite the fact that added or generated in situ metal species are indeed capable of activating the Ar–Cl bond under reasonably mild conditions. The proper ligand environment for C–Cl activation as the first key catalytic step may be poorly suitable or even detrimental for further transformations needed, such as ligand exchange, migratory insertion, and reductive elimination. It is often observed that the requirements for each of the elementary reactions in the catalytic cycle are in conflict, burying the entire idea of catalysis. Therefore, studying mechanistic aspects of organometallic reactions which are believed to participate in the proposed catalytic cycle is of great importance. Whether we like it or not, the process of creation of a new catalytic process nowadays is still based mostly on scouting and optimization of reaction conditions, rather than exhaustive knowledge of intimate reaction mechanisms. Under certain circumstances, this empirical approach to catalysis turns the research into a fascinating and enjoyable adventure for the chemist. On the other hand, those with sufficient laboratory experience in the field might admit that in many cases the "scouting and optimizing" methodology is monotonous, unimaginative, and poorly efficient. For this reason the description of the metal-catalyzed transformations below will be complemented with comments on mechanistic features of the key steps constituting the catalytic cycle.

3.1 Reductive Dechlorination of Chloroarenes

Replacing chlorine in ArCl for hydrogen (Eq. 12) is more important for synthesis than it might look upon initial consideration [10, 11, 87]. In particular, hydrogenolysis of the C–Cl bond with deuterium may be used for selectively labeling the corresponding position of the ring with D [88]. Furthermore, using chlorine as a protecting group offers the synthetic chemist a rare opportunity to alter the orientation rules of aromatic electrophilic substitution. This strategy has been used for the preparation of various cyclic compounds [89].

$$ArCl \xrightarrow{[H]} ArH \qquad (12)$$

As early as 1973, Love and McQuillin [73] reported that H_2 (P=1 atm) in DMF reduced chlorobenzene to benzene in the presence of $[Py_3RhCl_3]$ (5 mol%) and $NaBH_4$ at room temperature (50% conversion in 13 h). The dechlorination of PhCl was also carried out by indoline as a reducing agent in the presence of $PdCl_2$ in MeOH at 140°C [90]. Remarkably, the rate of reduction decreased in the order PhCl>PhBr>PhI, suggesting that oxidative addition of the C–Hal bond was not the rate limiting step of the process. This conclusion was strongly supported by the fact that the reaction was zero order in PhCl. A considerable number of catalytic systems have been developed for the reduction of Ar–Cl bonds, employing hydrides of main group elements in the presence of various d [66, 91] and f [92] block metal compounds. It is unclear whether these reactions are heterogeneously or homogeneously catalyzed by the metals. Soluble polymer anchored $PdCl_2$ [93] and genuine heterogeneous systems, Pd/C [94] and metallic Ni [95], have exhibited high catalytic activity in the reductive dechlorination of chloroarenes with various reducing agents under mild conditions. Radical anions of anthracene in conjunction with Ni(II) and Co(II) have been reported to reduce PhCl and other organic halides [96]. Polychlorinated compounds have been reduced in the presence of Ni [97], Pd [76a, 98], Rh [75, 76, 99], and Ru [100] soluble complexes. It is worth noting that although $[(Ph_3P)_4Pd]$ does oxidatively add PhCl under drastic conditions (Sect. 2.2) [44, 54], only activated aromatic C–Cl bonds (e.g., in chloropyrazines and their N-oxides [101]) can be efficiently reduced in the presence of triphenylphosphine Pd(0) complexes.

Only two systems have been developed for the reduction of ArCl, which (a) employ a genuinely homogeneous catalyst and (b) have proven broad functional group tolerance. Both techniques utilize bulky, electron-rich phosphines. Rhodium(III) complexes of the type $[L_2Rh(H)Cl_2]$, where L=Cy_3P or i-Pr_3P, efficiently catalyze hydrogenolysis of the C–Cl bond in various ArCl (Eq. 13) [77, 78]. The reaction occurs under exceedingly mild conditions (20–100°C and 1 atm H_2). We discussed some mechanistic features of the process in Sect. 2.4.

$$\mathrm{ArCl + H_2 + NaOH \xrightarrow{[(Cy_3P)_2Rh(H)Cl_2]} ArH + NaCl + H_2O}$$

ArCl = chlorobenzene, chlorotoluenes, 4-chloroanisole, 4-chloroaniline, 3-chlorobenzophenone, carprofen, 4-chlorobenzoic acid, 4-chlorophenylacetic acid, 5-chloro-1-ethyl-2-methylimidazole, 5-chlorobenzodioxole, 1-chloronaphthalene, tetrachloro-m-xylene

(13)

Palladium complexes containing basic, bulky phosphines (dippp, dippe, dippb, i-Pr_3P) have been demonstrated to efficiently catalyze the hydrodechlorination reaction of chloroarenes with methanol or sodium formate (Eq. 14) [102]. Of the ligands examined, dippp exhibited the highest catalytic activity.

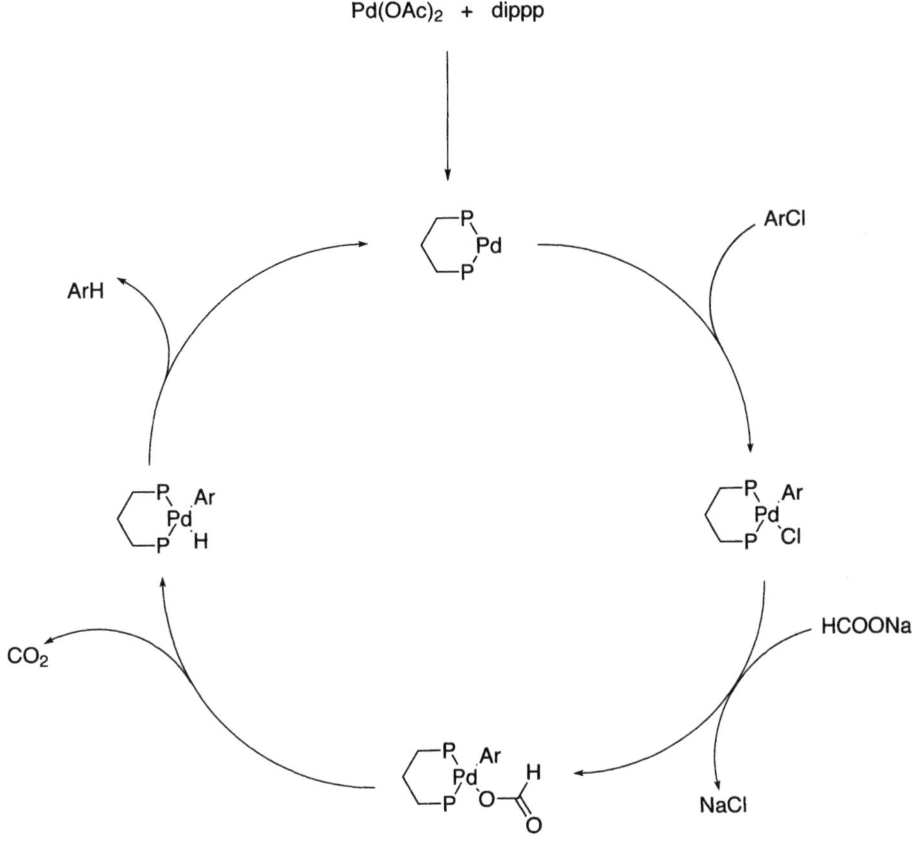

Scheme 3.

$$4\text{-}YC_6H_4Cl + HCOONa \text{ or } MeOH \xrightarrow{Pd(OAc)_2/dippp} C_6H_5Y \tag{14}$$

$Y = H, Me, MeO, NH_2, CN, CHO, MeCO, NO_2$

The mechanism of the Pd-catalyzed dechlorination likely involves oxidative addition of the C–Cl bond to electron-rich Pd(0) complexes generated in situ from $Pd(OAc)_2$ and a tertiary phosphine ligand [103]. The resulting organopalladium chloride, $[L_2Pd(Ar)Cl]$, where L=1/2dippb, 1/2dippp, 1/2dippe, or i-Pr_3P, undergoes ligand exchange with the reducing agent (MeO^- or $HCOO^-$) to give $[L_2Pd(Ar)OMe]$ or $[L_2Pd(Ar)O_2CH]$, respectively. Due to facile β-elimination [104], both of these complexes are then converted to an unstable [105] palladium hydride, $[L_2Pd(Ar)H]$, which reductively eliminates ArH with concomitant regeneration of the catalytically active Pd(0) species (Scheme 3).

3.2 The Heck Arylation of Olefins

As a unique method for the direct arylation of alkenes, the Heck reaction (Eq. 15) has been widely investigated, finding numerous elegant applications in organic synthesis [15, 36–38, 106].

$$ArX + CH_2{=}CHR \xrightarrow[-HX]{[Pd],\ base} ArCH{=}CHR \tag{15}$$

In recent years, many research groups have focussed their efforts on the development of new techniques for carrying out the Heck reaction of organic halides with olefins. The most interesting findings include the design of new homogeneous [107–109] and stabilized Pd or Pd/Ni cluster [110, 111] catalysts, as well as performing the reaction under high (10 kbar) pressure [112] or in superheated (260°C) and supercritical (400°C) water [113]. However, the palladium clusters stabilized by tetraalkylammonium salts or poly(vinylpyrrolidone) did not catalyze the Heck reaction of chlorobenzene [110]. Propylene carbonate stabilized nanostructured palladium clusters were more active, catalyzing the formation of stilbene from PhCl and styrene at moderate conversions and yields [111]. Surprisingly, PhI and PhBr were only slightly more reactive than PhCl when the Heck reaction was conducted in superheated/supercritical water [113]. Although a dienyl chloride was successfully olefinated with styrene under high pressure, an analogous reaction of PhCl was not mentioned [111]. The new organometallic palladacycles [107, 108] and Pd complexes of N-heterocyclic carbenes [109] exhibited exceptionally high catalytic activity in the Heck reaction of aryl bromides and activated chlorides but failed to catalyze the arylation of nonactivated chloroarenes, e.g., *p*-chloroanisole. Clearly, in spite of considerable efforts, very little progress has been made over the last few years in the arylation of alkenes with nonactivated aryl chlorides, such as chlorobenzene, chlorotoluenes, and chloroanisoles.

The classic Heck catalytic system, $Pd(OAc)_2/PPh_3$, normally exhibits poor activity in the olefination of chlorobenzene [114, 115], with Pd metal readily precipitating from the homogeneous mixtures even in the presence of a large excess of triphenylphosphine. Both intermolecular [116–118] and intramolecular [119] Heck-type reactions of nonactivated aryl chlorides have been carried out in the presence of nickel complexes which are normally much more reactive toward Ar–Cl bonds (Sects. 2.1, 2.2). However, triethylamine, which is commonly used as a base for the Pd-catalyzed Heck-type arylations [15, 36–38, 106], has been found to terminate the Ni-catalyzed reaction [117]. Using zinc metal instead of Et_3N gave positive results, although in a number of cases the nickel-catalyzed reaction between ArX and $RCH{=}CH_2$ gave rise to the saturated product, RCH_2CH_2Ar, in up to 45% yield [117]. This problem can be avoided and good yields of stilbenes (up to 82%) are obtained if the Ni-catalyzed arylation of styrene with chloroarenes is performed in MeCN in the presence of pyridine [118]. It is conceivable that the recently reported Ni-catalyzed electrochemical cyclization of *o*-chlorophenyl alkenyl or alkynyl ethers [120] and the Heck-type intramolecular cyclization of *o*-chlorophenyl alkenyl amines [119] occur via similar mechanisms. However, under the reaction conditions employed, the double bond emerging from the intramolecular arylation of the olefinic moiety on the ethers undergoes the electrochemical reduction in situ [120].

Bimetallic Pd/Ni [121] and Pd/Co [122] systems have exhibited considerable catalytic activity in the Heck reaction of nonactivated chloroarenes with ethyl acrylate, acrylonitrile, and acrylic acid. For instance, ethyl acrylate and acrylonitrile reacted smoothly with chlorobenzene in the presence of NaI and catalytic amounts of $NiBr_2$, $Pd_2(dba)_3$, and *o*-Tol_3P in DMF to give E-isomers of ethyl cinnamate and cinnamonitrile, respectively [121]. The reaction occurred via the nickel-catalyzed halogen exchange between ArCl and NaI, followed by the conventional palladium-catalyzed olefination of the iodoarene generated in situ.

Milstein and associates [123–125] have developed efficient methods for the Heck-type arylation of olefins with various chloroarenes, catalyzed by Pd complexes of dippb and dippp (Eq. 16). When a 1:2 mixture of $Pd(OAc)_2$ and dippb was used, the reaction was run in DMF at 150°C, in the presence of NaOAc as a base, giving predominantly E-isomers of the substituted stilbenes [123, 124]. Attempts to replace the solvent by MeCN resulted in no reaction, indicating that this catalytic process is very sensitive to the reaction medium. Choosing the right base seems to be as critical because when Et_3N was used instead of NaOAc the reaction was sluggish [123]. Palladium complexes of dippp exhibited very little activity under the reaction conditions optimized for the Pd/dippb catalyst. However, in the presence of zinc powder and no base the dippp Pd complexes catalyzed the Heck arylation of styrene with chloroarenes quite efficiently [125]. Remarkably, Z-isomers of the resulting stilbenes prevailed in this case (Eq. 16). The techniques developed by Milstein's group are not suitable for the

Heck reaction of electron-deficient olefins, which are reactive toward the strongly nucleophilic dippp and dippb ligands employed.

$$PhCl + CH_2{=}CHPh \xrightarrow[\text{HCl scavenger}]{Pd(OAc)_2/L} \textit{cis-}PhCH{=}CHPh + \textit{trans-}PhCH{=}CHPh \qquad (16)$$

	cis-PhCH=CHPh	trans-PhCH=CHPh
L = dippb; HCl scavenger = AcONa	4.4%	80%
L = dippp; HCl scavenger = Zn	81%	7%

The exceptional sensitivity of the Heck reaction to a wide variety of factors is intriguing, crying out for a detailed investigation of the intimate mechanism of the process. The generally accepted mechanism for the Heck reaction is presented in Scheme 4. Two reports [126, 127] have recently appeared, describing mechanistic studies of the Heck arylation of olefins.

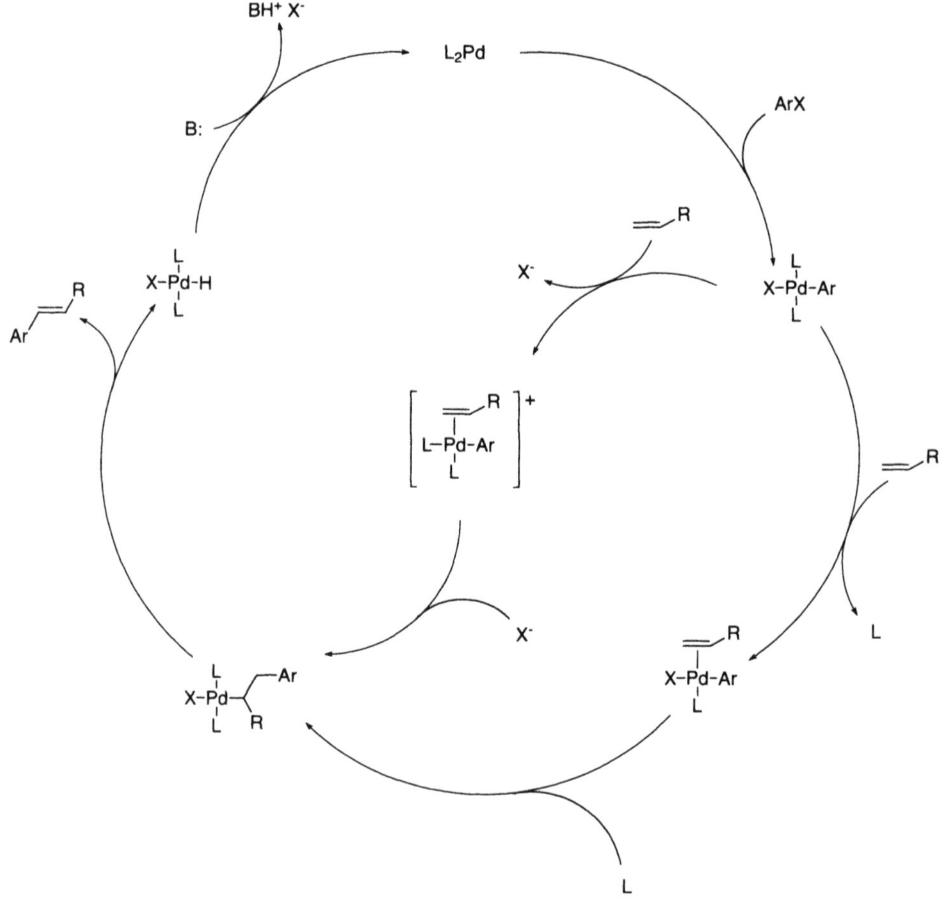

Scheme 4.

Herrmann et al. [126] undertook a detailed study of the Heck reaction between *n*-butyl acrylate and ArX (X=Br, Cl), catalyzed by Pd complexes of various phosphines. As it had been demonstrated [128] that $[(Ph_3P)_2Pd(Ph)Cl]$ can arylate olefins stoichiometrically, the poor reactivity of chloroarenes in the Heck reaction used to be rationalized by the very slow oxidative addition of the C–Cl bond to triphenylphosphine complexes of zero-valent palladium. It is clear now, however, that the reaction temperature required for the Heck reaction is considerably *higher* than that, at which the metal complex readily activates and cleaves the carbon-halogen bond [44, 54, 126]. Obviously, other steps in the catalytic cycle (Scheme 4) also require drastic conditions in order to occur. These high temperatures favor the Ar/Ph exchange in the intermediate arylpalladium complex, $[(Ph_3P)_2Pd\,(Ar)X]$ (Eq. 17) [54, 55, 126]. While readily accounting for the formation of side-products [126], the Ar/Ph exchange alone fails to provide a rationale for the catalyst deactivation observed. The loss of catalytic activity may be due to the Pd-mediated arylation of the phosphine ligand, resulting in the formation of tetraarylphosphonium salts [55b]. Of the numerous Ar_3P studied [126], only two (Ar=o-Tol and Mes) did not participate in the aryl/aryl exchange with aryl halides. However, both of these phosphines provided insufficient stabilization to the zero-valent Pd which precipitated in its metallic form, terminating the catalytic process. Trialkylphosphines, such as Cy_3P, Bu_3P, and i-Pr_3P, successfully stabilize Pd(0) toward precipitation and do not exchange their alkyl groups with aryls of the ArX substrates. Unfortunately, these phosphines are good promoters for the oxidative coupling of the olefinic substrates, which complicates and suppresses the desired Heck arylation [126].

$$Ph_3P\text{-}Pd(X)(Ar)\text{-}PPh_3 \longrightarrow ArPh_2P\text{-}Pd(X)(Ph)\text{-}PPh_3 \qquad (17)$$

The detailed study by Milstein and coworkers [127] revealed a number of mechanistic features of the Heck reaction catalyzed by electron-rich phosphine Pd complexes capable of activating the C–Cl bond under mild conditions. Originally it had been found that $[(dippp)_2Pd]$ oxidatively added the C–Cl bond in PhCl under mild conditions (Sect. 2.2), efficiently catalyzing carbonylation (Sect. 3.3) and reductive dechlorination (Sect. 3.1) reactions of chlorobenzene. Surprisingly, this complex exhibited practically no catalytic activity in the Heck reaction between styrene and PhCl [123, 127]. At the same time, good yields of stilbene were obtained when the dippp ligand was replaced by dippb containing one more methylene link in between the two phosphorus atoms. The reason for such a dramatic change in the catalytic activity is the exceptionally strong influence of the chelate effect on a number of steps of the catalytic cycle (Scheme 4). It was found that for complexes with strongly chelating phosphines, [(dippp) Pd(X)Ph] (X=Cl, Br) and [(dippe)Pd(Cl)Ph], halide dissociation, followed by the rate-limiting alkene insertion are involved. Rates of these processes are strongly solvent dependent, the fastest reaction being observed in DMF favoring ionization of the Pd–Cl bond. Addition of Cl^- slowed down the olefin insertion step. On the contrary, no cleavage of the Pd-X bond but rather phosphine dissociation is involved when complexes of monodentate phosphines, $[(i\text{-}Pr_2BuP)_2Pd(X)Ph]$ (X=Cl, Br), are reacted with olefins. The weak chelating properties of dippb placed it in the position between that of dippp and monodentate phosphines, allowing for "the lowest resistance pathway" and hence most efficient catalysis [127]. Various side-reactions were found to complicate the process and deactivate the catalyst, namely the formation of biphenyls and phosphonium cations, as well as β-carbon elimination occurring when norbornene was used as the olefin. A plausible explanation was also offered [127] for the catalytic activity of the dippp/$Pd(OAc)_2$/Zn system [125]. It is unfortunate that space limitations do not permit detailed analysis and discussion of the paper by Portnoy, Ben-David, Rousso, and Milstein [127]. Carefully studying this report would certainly be useful and instructive to those wishing to design an efficient catalytic system for the Pd-catalyzed Heck olefination of chloroarenes and/or use this reaction in synthesis. It has also been proposed that in some cases the catalytic cycle might involve Pd(II)/Pd(IV) rather than Pd(0)/ Pd(II) intermediates [108].

3.3 Carbonylation of Chloroarenes

As a building block, carbon monoxide is of special importance in organic synthesis on both the laboratory [15, 129, 130] and industrial [131, 132] scale. The exceedingly low cost of CO makes it especially attractive as a reagent [132]. Normally, however, organic substrates to be carbonylated are incomparably more expensive than carbon monoxide. This is especially true for organic iodides and bromides, which readily react with CO in the presence of transition metal catalysts and nucleophiles to give valuable aldehydes, ketones, carboxylic acids, anhydrides, esters, amides, lactones, and lactams [15, 129–132]. Replacing iodo and bromoarenes in the carbonylation reactions by considerably less costly aryl chlorides is highly desirable and challenging, given the notoriously poor reactivity of the C–Cl bond in nonactivated chloroarenes.

First patents on the carbonylation reactions of chloroarenes described processes requiring severe conditions [1] and will not be considered in this chapter. In the early 1970s, Cassar and Foà [26] succeeded in performing the catalytic carbonylation of both isomers of chloronaphthalene under mild conditions (P_{CO}=1 atm, T=110°C). The reaction occurred in polar solvents (DMF, dimethylacetamide, DMSO, and HMPA), in the presence of $Ca(OH)_2$ and catalytic quantities of $Ni(CO)_4$, furnishing the corresponding naphthoic acid in up to 95–97% yield. Since then, a number of electron-rich alkyl and alkoxycarbonylcobalt carbonyls, $[(CO)_4CoCH_2Y]$ (Y=H, COOMe, COOEt) [70, 133–135], and palladium complexes [136–141] have been used to carbonylate activated chloroarenes, such as chloronaphthalenes, 2-chlorofuran, 2-chlorothiophene, and various ClC_6H_4Y, where Y=SO_2NH_2, SO_2Ph, CN, CF_3, COR, etc. However, these catalytic systems were not suitable for the carbonylation of chlorobenzene, which was found to be 27, 500, and 325,000 times less reactive than 1-chloronaphthalene, bromobenzene, and iodobenzene, respectively [26]. Various chloroaromatic compounds, PhCl included, have been carbonylated to the corresponding carboxylic acids or their methyl esters in the presence of NaOH or NaOMe and cobalt carbonyl or acetate catalysts under $S_{RN}1$ conditions (photostimulation) [23, 67, 69, 142–145]. Interestingly, this approach suggested and realized by Caubère and coworkers [69] was a consequence of their previous original work in the field of the so-called "complex reducing agents" (CRAs), heterogeneous systems consisting of NaH, NaOR, and a transition metal halide or acetate [23, 66, 67]. It is worth mentioning that a mixture of sodium hydride, sodium neopentoxide, $Co(OAc)_2$, and CO, the so-called "CoCRACO," has been used for the stoichiometric carbonylation of chlorobenzene at 40% conversion [68]. An interesting heterogeneous catalytic system, Pd/C pretreated with $K_2Cr_2O_7$, has been found for the methoxycarbonylation reaction of chlorobenzene and some other aryl chlorides [146].

The first efficient, homogeneous, nonphotochemical catalytic carbonylation reactions of chlorobenzene and other nonactivated chloroarenes, proceeding under mild conditions, were reported only a decade ago. Ben-David, Portnoy, and Milstein [147–149] and Huser, Osborn, et al. [59, 150–153] discovered independently and simultaneously that palladium complexes of electron-rich bulky phosphines can catalyze the carbonylation of chlorobenzene and its derivatives. Milstein and associates [147–149] used bidentate dippp ligand, whereas Huser and Osborn [59, 150–153] employed monodentate Cy_3P and i-Pr_3P for their carbonylation reactions. In a few cases Et_3P [151], dippb [147], and bidentate phosphines containing 2-methoxyphenyl groups on the P atoms [154] also gave satisfactory results. In the presence of H_2 or sodium formate, chlorobenzene was catalytically carbonylated to benzaldehyde in nearly quantitative yield (Eq. 18) [59, 148, 149, 151]. Benzoic acid (Eq. 19) [147, 152], alkyl benzoates (Eq. 20) [59, 147, 150], and dialkylbenzamides (Eq. 21) [147, 153] were also synthesized, in high yields, from chlorobenzene and the corresponding nucleophile in the presence of Pd catalysts. The carbonylation reactions catalyzed by tricyclohexylphosphine palladium complexes were normally conducted at slightly higher temperatures and pressures (180°C, 15–30 atm) [59, 150–153] than those catalyzed by dippp (120–150°C, 4.8–5.5 atm) [147–149]. However, both the availability and lower cost of Cy_3P make it more attractive. It is not surprising, therefore, that other research groups [60, 155–157] have employed tricyclohexylphosphine complexes of palladium for the modification of the Huser-Osborn method. In particular, it has been reported that chloroarenes can be successfully converted to the corresponding acids under biphasic conditions in the presence of CO [60, 155, 157] or methyl formate [156] and $[(Cy_3P)_2PdCl_2]$. The carbonylation of nonactivated chloroarenes can be performed under as mild conditions as 100°C and an atmospheric pressure of CO [60, 155]. When HCOOMe was used instead of CO, the Pd-catalyzed reaction was promoted by $[Ru_3(CO)_{12}]$ and ammonium formate [156]. Miyawaki et al. [157] mentioned that $[(Cy_3P)_2Pd(AcO)_2]$,

$[(Cy_3P)_2Pd(acac)]$, and $[(dcpe)PdCl_2]$ also exhibited catalytic activity in the carbonylation of chloroarenes.

$$PhCl + CO + H_2 \text{ or } HCOONa \xrightarrow{[Pd]} PhCHO \quad (18)$$

$$PhCl + CO + OH^- \xrightarrow{[Pd]} PhCOO^- \quad (19)$$

$$PhCl + CO + RO^- \xrightarrow{[Pd]} PhCOOR \quad (20)$$

$$PhCl + CO + HNR_2 \xrightarrow{[Pd]} PhCONR_2 \quad (21)$$

Mechanistic aspects of the catalytic carbonylation reactions of chloroarenes merit comments. As mentioned above, $[Ni(CO)_4]$ efficiently catalyzes the hydroxycarbonylation of chloronaphthalenes [26] but not less reactive chlorobenzene because the complex is not electron-rich enough to activate the Ph–Cl bond. Much more nucleophilic tertiary phosphine complexes of Ni(0) oxidatively add the Ph–Cl bond under exceedingly mild conditions (Sect. 2.1). Moreover, the resulting organonickel compounds, $[(R_3P)_2Ni(Ph)Cl]$, readily form the carbonylated product, PhCOCl, upon treatment with carbon monoxide [45, 158]. However, the Ni(0) species emerging from this reaction are the notoriously inert carbonylphosphine complexes, $[(R_3P)_2Ni(CO)_2]$ and $[(R_3P)Ni(CO)_3]$, which are totally unreactive toward chloroarenes and even much stronger electrophiles. This is a good illustration of the disappointing absolute incompatibility of the ligand environment on the metal with only one single step of the desired catalytic cycle.

Carbonylphosphine complexes of zero-valent palladium are considerably less stable and more reactive than their Ni counterparts. Most common triphenylphosphine complexes of Pd are excellent catalysts for various carbonylation reactions of aryl iodides and bromides [15, 129–131]. It is conceivable that the palladium-catalyzed alkoxycarbonylation of ArCl proceeds via a mechanism similar to that proposed for the analogous reactions of bromo- and iodoarenes (Scheme 5) [45, 159, 160, 161].

It was recently established, however, that triphenylphosphine complexes of Pd do not catalyze the carbonylation of chlorobenzene at 180°C and 5 atm CO [157a]. Remarkably, the temperature employed for the experiments [157a] exceeded (by 40°C!) that required for the efficient oxidative addition of the Ph–Cl bond to $[(Ph_3P)_nPd]$ [44, 54]. Therefore, like in the high-temperature Heck olefination of PhCl with triphenylphosphine complexes of Pd (Sect. 3.2), the oxidative addition of the C–Cl to the metal is unlikely to be the impediment preventing the carbonylation. Let us consider how the nature of X in the substrate, ArX, might influence each step of the catalytic cycle presented in Scheme 5.

Once the Ar–X bond oxidatively adds to the Pd(0), the resulting complex, $[L_2Pd(Ar)X]$, is expected to react with CO to give $[L_2Pd(COAr)X]$. As far as the kinetics of the carbonylation is concerned, the rate constant was found to de-

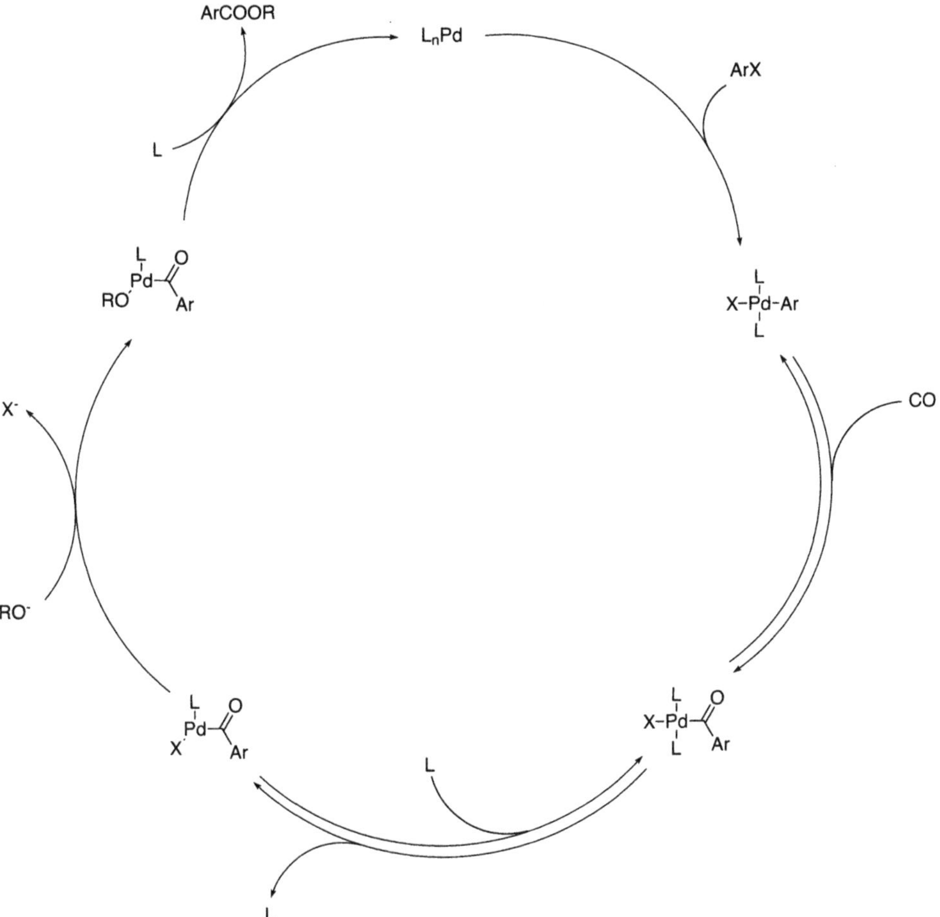

Scheme 5.

pend considerably on both L and Ar, and only slightly on the nature of X [45]. For L=PPh_3, Ar=p-$NO_2C_6H_4$, and X=Cl, Br, and I the observed rate constant ratio was found to be 1:1.6:9.3. Moreover, large amounts of halide anions did not seem to have a significant effect on the carbonylation. Remarkably, however, the thermodynamics of the reaction between $[(Ph_3P)_2Pd(Ph)X]$ and CO was strongly affected by the nature of X. Both the iodo and bromo palladium phenyls readily underwent quantitative conversion to the corresponding benzoyl complexes at room temperature and atmospheric pressure of CO. In contrast, the chloro analogue reacted with CO, under identical conditions, *reversibly*, with the conversion reached being only ca. 50% [45]! Likewise, the carbonylation of $[(Cy_3P)_2Pd(Ph)Cl]$ required 30 bar CO at room temperature and was *reversible* at 60°C under argon [59]. It is still poorly understood why the chloro complexes behaved so differently from their bromo and iodo counterparts. The diminished stability of the benzoyl palladium *chlorides* might be rationalized in terms of the chloro ligand being much stronger a π-base than Br and I. Therefore, filled/filled d_{π}/p_{π} repulsions [162] between the metal and X in $[(Ph_3P)_2Pd(COPh)X]$ should also be stronger for X=Cl, resulting in destabilization of the Pd-Cl bond and its ionization, followed by decarbonylation of the complex. This rationale is ideally consistent with the fact that the carbonylation of $[(Ph_3P)_2Pd(p\text{-}C_6H_4Y)Cl]$, where Y is a strong electron-withdrawing group (e.g., NO_2 or CN), occurred *quantitatively* and *irreversibly*, although at a slower rate, as compared to the analogous σ-phenyl chloro complex [45]. At the high temperatures required for the oxidative addition of PhCl to the $[(Ph_3P)_nPd]$, the equilibrium between $[(Ph_3P)_2Pd(COPh)Cl]$ and $[(Ph_3P)_2Pd(Ph)Cl]$ may be shifted entirely to the latter, shutting down the catalytic reaction. It is worth noting that the push-pull-type mechanism of stabilization of the Pd-Cl bond in $[(Ph_3P)_2Pd(Ph)Cl]$ [163] may not work for $[(Ph_3P)_2Pd(COPh)Cl]$, in which the phenyl ligand is separated from the metal by the inserted molecule of CO.

The last step of the catalytic cycle (Scheme 5), which may be influenced by the nature of the halogen, is the replacement of X by the alkoxide anion. It was established [159] that the reaction of the benzoyl intermediate, $[(Ph_3P)_2 Pd(COPh)X]$, with EtOH in the presence of Et_3N proceeded almost equally fast for X=Cl, Br, and I. Under identical conditions, the observed rate constant ratio was measured to be 3.75:3.46:3.25 for X=I, Br, and Cl, respectively. On the other hand, in anhydrous media of low polarity the affinity of Pd(II) for halide anions increases in the order I<Br<Cl<F [163], suggesting that under certain circumstances Pd(II) chloro complexes might be less prone to undergoing ligand exchange than their bromo and iodo analogues. It is clear from the mechanistic studies and considerations just described that some steps in the catalytic cycle (Scheme 5) might be even more problematic than the C–Cl activation step. Indeed, Huser, Youinou, and Osborn [59] found that the oxidative addition of the C–Cl bond is not the rate-limiting step in the carbonylation of chlorobenzene, catalyzed by tricyclohexylphosphine palladium complexes. Although considerable progress has been made toward elucidating the mechanism of the Pd-catalyzed haloarenes, ArCl included, important pieces of information are still missing, crying out for more research. For example, dinuclear complexes of palladium [8c, 60, 164] possibly play an important role in the catalytic carbonylation process, which may appear to be much more complex than the oversimplified model presented in Scheme 5.

3.4
Homocoupling and Cross-Coupling Reactions

The nickel-catalyzed reductive homocoupling of haloarenes is an excellent method for the synthesis of symmetrical biaryls (Eq. 22). The remarkable ability of $[Ni(cod)_2]$ to promote the stoichiometric formation of biaryls from aryl iodides and bromides was originally reported by Semmelhack, Helquist, and Jones [25a]. Unlike the Ullmann reaction [3], the Ni-promoted homocoupling of haloarenes occurs under exceedingly mild conditions, furnishing biaryls in excellent yields. The reaction can be rendered catalytic in nickel if conducted in the presence of a tertiary phosphine as a ligand and an active metal, usually zinc as the reductant.

$$2ArCl + Zn \xrightarrow{[L_nNi]} Ar\text{-}Ar + ZnCl_2 \tag{22}$$

An informative and concise review of the synthetic and mechanistic aspects of the Ni-catalyzed homocoupling of aromatic substrates has recently been published by Percec and Hill [41]. The first efficient procedure for the Ni-catalyzed coupling of chloroarenes was developed by Colon and associates [165]. To avoid the reduction of the substrate [166] only rigorously anhydrous solvents (DMF or dimethylacetamide) were used for the reaction, which was promoted by halide anions, especially I^- and Br^- [165]. Adding 2,2'-bipyridine was beneficial to the homocoupling, resulting in higher selectivity and yields. Iyoda et al. [167] later found that an efficient catalyst formed when $[(Ph_3P)_2NiBr_2]$, $[Et_4N]^+ I^-$, and Zn were mixed in anhydrous THF. The thus prepared system successfully catalyzed the homocoupling of various para- and meta-substituted chloroarenes and chlorinated N-heterocycles at 50°C. Aryl chlorides bearing a substituent ortho to Cl exhibited poor reactivity, although analogous bromides and iodides gave the corresponding 2,2'-disubstituted biphenyls in 56–90% yield [167]. The catalytic homocoupling of chloroarenes can also be performed electrochemically [168]. Various mechanisms involving Ni(0), Ni(I), Ni(II), and Ni(III) complexes have been reported for the catalytic reductive homocoupling of haloarenes [41, 56, 169].

A number of cross-coupling reactions of organic halides have been developed for the C–C bond formation [1, 15, 17, 36–42]. Although palladium compounds are widely used for such reactions of ArI and ArBr, the cross-coupling processes involving nonactivated chloroarenes are catalyzed by Ni complexes which are much more reactive toward the C–Cl bond (Sects. 2.1, 2.2). Since the early 1970s [1], tertiary phosphine complexes of nickel have been widely used to catalyze the cross-coupling of chloroarenes with Grignard reagents (Eq. 23). Although the reaction between ArCl and RMgX is well known and widely explored, basic research in this field has continued in the 1990s [170], resulting in the development of new Ni catalytic systems devoid of tertiary phosphines, as well as new exciting applications of the method. The Reformatsky reagent and other organozinc compounds have been reported to undergo the Ni-catalyzed cross-coupling with nonactivated chloroarenes, affording desired products in excellent yield [171].

$$ArCl + RMgX \xrightarrow{[L_nNi]} Ar\text{-}R + MgXCl \tag{23}$$

The Miyaura-Suzuki reaction (Eq. 24) [40] normally employs iodo and bromoarenes as substrates and Pd complexes as catalysts. Activated chloroarenes and electron-deficient chlorinated heterocycles also react [9c, 172], but the reaction of chlorobenzene with boronic acids is sluggish if it occurs at all. In particular, $[(Ph_3P)_4Pd]$ does not catalyze the coupling of PhCl with boronic acids. Even in the presence of $[(dppb)PdCl_2]$ (dppb=1,4-bis(diphenylphosphino)butane), a much more active catalyst, the reaction between chlorobenzene and $PhB(OH)_2$ affords biphenyl in only 28% yield [172a].

$$ArX + PhB(OH)_2 \xrightarrow{[L_nPd]} Ar\text{-}Ph \tag{24}$$

Only very recently, Miyaura and associates [173a,b] and later Indolese [173c] reported that monochlorinated benzene, toluene, anisole, aniline, phenylacetamide, and other chloroarenes can be smoothly arylated with arylboronic acids to give corresponding biphenyls in high yield. The reaction is catalyzed by $[(dppf)NiCl_2]$ (dppf=1,1'-bis(diphenylphosphino)ferrocene) in the presence of potassium phosphate, and occurs in dioxane at 80–95°C. Interestingly, both groups [173] adopted the catalytic conditions (dioxane, $[(dppf)NiCl_2]$, K_3PO_4, Zn) originally developed by Percec, Bae, and Hill for the coupling of arenesulfonates with arylboronic acids [174]. To activate the catalyst, Miyaura's group [173a,b] successfully used BuLi instead of Zn. Indolese [173c] found that the Ni catalyst did not need a special activator at all. Moreover, in terms of catalytic turnover numbers the catalyst performed an order of magnitude better when neither Zn nor BuLi was used [173c].

Bykov, Bumagin and Beletskaya [175] recently reported that so-called "ligand-free" palladium efficiently catalyzed the cross-coupling reaction of tetraphenylborate anion with a variety of chloroarenes, such as chlorophenols, chlorobenzoic acids, *p*-chloroacetophenone, and *p*-chloroanisole. These reactions smoothly occurred in water or aqueous DMF at 100–140°C in the presence of NaOH and 1–5% $PdCl_2$, to give the corresponding biaryls in 40–95% yield. Under optimized conditions, each BPh_4^- anion donated ca. two phenyl groups for the coupling, which was normally accompanied by precipitation of Pd black. To suppress the formation of Pd metal some of the reactions were run in the presence of $K_2Cr_2O_7$ (10 mol %) as the reoxidant. The "ligand-free" palladium systems were inefficient in the Stille coupling of $PhSnMe_3$ with ArCl [175]. At the same time, tertiary phosphine complexes of Ni and Pd catalyzed cross-coupling reactions of activated chloroarenes with organotin [176] and organosilicon [177] compounds. Although palladium complexes of electron-rich $i\text{-}Pr_3P$ and dcpe were used as the catalysts for the Si-coupling, *p*-chlorotoluene and *p*-chloroanisole remained quite unreactive, failing to give satisfactory yields of the desired products [177].

3.5 Nucleophilic Substitution

Finding ways to make nonactivated haloarenes susceptible to nucleophilic attack has always been a great challenge for chemists. Catalysis with transition metal complexes has proven to be efficient in activating inert aryl-halogen bonds and performing various nucleophilic displacement reactions (Eq. 25) [178]. The coupling reactions of ArX (Sect. 3.4) are one of the types of metal-assisted aromatic nucleophilic displacement reactions employing various organometallic compounds as precursors or synthons of carbanionic nucleophiles. Copper reagents have been widely used to promote S_NAr reactions of aryl bromides and iodides [3]. In recent years, considerable progress has been made toward the development of copper-catalyzed processes involving chlorobenzene, such as the Cu-catalyzed phenylation of alkoxide anions [179], aryloxide anions [180], and ammonia [181] to give aryl ethers and aniline, respectively. In most instances, however, nickel catalysts have been utilized for the homogeneous or phase-transfer promoted S_N-type reactions of chloroarenes.

$$\mathrm{ArX} + \mathrm{Nu} \xrightarrow{[\mathrm{L_nM}]} \mathrm{ArNu} \tag{25}$$

$$\mathrm{Nu} = \mathrm{RO^-, RS^-, CN^-, NH_3, RNH_2, R_2NH,\ etc.}$$

The most extensively explored nucleophilic displacement reaction of haloarenes is the substitution by CN^- [182]. Since 25 years ago, when Cassar [183] communicated the synthesis of ArCN from the corresponding ArCl and NaCN in the presence of $[(Ph_3P)_4Ni]$ as the catalyst, this reaction has been intensively studied under both homogeneous [184] and phase-transfer [137, 185] conditions. Sakakibara and coworkers [186] reported that performing the Ni-catalyzed cyanation of chloroarenes in polar aprotic solvents (MeCN or HMPA) is advantageous in terms of both yield and reproducibility. As anticipated, cobalt catalysts were considerably less active in the cyanation of chloroarenes. In fact, the cobalt-catalyzed reaction of 1-chloronaphthalene with CN^- furnished 1-naphthyl cyanide in only 18% yield [187], despite the fact that chloronaphthalenes are much more reactive than chlorobenzene. Although very efficient in the cyanation of aryl iodides, bromides, and triflates [188], palladium complexes normally do not catalyze the reaction of ArCl with CN^-, unless the carbon-chlorine bond is activated [189]. However, Andersson and Långström [8f] recently reported that ^{11}C–labeled benzonitrile formed in 45% radiochemical yield from the reaction of chlorobenzene with $K^{11}CN$ in THF, catalyzed by $[(Ph_3P)_4Pd]$. The reaction was surprisingly efficient and fast, giving the product within 5 min at 90°C, whereas normally the oxidation addition of PhCl to $[(Ph_3P)_4Pd]$ requires hours at 140°C [44, 54] (Sect. 2.2). An alternative way for the synthesis of benzonitrile from chlorobenzene is the Pd- or Ni-catalyzed reaction between PhCl and organic or inorganic cyanates in the presence of CO [190].

A variety of other nucleophiles have been used for the metal-catalyzed displacement of chlorine in nonactivated chloroarenes, including arylthiolate [191–193] and iodide [194, 195] anions, primary and secondary amines [196, 197], tertiary phosphines [198, 199], and aminophosphines [200]. All these reactions are catalyzed by either preformed or generated in situ Ni(0) complexes. Very recently, however, Reddy and Tanaka [201] and Koie et al. [202] reported the arylation of secondary amines with chlorobenzene and other chloroarenes, catalyzed by palladium complexes containing bulky, electron-rich phosphines,

Cy_3P, i-Pr_3P [201], and t-Bu_3P [202]. These findings are of exceptional importance since electron-rich complexes of palladium have never been successfully used before for catalysis of S_N reactions of nonactivated chloroarenes. It is also remarkable that the Pd/t-Bu_3P system exhibited catalytic activity in the amination reaction [202], while failing to catalyze the carbonylation of chlorobenzene [59].

A totally different approach to metal-catalyzed S_NAr reactions of chlorobenzene involves reversible π-coordination of the metal to PhCl, leading to the increase in electron deficiency of the benzene ring, sufficient for nucleophilic displacement of chlorine via the Meisenheimer-type path (Sect. 2.4) [85, 86, 203]. This way, anisole can be prepared from chlorobenzene in a catalytic manner [86, 203], although with very low catalytic turnover numbers of 2–6 (Eq. 26).

$$\text{PhCl} + \text{MeOH} \xrightarrow[\text{-HCl}]{[(\text{MeOH})_n\text{Rh}(\text{C}_5\text{Me}_4\text{Et})]^{2+}} \text{PhOMe} \tag{26}$$

4 Recent Progress, Conclusions, and Perspectives

The area of catalytic activation of most unreactive C–Cl bonds has flourished tremendously over the last decade. Because of its considerable practical importance the field keeps growing at an impressive pace. Numerous novel techniques have been developed for the synthesis of various functionalized aromatic compounds from the corresponding chloroarenes. A series of new palladium, nickel, and rhodium catalysts have been synthesized for C–Cl activation and much information has been accrued on the mechanism of catalysis with these complexes.

Most recently, already after completion of our work on this Chapter, a number of new interesting reports appeared in the literature. Novel Fe [204] and Re [205] systems for stoichiometric Ar–Cl activation were reported. Reetz and associates [206] claimed a simple catalytic system, $[Pd(MeCN)_2Cl_2]/[Ph_4P]^+X^-$, for the Heck phenylation of styrene with PhCl. This work and the communication by Andersson and Långström [8f] are the only two reports that claim an uncommonly efficient catalytic C–Cl functionalization of nonactivated chloroarenes with Pd complexes devoid of bulky, basic phosphine ligands. Beller and his group [207] successfully applied triphenylphosphine and even less electron-rich phosphite complexes of Pd for various catalytic reactions of *activated* chloroarenes, such as 4-chlorobenzotrifluoride. The oxygenative cleavage of chlorocatechols with O_2 in the presence of tris(2-pyridylmethyl)amine complexes of Fe (III) was described by Funabiki et al. [208]. Cheng's group [209] reported the unprecedented synthesis of 2-methylbenzonitrile from 2-chlorotoluene, MeCN, and Zn in the presence of a Ni catalyst and similar reactions of aryl bromides, catalyzed by Pd complexes [209]. Palladium-grafted molecular sieves were found to catalyze the Heck phenylation of butyl acrylate with PhCl, albeit both selectivity and conversion were low [210]. Following their original finding of the active t-Bu_3P/Pd system [202], Nishiyama, Yamamoto and Koie reported a number of useful amination reactions of nonactivated chloroarenes [211]. Palladium complexes of electron-rich, bulky monophosphine [212], diphosphine [213] and aminophosphine [214] ligands were used for the Suzuki-Miyaura coupling and amination reactions of electron-rich aryl chlorides. Most recently, the Koie t-Bu_3P/Pd system [202, 211] was adopted, with much success, for catalysis of the Suzuki-Miyaura [215a] and Heck [215b] reactions. A "NiCRACO"-type [23, 66, 67] system was used for the amination of nonactivated chloroarenes [216]. Chlorobiphenyls were obtained via photolysis of PhCl in the presence of a Pd complex [217]. A novel $PdCl_2/MCl_3$ (M = Al or Ga) system was used to carbonylate chlorobenzene to 4-chlorobenzophenone and/or benzoyl chloride [218].

The rapidly growing area of catalytic activation of chloroarenes is still full of challenging problems to be solved in the years to come. Among them are the arylation of terminal acetylenes with nonactivated chloroarenes, finding new efficient catalysts for and studying the mechanism of the Heck olefination of chloroarenes, and the development of new Pd- and Rh-catalyzed S_NAr processes, to name just a few. Regarding chloro-derivatives of nonaromatic hydrocarbons, activation of the poorly reactive C–Cl bond [219] of readily available gem-dichlorocyclopropanes would certainly be an important and likely rewarding area for organometallic and catalytic research.

References

1. Grushin VV, Alper H (1994) Chem Rev 94:1047
2. Rossi RA, de Rossi RH (1983) Aromatic substitution by the $S_{RN}1$ mechanism. ACS, Washington DC
3. (a) Lindley J (1984) Tetrahedron 40:1433. (b) Sainsbury M (1980) Tetrahedron 36:3327. (c) Fanta PE (1974) Synthesis 9. (d) Couture C, Paine AJ (1985) Can J Chem 63:111
4. Miller J (1968) Aromatic nucleophilic substitution. Elsevier, New York
5. March J (1985) Advanced organic chemistry, 3rd edn. John Wiley and Sons, New York
6. (a) Grushin VV (1992) Acc Chem Res 25:529. (b) Grushin VV, Demkina II, Tolstaya TP (1991) Inorg Chem 30:1760. (c) Tolstaya TP, Demkina II, Grushin VV, Vanchikov AN (1989) J Org Chem USSR 25:2305. (d) Grushin VV, Kantor MM, Tolstaya TP, Shcherbina TM (1984) Bull Acad Sci USSR Div Chem Sci 33:2130
7. (a) Kalinin VN (1987) Russ Chem Rev 56:682. (b) Balas L, Jhurry D, Latxague L, Grelier S, Morel Y, Hamdani M, Ardoin N, Astruc D (1990) Bull Soc Chim Fr 401
8. (a) Mutin R, Lucas C, Thivolle-Cazat J, Dufaud V, Dany F, Basset J-M (1988) J Chem Soc Chem Commun 896. (b) Dany F, Mutin R, Lucas C, Dufaud V, Thivolle-Cazat J, Basset J-M (1989) J Mol Catal 51:L15. (c) Dufaud V, Thivolle-Cazat J, Basset J-M, Mathiew R, Jaud Z, Waissermann J (1991) Organometallics 10:4005. (d) Carpentier JF, Petit F, Mortreux A, Dufaud V, Basset J-M, Thivolle-Cazat J (1993) J Mol Catal 81:1. (e) Carpentier J-F, Finet E, Castanet Y, Brocard J, Mortreux A (1994) Tetrahedron Lett 35:4995. (f) Andersson Y, Långström B (1994) J Chem Soc Perkin Trans 1 1395. (g) Uemura M, Nishimura H, Kamikawa K, Nakayama K, Hayashi Y (1994) Tetrahedron Lett 35:1909. (h) Uemura M, Nishimura H, Hayashi T (1994) J Organomet Chem 473:129. (i) Caldirola P, Chowdhury R, Johansson AM, Hacksell U (1995) Organometallics 14:3897. (j) Kiji J (1996) Macromol Symp 105:167
9. (a) Illuminati G (1964) Adv Heterocycl Chem 3:285. (b) Oae S, Furukawa N (1989) Adv Heterocycl Chem 48:1. (c) Zoltewicz JA, Cruskie, Jr MP, Dill CD (1995) J Org Chem 60:264

10. Krespan CG (1995) In: Hudlický M, Pavlath AE (eds) Chemistry of organic fluorine compounds II. ACS, Washington DC, p 297
11. Lunin VV, Lokteva ES (1996) Russ Chem Bull 45:1519
12. Dehmlow EV, Dehmlow SS (1983) Phase transfer catalysis. Verlag Chemie, Weinheim
13. (a) Bekkevol S, Svorstoel I, Hoeiland H, Songstad J (1983) Acta Chem Scand B 37:935. (b) Fanning JC, Keefer LK (1987) J Chem Soc Chem Commun 955. (c) Grushin VV (1998) Angew Chem Int Ed Engl 37:994
14. Zefirov NS, Makhon'kov DA (1982) Chem Rev 82:615
15. Tsuji J (1995) Palladium reagents and catalysis: innovations in organic synthesis. Wiley, Chichester
16. (a) Reetz MT, Wanninger K, Hermes M (1997) Chem Commun 535. (b) Voigt K, Schik U, Meyer FE, de Meijere A (1994) Synlett 189. (d) Horino H, Inone N, Asao T (1981) Tetrahedron Lett 22:741
17. Rossi R, Carpita A, Bellina F (1995) Org Prep Proc Int 27:129
18. (a) Ciriano MA, Tena MA, Oro LA (1992) J Chem Soc Dalton Trans 2123. (b) Tejel C, Ciriano MA, Oro LA, Tiripicchio A, Ugozzoli F (1994) Organometallics 13:4153. (c) Kiplinger JL, Richmond TG (1996) Polyhedron 16:409. (d) Aulwurm UR, Knoch F, Kisch H (1996) Z Naturforsch B Chem Sci (1996) 51:1555. (e) Crespo M, Solans X, Font-Bardia M (1996) J Organomet Chem 518:105. (f) Crespo M, Grande C, Klein A, Font-Bardia M, Solans X (1998) J Organomet Chem 563:179. (g) Santra BK, Lahiri GK (1998) J Chem Soc Dalton Trans 1613
19. (a) Portnoy M, Ben-David Y, Milstein D (1995) J Organomet Chem 503:149. (b) Baar CR, Hill GS, Vittal JJ, Puddephatt RJ (1998) Organometallics 17:32
20. (a) Haarman HF, Ernsting JM, Kranenburg M, Kooijman H, Veldman N, Spek AL, van Leeuwen PWNM, Vrieze K (1997) Organometallics 16:887. (b) Le Bras J, Amouri H, Vaissermann J (1997) J Organomet Chem 548:305
21. (a) Rondon D, He X-D, Chaudret B (1992) J Organomet Chem 433:C18. (b) Rondon D, Delbeau J, He X-D, Sabo-Etienne S, Chaudret B (1994) J Chem Soc Dalton Trans 1895
22. Peng T-S, Winter CH, Gladysz JA (1994) Inorg Chem 33:2534
23. Caubère P (1991) Rev Heteroatom Chem 4:78
24. Kunishima M, Hioki K, Kono K, Sakuma T, Tani S (1994) Chem Pharm Bull 42:2190
25. (a) Semmelhack MF, Helquist PM, Jones LD (1971) J Am Chem Soc 93:5908. (b) Wenschuh E, Zimmering R (1987) Z Chem 27:448
26. Cassar L, Foà M (1973) J Organomet Chem 51:381
27. Gerlach DH, Kane AR, Parshall GW, Jesson JP, Muetterties EL (1971) J Am Chem Soc 93:3543
28. Morvillo A, Turco A (1981) J Organomet Chem 208:103
29. Fahey DR (1970) J Am Chem Soc 92:402
30. Fahey DR, Mahan JE (1977) J Am Chem Soc 99:2501
31. Hidai M, Kashiwagi T, Ikeuchi T, Uchida Y (1971) J Organomet Chem 30:279
32. Fóa M, Cassar L (1975) J Chem Soc Dalton Trans 2572
33. Tsou TT, Kochi JK (1979) J Am Chem Soc 101:6319
34. Osborn JA (1975) In: Ishii Y, Tsutsui M (eds) Prospects in organotransition-metal chemistry. Plenum Press, New York, p 65
35. Portnoy M, Milstein D (1993) Organometallics 12:1665
36. Kalinin VN (1992) Synthesis 413
37. de Meijere A, Meyer F (1994) Angew Chem Int Ed Engl 33:2379
38. Cabri W, Caudiani I (1995) Acc Chem Res 28:2
39. Tsuji J, Mandai T (1995) Angew Chem Int Ed Engl 34:2589
40. Miyaura N, Suzuki A (1995) Chem Rev 95:2457
41. Percec V, Hill HH (1996) ACS Symp Ser 624:2

42. (a) Beletskaya IP, Cheprakov AV (1998) In: Grieco PA (ed) Organic synthesis in water. Blackie Academic & Professional, London, p 141. (b) Beletskaya IP (1997) Pure Appl Chem 69:471
43. Coulson DR (1968) J Chem Soc Chem Commun 1530
44. Fitton P, Rick EA (1971) J Organomet Chem 28:287
45. Garrou PE, Heck RF (1976) J Am Chem Soc 98:4115
46. Fauvarque J-F, Pflüger F, Troupel M (1981) J Organomet Chem 208:419
47. Caspar JV (1985) J Am Chem Soc 107:6718
48. Amatore C, Pflüger F (1990) Organometallics 9:2276
49. Amatore C, Azzabi M, Jutand A (1991) J Am Chem Soc 113:8375
50. Amatore C, Jutand A, Khalil F, M'Barki MA, Mottier L (1993) Organometallics 12:3168
51. Amatore C, Jutand A, Suarez A (1993) J Am Chem Soc 115:9531
52. (a) Grushin VV, Alper H (1993) Organometallics 12:3846. (b) Wallow TI, Goodson FE, Novak BM (1996) Organometallics 15:3708
53. Amatore C, Carré E, Jutand A, Tanaka H, Ren Q, Torii S (1997) Chem Eur J 2:957
54. Herrmann WA, Broßmer C, Priermeier T, Öfele K (1994) J Organomet Chem 481:97
55. (a) Kong K-C, Cheng C-H (1991) J Am Chem Soc 113:6313. (b) Goodson FE, Wallow TI, Novak BM (1997) J Am Chem Soc 119:12441 and references cited therein
56. (a) Amatore C, Jutand A (1988) Organometallics 7:2203. (b) Amatore C, Jutand A (1990) Acta Chem Scand 44:755
57. Amatore C, Jutand A, Khalil F, Nielsen MF (1992) J Am Chem Soc 114:7076
58. Parshall GW (1974) J Am Chem Soc 96:2360
59. Huser M, Youinou M-T, Osborn JA (1989) Angew Chem Int Ed Engl 28:1386
60. Grushin VV, Alper H (1993) Organometallics 12:1890
61. Grushin VV, Alper H (1995) J Am Chem Soc 117:4305
62. Portnoy M, Milstein D (1993) Organometallics 12:1655
63. Pan Y, Mague T, Fink MJ (1993) J Am Chem Soc 115:3842
64. Widenhoefer RA, Zhong HA, Buchwald SL (1996) Organometallics 15:2745
65. Harvey PD, Provencher R, Gagnon J, Zhang T, Fortin D, Hierso K, Drouin M, Socol SM (1996) Can J Chem 74:2268
66. Caubère P (1983) Angew Chem Int Ed Engl 22:599
67. Caubère P (1985) Pure Appl Chem 57:1875
68. Brunet J-J, Sidot C, Loubinoux B, Caubère P (1979) J Org Chem 44:2199
69. Brunet J-J, Sidot C, Caubère P (1983) J Org Chem 48:1166
70. Foà M, Francalanci F (1987) J Mol Catal 41:89
71. Pilloni G, Valcher S, Martelli M (1972) J Electroanal Chem 40:63
72. Zecchin S, Schiavon G, Pilloni G, Martelli M (1976) J Organomet Chem 110:C45
73. Love CJ, McQuillin FJ (1973) J Chem Soc Perkin Trans 1 2509
74. Navazio G, Sandrini PL, Troilo G (1977) Atti Ist Veneto Sci Lett Arti Cl Sci Mat Nat 135:67; Chem Abstr 90:87639v
75. (a) Baker RT (1994) US Patent 5,300,712; Chem Abstr 121:56984 h. (b) Baker RT (1994) US Patent 5,326,914; Chem Abstr 123:55371 k
76. (a) Cho O-J, Lee I-M, Park K-Y, Kim H-S (1995) J Fluorine Chem 71:107. (b)Kim H-S, Cho O-J, Lee I-M, Hong S-P, Kwag C-Y, Ahn B-S (1996) J Mol Catal A 111:49
77. Grushin VV, Alper H (1991) Organometallics 10:1620
78. Grushin VV (1993) Acc Chem Res 26:279
79. Grushin VV, Vymenits AB, Vol'pin ME (1990) J Organomet Chem 382:185
80. Gusev DG, Bakhmutov VI, Grushin VV, Vol'pin ME (1990) Inorg Chim Acta 175:19
81. Gusev DG, Bakhmutov VI, Grushin VV, Vol'pin ME (1990) Inorg Chim Acta 177:115
82. Albinati A, Bakhmutov VI, Caulton KG, Clot E, Eckert J, Eisenstein O, Gusev DG, Grushin VV, Hauger BE, Klooster W, Koetzle TF, McMullan RK, O'Loughlin TJ, Pelissier M, Ricci JS, Sigalas MP, Vymenits AB (1993) J Am Chem Soc 115:7300

83. (a) Grushin VV, Vymenits AB, Vol'pin ME (1990) Metalloorg Khim 3:702. (b) Young RJ Jr, Grushin VV (1999) Organometallics 18:294
84. Grushin VV, Gusev DG (1989) unpublished observations
85. Goryunov LI, Shteingarts VD (1991) Zh Org Khim 27:1144
86. Goryunov LI, Shteingarts VD (1993) Zh Org Khim 29:2230
87. Pinder AR (1980) Synthesis 425
88. Bosin TR, Raymond MG, Buckpitt AR (1973) Tetrahedron Lett 14:797
89. (a) Surrey AR, Hammer HF (1946) J Am Chem Soc 68:1244. (b) Huffman JW (1959) J Org Chem 24:1759. (c) Newman MS, Powell WH (1969) J Org Chem 34:3923. (d) Fox BA, Threlfall (1973) Org Synth 5:346
90. (a) Imai H, Nishiguchi M, Tanaka M, Fukuzumi K (1977) Chem Lett 855. (b) Imai H, Nishiguchi M, Tanaka M, Fukuzumi K (1977) J Org Chem 42:2309
91. (a) Guillaumet G, Mordenti L, Caubére P (1975) J Organomet Chem 92:43. (b) Carfagna C, Musco A, Pontellini R, Terzoni G (1989) J Mol Catal 54:L23. (c) Carfagna C, Musco A, Pontellini R, Terzoni G (1989) J Mol Catal 57:23. (d) Davydov DV, Beletskaya IP (1993) Metalloorg Khim 6:111. (e) Davydov DV, Beletskaya IP (1993) Russ Chem Bull 42:575. (f) Včelák J, Hetfleiš J (1994) Collect Czech Chem Commun 59:1645. (g) Li H, Liao S, Xu Y (1996) Chem Lett 1059. (h) Boukherroub R, Chatgilialoglu C, Manuel G (1996) Organometallics 15:1508
92. (a) Qian C, Zhu D, Gu Y (1990) J Mol Catal 63:L1. (b) Qian C, Zhu D, Gu Y (1991) J Organomet Chem 401:23. (c) Deng DL, Qian CT, Penn JH (1994) Chin Chem Lett 5:303; Chem Abstr 121:116732 m. (d) Penn JH, Deng DL, Chang TQ (1996) Chin Chem Lett 7:845; Chem Abstr 125:275307a
93. (a) Zhang Y, Liao S, Xu Y (1994) Tetrahedron Lett 35:4599 b) Kang R, Yu H (1998) Huanjing Huaxue 17:159; Chem Abstr 129:148680
94. (a) Wiener H, Blum J, Sasson Y (1991) J Org Chem 56:6145. (b) Pews RG, Hunter JE, Wehmeyer RM (1993) Tetrahedron 49:4809. (c) Balko EN, Przybylski E, Von Trentini F (1993) Appl Catal B 2:1. (d) Marques CA, Selva M, Tundo P (1993) J Chem Soc Perkin Trans 1 529. (e) Marques CA, Selva M, Tundo P (1993) J Org Chem 58:5256. (f) Marques CA, Selva M, Tundo P (1994) J Org Chem 59:3830. (g) Marques CA, Selva M, Tundo P (1995) J Org Chem 60:2430. (h) Marques CA, Rogozhnikova O, Selva M, Tundo P (1995) J Mol Catal A 96:301. (i) Sakra T, Kacetl L, Skoloudova A (1995) Chem Prum 45:118; Chem Abstr 123:313446p. (j) Rajagopal S, Spatola AF (1995) J Org Chem 60:1347. (k) Simagina VI, Litvak VV, Stoyanova IV, Yakovlev VA, Mastikhin VM, Afanasenkova IV, Likholobov VA (1996) Izv Akad Nauk Ser Khim 1391
95. (a) Sakai M, Lee M-S, Yamaguchi K, Kawai Y, Sasaki K, Sakakibara Y (1992) Bull Chem Soc Jpn 65:1739. (b) Birke P, Herda WR, Heubner U, Koppe J, Neumann U, Schoedel R (1995) Ger DE 4,320,462
96. Yanilkin VV, Maksimyuk NI, Kargin YuM (1994) Russ Chem Bull 43:957
97. (a) Stiles M (1994) J Org Chem 59:5381. (b) Scrivanti A, Vicentini B, Beghetto V, Chessa G, Matteoli U (1998) Inorg Chem Commun 1:246
98. Huser M, Osborn JA (1990) Eur Pat Appl EP 352,164; Chem Abstr 113:39527d
99. Ferrughelli DT, Horváth IT (1992) J Chem Soc Chem Commun 806
100. Bényei AC, Lehel S, Joó F (1977) J Mol Catal A 116:349
101. Akita Y, Ohta A (1981) Heterocycles 16:1325
102. Ben-David Y, Gozin M, Portnoy M, Milstein D (1992) J Mol Catal 73:173
103. (a) Amatore C, Jutand A, M'Barki MA (1992) Organometallics 11:3009. (b) Ozawa F, Kubo A, Hayashi T (1992) Chem Lett 2177. (c) Mandai T, Matsumoto T, Tsuji J, Saito S (1993) Tetrahedron Lett 34:2513. (d) Amatore C, Carré E, Jutand A, M'Barki MA (1995) Organometallics 14:1818
104. Grushin VV, Bensimon C, Alper H (1995) Organometallics 14:3259
105. Grushin VV (1996) Chem Rev 96:2011

106. (a) Heck RF (1979) Acc Chem Res 12:146. (b) Heck RF (1982) Org React 27:345. (c) Heck RF (1985) Palladium reagents in organic synthesis. Academic Press, New York. (d) Heck RF (1991) Vinyl substitutions with organopalladium intermediates. In: Trost BM, Flemming I (eds) Comprehensive organic synthesis. Pergamon Press, Oxford, vol 4, chap 4.3, p 833
107. (a) Herrmann WA, Brossmer C, Öfele K, Reisinger C-P, Priermeier T, Beller M, Fischer H (1995) Angew Chem Int Ed Engl 34:1844. (b) Beller M, Riermeier TH, Haber S, Kleiner H-J, Herrmann WA (1996) Chem Ber 129:1259. (c) Herrmann WA, Brossmer C, Reisinger C-P, Priermeier T, Öfele K, Beller M (1997) Chem Eur J 3:1357
108. Ohff M, Ohff A, van der Boom ME, Milstein D (1997) J Am Chem Soc 119:11687
109. Herrmann WA, Elison M, Fischer J, Köcher C, Artus GRJ (1995) Angew Chem Int Ed Engl 34:2371
110. Reetz MT, Breinbauer R, Wanninger K (1996) Tetrahedron Lett 37:4499
111. Reetz MT, Lohmer G (1996) Chem Commun 1921
112. Voigt K, Schik U, Meyer FE, de Meijere A (1994) Synlett 189
113. Reardon P, Metts S, Crittendon C, Daugherity P, Parsons EJ (1995) Organometallics 14:3810
114. Davison JB, Simon NM, Sojka SA (1984) J Mol Catal 22:349
115. Spencer A (1984) J Organomet Chem 270:115
116. Colon I (1982) U.S. Patent 4,334,081; Chem Abstr 97:72052u
117. Boldrini GP, Savoia D, Tagliavini E, Trombini C, Umani Ronchi A (1986) J Organomet Chem 301:C62
118. Lebedev SA, Pedchenko VV, Lopatina VS, Berestova SS, Petrov ES (1990) Zh Org Khim 26:1520; Chem Abstr 114:81104j
119. (a) Mori M, Ban Y (1976) Tetrahedon Lett 1803. (b) Mori M, Kudo S, Ban Y (1979) J Chem Soc Perkin Trans 1 771
120. (a) Olivero S, Duñach E (1994) Synlett 531. (b) Olivero S, Clinet JC, Duñach E (1995) Tetrahedron Lett 36:4429
121. Bozell JJ, Vogt CE (1988) J Am Chem Soc 110:2655
122. Mitra J, Mitra AK (1997) J Indian Chem Soc 74:146; Chem Abstr 126:250966
123. Ben-David Y, Portnoy M, Gozin M, Milstein D (1992) Organometallics 11:1995
124. Milstein D, Ben David Y (1993) PCT Int Appl WO 9301,173; Chem Abstr 118:254539 t
125. Portnoy M, Ben-David Y, Milstein D (1993) Organometallics 12:4734
126. (a) Herrmann WA, Broßmer C, Öfele K, Beller M, Fischer H (1995) J Mol Catal A 103:133. (b) Herrmann WA, Broßmer C, Öfele K, Beller M, Fischer H (1995) J Organomet Chem 491:C1
127. Portnoy M, Ben-David Y, Rousso I, Milstein D (1994) Organometallics 13:3465
128. Dieck HA, Heck RF (1974) J Am Chem Soc 96:1133
129. Colquhoun HM, Thompson DJ, Twigg MV (1991) Carbonylation. Direct synthesis of carbonyl compounds. Plenum Press, New York
130. Falbe J (1980) New synthesis with carbon monoxide. Springer, Berlin
131. Parshall GW, Ittel SD (1992) Homogeneous catalysis. The applications and chemistry of catalysis by soluble transition metal complexes. John Wiley and Sons, New York
132. Weissermel K, Arpe H-J (1997) Industrial organic chemistry, 3rd edn. VCH, New York
133. Foà M, Francalanci F, Bencini E, Gardano A (1985) J Organomet Chem 285:293
134. Foà M, Francalanci F (1986) J Organomet Chem 301:C27
135. Zhesko TE, Boyarskii VP, Beletskaya IP (1989) Metalloorg Khim 2:385; Chem Abstr 111:214202b
136. Cassar L Foà M, Gardano A (1976) J Organomet Chem 121:C55
137. Cassar L (1980) Ann NY Acad Sci 333:208
138. Sudo K, Nakasa K, Kudo M, Yamamoto M (1988) Eur Pat Appl EP 283,194; Chem Abstr 110:172884x

139. (a) Sudo K, Kudo M, Yamamoto M (1988) Jpn Kokai Tokkyo Koho JP 63,227,547 [88,227,547]; Chem Abstr 111:57304 s. (b) Sudo K, Kudo M, Yamamoto M (1988) Eur Pat Appl EP 282,266; Chem Abstr 111:133974c
140. Milstein D (1991) US Pat 5,034,534; Chem Abstr 116:6544r
141. Perry RJ, Wilson BD (1996) J Org Chem 61:7482
142. Kashimura T, Kudo K, Mori S, Sugita N (1986) Chem Lett 299
143. Kashimura T, Kudo K, Mori S, Sugita N (1986) Chem Lett 483
144. Kashimura T, Kudo K, Mori S, Sugita N (1986) Chem Lett 851
145. Kudo K, Shibata T, Kashimura T, Mori S, Sugita N (1987) Chem Lett 577
146. Dufaud V, Thivolle-Cazat J, Basset J-M (1990) J Chem Soc Chem Commun 426
147. Ben-David Y, Portnoy M, Milstein D (1989) J Am Chem Soc 111:8742
148. Ben-David Y, Portnoy M, Milstein D (1989) J Chem Soc Chem Commun 1816
149. Milstein D, Ben-David Y (1991) Eur Pat Appl EP 406,848; Chem Abstr 115:135677q
150. Huser M, Osborn JA (1990) Eur Pat Appl EP 352,167; Chem Abstr 113:40170p
151. Huser M, Osborn JA (1990) Eur Pat Appl EP 352,166; Chem Abstr 113:58687x
152. Huser M, Metz F, Osborn JA (1990) Fr Demande FR 2,637,281; Chem Abstr 113:152035e
153. Huser M, Metz F, Osborn JA (1990) Fr Demande FR 2,637,283; Chem Abstr 114:121757 h
154. Drent E (1993) Brit UK Pat Appl GB 2,261,662; Chem Abstr 119:159881 k
155. Grushin VV, Alper H (1992) J Chem Soc Chem Commun 611
156. Jenner G, Ben Taleb A (1994) J Organomet Chem 470:257
157. (a) Miyawaki T, Nomura K, Hazama M, Suzukamo G (1997) J Mol Catal A 120:L9. (b) Nomuro K, Miyawaki T (1996) Jpn Kokai Tokkyo Koho JP 08,104,661 [96,104,661]; Chem Abstr 125:86311 m
158. Corain B, Favevo G (1975) J Chem Soc Dalton Trans 283
159. Ozawa F, Kawasaki N, Okamoto H, Yamamoto T, Yamamoto A (1987) Organometallics 6:1640
160. Moser WR, Wang AW, Kildahl NK (1988) J Am Chem Soc 110:2816
161. Ozawa F, Soyama H, Yanagihara H, Aoyama I, Takino H, Izawa K, Yamamoto T, Yamamoto A (1985) J Am Chem Soc 107:3235
162. Caulton KG (1994) New J Chem 18:25
163. (a) Flemming JP, Pilon MC, Borbulevitch OYa, Antipin MYu, Grushin VV (1998) Inorg Chim Acta 280:87. (b) Amatore C, Carré E, Jutand A (1998) Acta Chem Scand 52:100
164. (a) Portnoy M, Frolow F, Milstein D (1991) Organometallics 10:3960. (b) Portnoy M, Milstein D (1994) Organometallics 13:600
165. (a) Colon I, Kelsey DR (1986) J Org Chem 51:2627. (b) Colon I, Maresca LM, Kwiatkowaski GT (1981) US Patent 4,263,466
166. Colon I (1982) J Org Chem 47:2622
167. Iyoda M, Otsuka H, Sato K, Nisato N, Oda M (1990) Bull Chem Soc Jpn 63:80
168. (a) Rollin Y, Troupel M, Tuck DG, Perichon J (1986) J Organomet Chem 303:131. (b) Mabrouk S, Pelligrini S, Folest JC, Rollin Y, Perichon J (1986) J Organomet Chem 301:391. (c) Meyer G, Rollin Y, Perichon J (1987) J Organomet Chem (1987) 333:263. (d) Meyer G, Troupel M, Perichon J (1990) J Organomet Chem 393:137
169. Yamamoto T, Wakabayashi S, Osakada K (1992) J Organomet Chem 428:223
170. (a) Bochmann M, Creaser CS, Wallage L (1990) J Mol Catal 60:344. (b) Ikoma Y, Taya F, Ozaki E-i, Higuchi S, Naoi Y, Fuji-i K (1990) Synthesis 147. (c) Qiu C, Lan Z (1993) Lanzhou Daxue Xuebao, Ziran Kexueban 29:176; Chem Abstr 122:55649m. (d) Poetsch E, Meyer V (1995) Ger Offen DE 4,326,169; Chem Abstr 122:239312c. (e) Lan Z (1996) Huaxue Shiji 18:11; Chem Abstr 125:58005w. (f) Ikoma Y, Naoi Y, Otani M, Ando K, Akiyama T, Sugimori A (1997) Nippon Kagaku Kaishi 119; Chem Abstr 126:250937
171. (a) Fauvarque JF, Jutand A (1979) J Organomet Chem 177:273. (b) Lebedev SA, Sorokina RS, Berestova SS, Petrov ES (1986) Izv Akad Nauk SSSR Ser Khim 679; Chem Abstr 106:84760r

172. (a) Mitchell MB, Wallbank PJ (1991) Tetrahedron Lett 32:2273. (b) Ali NM, McKillop A, Mitchell MB, Rebelo RA, Wallbank PJ (1992) Tetrahedron 48:8117. (c) Alcock NW, Brown JM, Hulmes DI (1993) Tetrahedron: Asymmetry 4:743. (d) Achab S, Guyot M, Potier P (1993) Tetrahedron Lett 34:2127. (e) Janietz D, Bauer M (1993) Synthesis 33. (f) Beller M, Fischer H, Herrmann WA, Öfele K, Brossmer C (1995) Angew Chem Int Ed Engl 34:1848. (g) Shen W (1997) Tetrahedron Lett 38:5575
173. (a) Saito S, Sakai M, Miyaura N (1996) Tetrahedron Lett 37:2993. (b) Saito S, Oh-tani S, Miyaura N (1997) J Org Chem 62:8024. (c) Indolese AF (1997) Tetrahedron Lett 38:3513
174. Percec V, Bae J-Y, Hill DH (1995) J Org Chem 60:1060
175. (a) Bumagin NA, Bykov VV (1996) Russ J Gen Chem 66:1925. (b) Bykov VV, Bumagin NA, Beletskaya IP (1995) Doklady Chem 340:52. (c) Bumagin NA, Bykov VV (1997) Tetrahedron 53:14437
176. Shirakawa E, Yamasaka K, Hiyama T (1997) J Chem Soc Perkin Trans 1 2449
177. (a) Gouda K-i, Hagiwara E, Hatanaka Y, Hiyama T (1996) J Org Chem 61:7232. (b) Hagiwara E, Gouda K-i, Hatanaka Y, Hiyama T (1997) Tetrahedron Lett 38:439
178. Cristau H-J, Desmurs J-R, Ratton S, Rignol S, Taillefer M (1996) Ind Chem Libr 8:90
179. (a) Keegstra MA, Brandsma L (1991) Rec Trav Chim Pays-Bas 110:299. (b) Keegstra MA, Peters THA, Brandsma L (1992) Tetrahedron 48:3633
180. (a) Muganlinskii FF, Sandler OL, Kakhramanov VB, Guseinova DD (1990) Zh Prikl Khim 63:914. (b) Shankaraiah B (1991) Indian Patent 167,934; Chem Abstr 116:214136m
181. (a) Kawahara S (1995) Jpn Kokai Tokkyo Koho JP 07,330,688 [95,330,688]; Chem Abstr 124:288966u. (b) Kawahara S (1995) Jpn Kokai Tokkyo Koho JP 07,330,689 [95,330,689]; Chem Abstr 124:288967v
182. Ellis GP, Romney-Alexander TM (1987) Chem Rev 87:779
183. Cassar L (1973) J Organomet Chem 54:C57
184. Cassar L, Ferrara S, Foà M (1974) Adv Chem Ser 132:252
185. Cassar L, Foà M, Montanari F, Marinelli GP (1979) J Organomet Chem 173:335
186. Sakakibara Y, Okuda F, Shimobayashi A, Kirino K, Sakai M, Uchino N, Takagi K (1988) Bull Chem Soc Jpn 61:1985
187. Funabiki T, Nakamura H, Yoshida S (1983) J Organomet Chem 243:95
188. Takagi K, Sasaki K, Sakakibara Y (1991) Bull Chem Soc Jpn 64:1118 and references cited therein
189. Akita Y, Shimazaki M (1981) Synthesis 974
190. Harris JF (1973) US Patent 3,755,409; Chem Abstr 79:104970p
191. Foà M, Santi R, Garavagila F (1981) J Organomet Chem 206:C29
192. Cristau HJ, Chabaud B, Chêne A, Christol H (1981) Synthesis 892
193. Takagi K (1987) Chem Lett 2221
194. Takagi K (1978) Chem Lett 191
195. Tsou TT, Kochi JK (1980) J Org Chem 45:1930
196. (a) Cramer R, Coulson DR (1975) J Org Chem 40:2267. (b) Coulson DR (1975) US Patent 3,914,311; Chem Abstr 84:43584v
197. Wolfe JP, Buchwald SL (1997) J Am Chem Soc 119:6054
198. Cassar L, Foà M (1974) J Organomet Chem 74:75
199. Allen DW, Nowell IW, March LA (1982) Tetrahedron Lett 5479
200. Krasil'nikova EA, Sentemov VV, Gavrilova EL (1993) Zh Obshch Khim 63:848
201. Reddy NP, Tanaka M (1997) Tetrahedron Lett 38:4807
202. (a) Nishiyama M, Yamamoto T, Koie Y (1997) 72nd Spring annual meeting of the Chemical Society of Japan. Tokyo, Japan. Abstract 2PA052. (b) Yamamoto T, Nishiyama M, Koie Y (1997) 72nd Spring annual meeting of the Chemical Society of Japan. Tokyo, Japan. Abstract 2PA053. (c) Nishiyama M, Koie Y (1997) Eur Pat Appl EP 802,173; Chem Abstr 127:346419 t
203. Goryunov LI, Shteingarts VD (1995) Zh Org Khim 31:472

204. Poignant G, Sinbandhit S, Toupet L, Guerchais V (1998) Angew Chem Int Ed Engl 37:963
205. Leiva C, Sutton D (1998) Organometallics 17:4568
206. Reetz MT, Lohmer G, Schwickardi R (1998) Angew Chem Int Ed Engl 37:481
207. (a) Riermeier TH, Zapf A, Beller M (1997) Top Catal 4:301. (b) Beller M, Zapf A (1998) Synlett 793
208. Funabiki T, Yamazaki T, Fukui A, Tanaka T, Yoshida S (1998) Angew Chem Int Ed Engl 37:513
209. Luo F-H, Chu C-I, Cheng C-H (1998) Organometallics 17:1025
210. Mehnert CP, Weaver DW, Ying JY (1998) J Am Chem Soc 120:12289
211. (a) Yamamoto T, Nishiyama S, Koie Y (1997) Jpn Kokai Tokkyo Koho JP 10,310,561 [97,119,477]; Chem Abstr 130:52227. (b) Yamamoto T, Nishiyama M, Koie Y (1998) Tetrahedron Lett 39:2367. (c) Nishiyama M, Yamamoto T, Koie Y (1998) Tetrahedron Lett 39:617
212. (a) Firooznia F, Gude C, Chan K, Satoh Y (1998) Tetrahedron Lett 39:3985. (b) Shen W (1997) 38:5575
213. Hamann BC, Hartwig JF (1998) J Am Chem Soc 120:7369, 12706
214. Old DW, Wolfe JP, Buchwald SL (1998) J Am Chem Soc 120:9722
215. (a) Littke AF, Fu GC (1998) Angew Chem Int Ed Engl 37:3387. (b) Littke AF, Fu GC (1999) J Org Chem 64:10
216. Brenner E, Fort Y (1998) Tetrahedron Lett 39:5359
217. Tsubomura T, Ishikura A, Hoshino K, Narita H, Sakai K (1997) Chem Lett 1171
218. Noskov YuG, Petrov ES (1998) React Kinet Catal Lett 64:359
219. Reyne F, Waegell B, Brun, P (1995) Bull Chem Soc Jpn 68:1162

Activation of the N–N Triple Bond in Molecular Nitrogen: Toward its Chemical Transformation into Organo-Nitrogen Compounds

Masanobu Hidai[a] and Yasushi Mizobe[b]

[a]Department of Chemistry and Biotechnology, Graduate School of Engineering, The University of Tokyo, Hongo, Bunkyo-ku, Tokyo 113-8656, Japan
E-mail: hidai@chembio.t.u-tokyo.ac.jp

[b]Institute of Industrial Science, The University of Tokyo, Roppongi, Minato-ku, Tokyo 106-8558, Japan
E-mail: ymizobe@cc.iis.u-tokyo.ac.jp

The reactivities of coordinated N_2 in transition metal complexes are outlined. Emphasis is placed upon the transformation of coordinated dinitrogen into organo-nitrogen compounds such as amines, azines, and pyrroles under ambient conditions, most of which are attainable for Mo and W dinitrogen complexes with tertiary phosphine coligands, e.g., *trans*-$[M(N_2)_2(Ph_2PCH_2CH_2PPh_2)_2]$ and *cis*-$[M(N_2)_2(PMe_2Ph)_4]$ (M=Mo, W). Recent findings about direct cleavage of the N–N triple bond of the bridging N_2 ligand in Mo and Nb complexes are also given. A remarkable W–Ru bimetallic system, which is capable of converting N_2 into either NH_3 by the use of H_2 gas or acetone azine by treatment with acetone under H_2, is highlighted finally as a clue to developing novel homogeneous systems for catalytic N_2 fixation.

Keywords: Dinitrogen complexes, Organo-nitrogen compounds, Nitrogen fixation, Molybdenum complexes, Tungsten complexes

1 Introduction 228

2 Reactions of Coordinated Dinitrogen 229

2.1 Electrophilic Attack on the Terminal Nitrogen 229
2.2 Radical Attack on the Terminal Nitrogen 230
2.3 Nucleophilic Attack on the Nitrogen Adjacent to Metal 232
2.4 Reactions of the Dinitrogen Bridging High-Valent Metals 232

3 C–N Bond Formation at Coordinated Dinitrogen via Hydrazido(2-) Species 233

4 Direct Cleavage of the N–N Triple Bond 236

5 Future Prospects 237

References 239

Topics in Organometallic Chemistry, Vol. 3
Volume Editor: S. Murai

1
Introduction

Nitrogen is an essential element for life. To supply the increasing demand of nitrogenous compounds, the Haber process has long been used industrially to reduce N_2 with H_2 into NH_3; the nitrogen source in all other nitrogen-containing compounds is the NH_3 produced by this process. Although N_2 is readily available in plenty from the atmosphere, the synthesis of NH_3 by the Haber process requires quite drastic conditions due to the extreme chemical inertness of N_2. Development of the alternative to this energy-consuming process, which includes the direct synthesis of organo-nitrogen compounds from N_2, has therefore long been awaited.

Since the discovery of the first stable N_2 complex $[Ru(NH_3)_5(N_2)]^{2+}$ [1], a number of N_2 complexes have been isolated and the reactivities of the N_2 ligand have been studied extensively with the aim of exploiting novel homogeneous catalysts capable of transforming N_2 into nitrogenous compounds under mild conditions [2]. Almost all of the d-block transition metals are now known to bind molecular N_2 to give fully characterized N_2 complexes if the valence state of the metal and the ligands around the metal are appropriately chosen. However, the N_2 ligand in most of these complexes tends to dissociate under certain reaction conditions and well-defined reactions converting the coordinated N_2 into nitrogen-containing compounds are still limited, although the N_2 bound to transition metal(s) seems to be more or less activated. With respect to the reactions of the N_2 ligand, the C–N bond formation leading to organo-nitrogenous compounds may be of greater importance than the N–H bond formation yielding NH_3, but has been observed less commonly than the latter [2, 3]. In this regard, the Mo and W complexes *trans*-$[M(N_2)_2(dppe)_2]$ (**1**; M=Mo, W; dppe=$Ph_2PCH_2CH_2PPh_2$] [4] and *cis*-$[M(N_2)_2(PMe_2Ph)_4]$ (**2**; M=Mo, W) [4c, 5] are quite outstanding, since these N_2 complexes readily undergo not only the N–H but also the N–C and N–Si bond formation reactions at the coordinated N_2 to give a variety of nitrogenous ligands and compounds. In this short review, recent advances in the chemistry of dinitrogen complexes are outlined, where the reactions leading to organo-nitrogen compounds will be emphasized. Extensive studies on the protonation of coordinated dinitrogen yielding ammonia or sometimes hydrazine [2, 6] are outside the scope of this article and are mostly omitted.

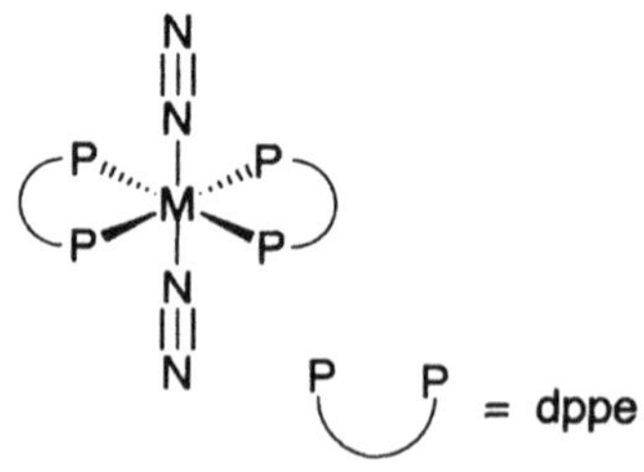

P⌒P = dppe

1 (M = Mo, W)

P = PMe_2Ph

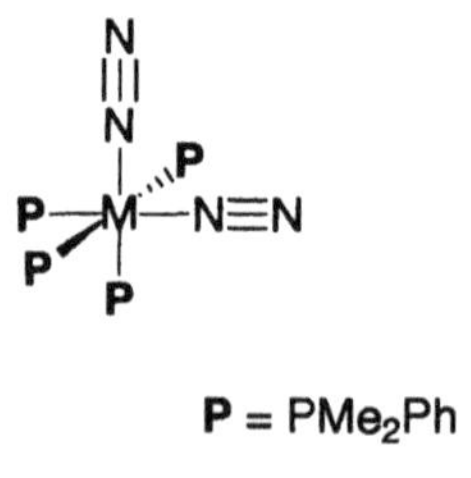

2 (M = Mo, W)

2 Reactions of Coordinated Dinitrogen

In transition metal complexes, N_2 is known to coordinate with one or more metals in several fashions, among which the terminal end-on mode is most ubiquitous as observed in **1** and **2**. Thus, the reactions of the N_2 ligand of the linear end-on type have extensively been investigated, which may be classified into three categories depending upon the nature of the attacking reagents, viz., reactions with electrophiles, radicals, or nucleophiles. These are outlined in this section, together with some examples of the formation of nitrogenous organic compounds from a bridging N_2 ligand.

2.1 Electrophilic Attack on the Terminal Nitrogen

Of the end-on N_2 ligand bound to a low-valent metal center, the terminal N atom generally carries more negative charge than the inner N atom due to the interaction of N_2 with the metal, which is interpreted in terms of both the σ-donation of the lone-pair electrons at the inner N atom to the metal and the concomitant back-donation of the metal d-electrons into the π^* orbital equally distributed over the two N atoms. Thus, the N_2 ligand binding to a highly electron-rich metal site is susceptible to electrophilic attack at the terminal N atom. Typical examples displaying such reactivities are found in Mo and W complexes **1** and **2**, which react with inorganic acids to form either a hydrazido(2-) species ($MNNH_2$) from **1** or NH_3 and less commonly N_2H_4 via an analogous hydrazido(2-) species from **2** [7].

In addition to a proton, several organic and organometallic electrophiles are also known to attack the terminal N atom to give numerous organo-nitrogen ligands. As illustrated in Scheme 1, the complexes containing the organo-diazenido (MNNR) and sometimes the organo-hydrazido(2-) (MNNRR') ligands are available by treatment of **1** and/or **2** with acid chlorides [8], silyl iodides [9], germyl iodides [9c] and $R_3SiCo(CO)_3$ [10] (Scheme 1). Analogous reactions leading to organo-nitrogen ligands have not been observed for the other N_2 complexes except the acylation and aroylation of $[ReCl(N_2)(PMe_2Ph)_4]$ [11].

As expected, arylation of the coordinated N_2 in **1** and **2** by aryl halides does not proceed. However, it has been recently found that the anionic complex having a more negatively charged N_2 ligand $[W(N_2)(NCS)(dppe)_2]^-$ undergoes arylation at the terminal N atom with the aryl fluorides activated by coordination to the $Cr(CO)_3$ or $RuCp^+$ fragment (Scheme 2) [12]. It is noteworthy that the direct arylation by aryl halides such as PhBr, PhI, and p-$MeOCOC_6H_4I$ takes place in the case of the tetrathioether complex *trans*-$[Mo(N_2)_2(L)]$ (**4**; L=3,3,7,7,11,11,15,15-octamethyl-1,5,9,13-tetrathiacyclohexadecane) [13], although the mechanism is not yet known.

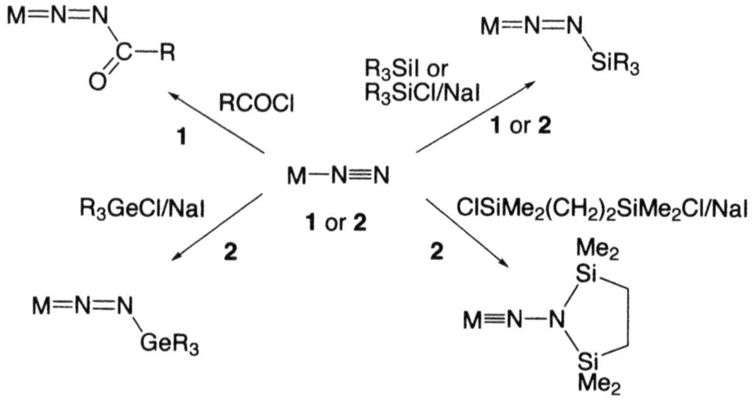

Scheme 1.

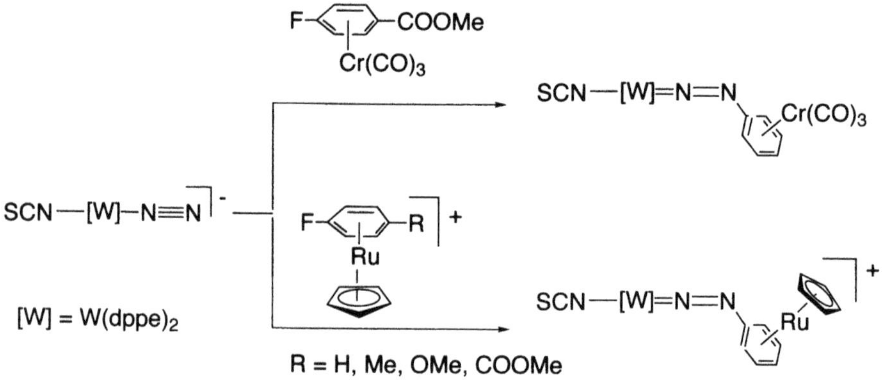

Scheme 2.

2.2
Radical Attack on the Terminal Nitrogen

The C–N bond formation reactions occur at the coordinated N_2 in the diphosphine complexes 1 by treatment with alkyl halides RX under irradiation with the W lamp, affording various diazenido complexes *trans*-[MX(NNR)(dppe)$_2$] [8, 14]. This alkylation reaction proceeds by the radical mechanism; the alkyl radical ·R is first generated by homolysis of RX around the coordination sphere and subsequently attacks at the terminal N atom. Reactivities have been investigated for some alkyldiazenido complexes thus obtained and, for example, degradation of *trans*-[MoBr(NNBu^n)(dppe)$_2$] by treatment with $NaBH_4$ or NaOMe in benzene/MeOH gives rise to the formation of butylamines together with NH_3 [14c]. Since the remote N atom in the diazenido ligand is significantly nucleophilic, reactions of 1 with α,ω-dibromoalkanes $Br(CH_2)_nBr$ (n=4, 5) result in the forma-

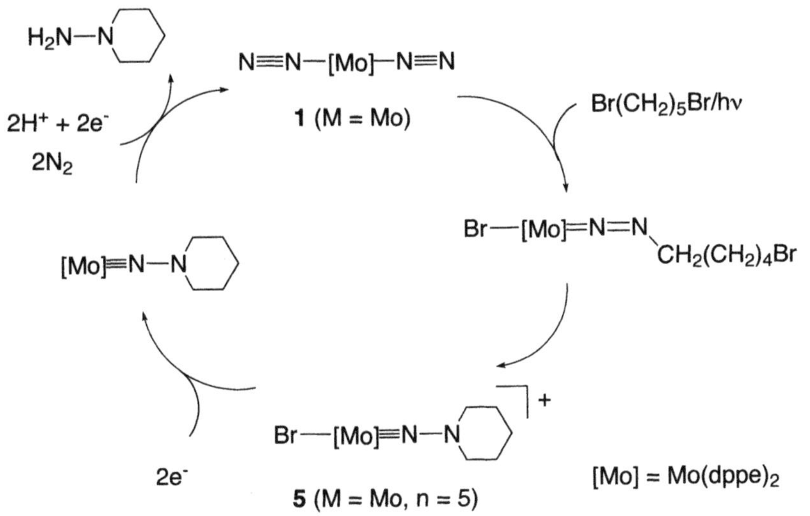

Scheme 3.

tion of dialkylhydrazido(2-) complexes *trans*-$[MBr\{N\overline{N(CH_2)_{n-1}C}H_2\}(dppe)_2]Br$ (**5**) via the initial attack of the $\cdot CH_2(CH_2)_{n-1}Br$ radical on the N_2 ligand to form a diazenido ligand and the subsequent intramolecular nucleophilic substitution of the diazenido ligand at the remote N atom (see Scheme 3) [15]. Complexes **5** (M=Mo, W; n=4) produce pyrrolidine by treatment with $LiAlH_4$ followed by workup with MeOH and then HBr [16]. More interestingly, the electroreduction of complex **5** (M=Mo; n=5) in THF releases *N*-aminopiperidine, accompanied by the recovery of the parent **1** (M=Mo) (Scheme 3) [17].

The detailed study on the silylation reaction of the N_2 ligand in **1** and **2** (see above) has led to the exploitation of the system transforming N_2 into silylamines catalytically [18]. To our knowledge, this provides the sole example of the catalytic N_2 conversion into nitrogenous compounds promoted by a well-defined N_2 complex. Thus, in the presence of **1** or **2**, reactions of the equimolar amounts of Me_3SiCl and Na under N_2 (1 atm) afford $N(SiMe_3)_3$ together with some $HN(SiMe_3)_2$. Among these complexes, **2** (M=Mo) shows the highest catalytic activity and ca. 25 mol silylamines per Mo atom has been obtained under certain conditions (Eq. 1). Although the precise mechanism of this reaction is still uncertain, the catalytic cycle might be initiated by the attack of the silyl radical on the terminal N atom, which is formed at the Mo site from Me_3SiCl and Na. As expected, the major by-product from this reaction is $Me_3SiSiMe_3$ resulting from the Wurtz type coupling.

$$Me_3SiCl + Na \xrightarrow[\text{catalyst: } \mathbf{2} \text{ (M = Mo)}]{N_2 \text{ (1 atm)}} N(SiMe_3)_3 + HN(SiMe_3)_2 + Me_3SiSiMe_3 \quad (1)$$

2.3
Nucleophilic Attack on the Nitrogen Adjacent to Metal

The significantly electron-deficient N_2 ligand in $[CpMn(CO)_2(N_2)]$ (**6**) is susceptible to nucleophilic attack at the inner N atom. Thus, treatment of **6** with MeLi affords the isolable $[CpMn(CO)_2(MeN{=}N^-Li^+)]$ in low yield, which reacts subsequently with $[Me_3O][BF_4]$ to form $[CpMn(CO)_2(MeN{=}NMe)]$ [19]. This reaction of the coordinated N_2, being reminiscent of the synthesis of Fischer carbenes from the isoelectronic CO, is quite unique in that the N_2 ligand does react with a nucleophilic reagent to give a characterizable metal species. Under pressurized N_2, the MeN=NMe ligand can be replaced by N_2 to regenerate **6** (Scheme 4) [19].

2.4
Reactions of the Dinitrogen Bridging High-Valent Metals

In contrast to the terminal end-on N_2 complexes for which the N–N bond length of around 1.12 Å is unexceptionally observed in solid state, the N–N separation clarified by the X-ray crystallography for the N_2 ligand bound to two or more metals varies significantly with the nature of the complexes. Except for the unusually short N–N distance at 1.088(12) Å found in the Sm complex $[\{(C_5Me_5)_2Sm\}_2(\mu_2\text{-}\eta^2{:}\eta^2\text{-}N_2)]$ [20], the N–N bond lengths of this type are generally close to or longer than those of the terminal end-on N_2. The longest N–N distance to date at 1.548(7) Å is observed in $[\{((Pr^i{}_2PCH_2SiMe_2)_2N)ZrCl\}_2(\mu_2\text{-}$

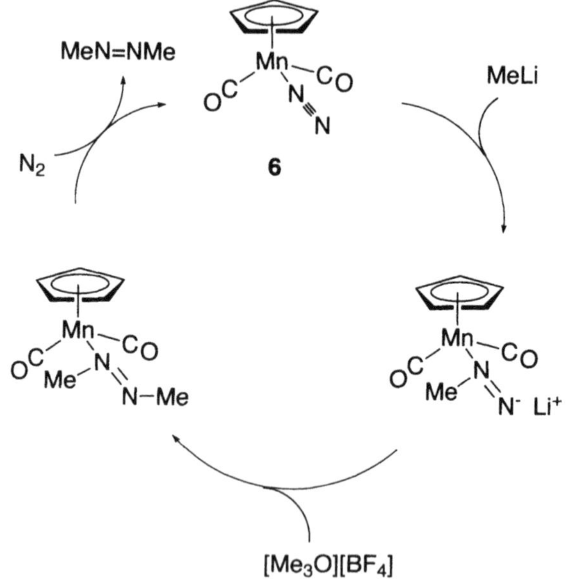

Scheme 4.

η^2:η^2-N_2)] [21], which exceeds even that of hydrazine (1.47 Å). It has been demonstrated that a substantial number of these complexes containing a bridging N_2 ligand are capable of producing NH_3 and/or N_2H_4 upon protolysis, although the mechanisms are ambiguous for most of these reactions [2a, 22]. In contrast, formation of the organo-nitrogen compounds from these complexes has still been poorly explored.

The μ_2-η^1:η^1-N_2 ligand (MNNM) may be classified into three types represented by the formalism: M–N≡N–M (**i**), M=N=N=M (**ii**), and M≡N–N≡M (**iii**), on the basis of the observed N–N and M–N distances in the essentially linear MNNM unit. In general, type **i** is observed in the case of low-valent complexes, whereas type **iii**, featured by the short M–N and long N–N bond distances, is found in the complexes containing relatively high-valent metal centers. The mixed-metal complex [WI(PMe_2Ph)$_3$(py)(μ-N_2)ZrCp_2Cl] (py=pyridine) has been assigned to type **ii** [23]. Interesting reactivities other than protonation have been manifested only for the type **iii** ligand, which include the formation of azines from [Nb_2{calix[4]-(O)$_4$}$_2$(μ-N_2)Na_2(THF)$_6$] (**7**) and PhCHO [24] and from ditantalum complexes such as [{Ta(L)$_3$(THF)}$_2$(μ-N_2)] (L=O-2,6-$Pr^i_2C_6H_3$, OBu^t, neopentyl) and PhCHO or Me_2CO [25]. These reactions may be interpreted as resulting from the metathesis of the C–O and M-N multiple bonds.

3
C–N Bond Formation at Coordinated Dinitrogen via Hydrazido(2-) Species

Hydrazido(2-) complexes of Mo and W readily derived from **1** and **2** are amenable to the electrophilic attack at the terminal N atom by organic compounds containing carbonyl or related functional groups to give the complexes with a variety of organo-nitrogen ligands. The importance of these reactions lies not only in the accessibility of a wide range of diazoalkane ligands by the use of various aldehydes and ketones but also in that these are applicable to both dppe and PMe_2Ph complexes **1** and **2**. At present the C–N bond formation at the N_2 ligand in **2** is attainable only through these indirect routes via the hydrazido(2-) species.

At the terminal N atom in the hydrazido(2-) ligands, nucleophilic substitution, addition, and condensation are now known to occur, among which the condensation reactions have been investigated most elaborately. Thus, the hydrazido(2-) complexes *trans*-[MF(NNH_2)(dppe)$_2$]$^+$ (**8**) and *mer*-[MX_2(NNH_2) (PMe_2 Ph)$_3$](**9**; X=Cl, Br, I) react with aldehydes or ketones RR'C=O to give various diazoalkane complexes *trans*-[MF(NN=CRR')(dppe)$_2$]$^+$ and *mer*-[MX_2(NN= CRR')(PMe_2Ph)$_3$], respectively [26, 27]. For *mer*-[WBr_2(NN=CMe_2)(PMe_2Ph)$_3$], liberation of Pr^iNH_2 together with NH_3 or formation of a mixture of N_2H_4 and Me_2C=NN=CMe_2 is observed by treatment with either $LiAlH_4$ or HBr gas [26b] (Scheme 5). The latter reaction is extended to the synthesis of a series of ketazines RR'C=NN=CRR' from the reactions of **2** (M=W) or *trans*-[W(N_2)$_2$ (PPh_2Me)$_4$] with MeOH/RR'C=O mixtures (Eq. 2). In these reactions, hydrazido(2-) complexes are initially formed by protonation of the N_2 ligand with MeOH and subsequently react with ketones to give diazoalkane complexes. Hy-

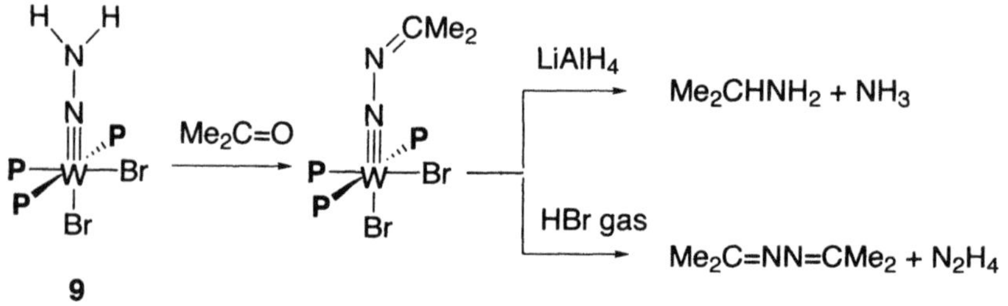

9 $\xrightarrow{Me_2C=O}$ $[WBr_2(NN{=}CMe_2)(P)_3]$ $\xrightarrow{LiAlH_4}$ $Me_2CHNH_2 + NH_3$

$\xrightarrow{HBr\ gas}$ $Me_2C{=}NN{=}CMe_2 + N_2H_4$

Scheme 5.

drazones ($RR'C{=}NNH_2$) are then liberated from the metal by their reactions with MeOH, which further react with ketones to give finally ketazines [28].

$$\underset{\mathbf{2}\ (M = W)}{cis\text{-}[W(N_2)_2(PMe_2Ph)_4]} \xrightarrow[RR'C=O]{MeOH} RR'C{=}NN{=}CRR' + 2H_2O + W(IV)\ species \quad (2)$$

Quite interestingly, the hydrazido(2-) ligands derived from the ligating N_2 in complexes **1** and **2** are transformed into N-heterocyclic compounds by application of the condensation and related methods (Scheme 6). Thus, their reactions with 2,5-dimethoxytetrahydrofuran, pyrylium salts, and phthalaldehyde, followed by workup of the complexes containing N-heterocyclic ligands with $LiAlH_4$ or KOH/alcohol, result in the formation of pyrroles [29], pyridines [30], and phthalimidines [31], respectively.

The remarkable feature of the synthesis of pyrrole from coordinated dinitrogen is that the original N_2 compounds **1** are recovered in moderate yield when the metal-containing products $[MH_4(dppe)_2]$ in the reaction of the pyrrolylimido complexes with $LiAlH_4$ are treated with N_2 under irradiation. This completes the cyclic reaction pathway for the synthesis of pyrrole from N_2 as illustrated in Scheme 7.

As commented above, nucleophilic substitution and addition reactions take place at the terminal N atom of the hydrazido(2-) ligand to give the organo-nitrogen ligands; the reaction of **8** (M=W) with succinyl chloride affords a diacylhydrazido(2-) complex *trans*-$[WF(NNCOCH_2CH_2CO)(dppe)_2]^+$ [32], while treatment of *trans*-$[MBr(NNH_2)(dppe)_2]^+$ with PhN=C=O [32] and **9** with $Ph_2C{=}C{=}O$ [33] both yield the acylhydrazido(2-) type complexes *trans*-$[MBr(NNHCONHPh)(dppe)_2]^+$ and mer-$[MX_2(NNHCOCHPh_2)(PMe_2Ph)_3]$, respectively.

Scheme 6.

Scheme 7.

4
Direct Cleavage of the N–N Triple Bond

In the Haber process, the metallic iron is primarily responsible for the catalytic activity and the formation of NH_3 proceeds through a rate-determining step in which the N–N triple bond of the adsorbed N_2 is cleaved at the Fe surface. However, direct N–N bond scission by using transition metal N_2 complexes has rarely been observed. Recently, Cummins et al. have found that the N–N triple bond can be cleaved by the three-coordinate Mo(III) amide complexes $[Mo(NRAr)_3]$ ($R=C(CD_3)_2CH_3$, $Ar=3,5\text{-}Me_2C_6H_3$). The reaction proceeds through an observable dinuclear intermediate $[(\mu\text{-}N_2)\{Mo(NRAr)_3\}_2]$ to give finally the monomeric nitrido complex $[NMo(NRAr)_3]$ (**10**) at as low as –35° to –30°C (Scheme 8) [34]. This attractive finding might give an insight into the mechanism for the formation of NH_3 from N_2 at the metallic surface as well as the $MoFe_7S_9$ cluster core demonstrated for the FeMo-cofactor [35]. Development of new reaction pathways for formation of nitrogenous compounds involving the N–N bond cleavage as the initial step might also become possible based on this observation, although the nitride ligand in **10** seems to be so inert that an appropriate N atom-transfer reaction should be coupled to give nitrogenous compounds. Quite recently, a similar direct cleavage of $\mu\text{-}N_2$ ligand has been observed for the dinuclear Nb calix[4]arene N_2 complex **7**, when treated with Na metal [24].

Several systems containing Ti are known, in which molecular nitrogen is incorporated into metal compounds with concurrent N–N bond scission, although in most cases the resultant nitride species are only poorly characterized. Thus, the reaction of N_2 with a $TiCl_3$/Mg/THF system is believed to afford a nitride species having a composition of $[TiNMg_2Cl_2(THF)]$ (**11**) [36], which is further treated with CO_2 to give another Ti compound formulated as $[Ti(NCO)Mg_2Cl_2O(THF)_3]$ (**12**) [37]. Recently synthesis of a range of organo-nitrogen compounds using in situ-generated **11** and **12** has been reported by Mori and coworkers.

Treatment of **11** with excess aroyl chloride ArCOCl followed by hydrolysis gives a mixture of $ArCONH_2$ and $(ArCO)_2NH$, while the reaction of **12** with phthaloyl dichloride or phthalic anhydride results in the formation of phthalimide. Aroylpalladium complexes, prepared from ArX (X=Br, I) and $[Pd(PPh_3)_4]$ under CO, also react with **12** to yield the aroylimides and/or aroylamides after

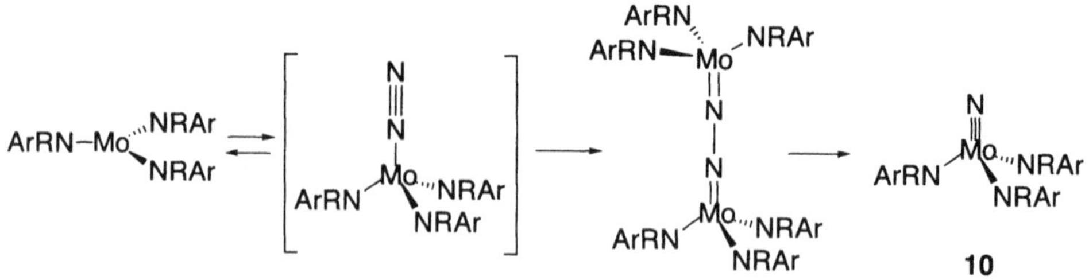

Scheme 8.

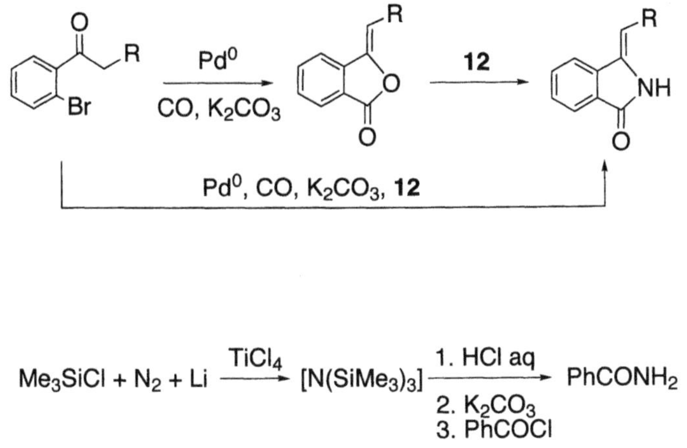

Pd^0; CO, K_2CO_3; **12**; Pd^0, CO, K_2CO_3, **12**

$$Me_3SiCl + N_2 + Li \xrightarrow{TiCl_4} [N(SiMe_3)_3] \xrightarrow[\text{2. } K_2CO_3 \text{ 3. } PhCOCl]{\text{1. HCl aq}} PhCONH_2$$

Scheme 9.

hydrolysis [38]. These findings have led to the exploitation of the new synthetic route towards N-heterocycles from α-halophenyl alkylketones, which involves both catalytic carbonylation and stoichiometric nitrogenation (Scheme 9) [39]. More recently, catalytic synthesis of $PhCONH_2$ from PhCOCl has been attained by the use of a $TiCl_4/Li/Me_3SiCl$ mixture under N_2, whose mechanism presumably involves the initial formation of $N(SiMe_3)_3$ (Scheme 9). This catalytic system has also been applied to the syntheses of a variety of heterocyclic compounds such as indoles and quinolines [40].

5 Future Prospects

The quest for the novel complexes with a highly reactive N_2 ligand is still continuing. Stimulated by the elucidation of the FeMo-cofactor core structure, syntheses of transition metal-sulfur clusters are currently being investigated extensively. However, metal sulfide clusters which can coordinate N_2 have not yet been isolated. One of the attractive N_2 complexes isolated recently is a series of gold clusters $[\{(LAu)_3\}_2(N_2)]^{2+}$(**13**; L=tertiary phosphines) with a novel N_2 ligand bridging the two trimetallic clusters, although these complexes are only accessible by using N_2H_4 as the N_2 source (Scheme 10) [41]. Interestingly, the novel coordination mode of N_2 in **13** is suggested to occur at the active site in the FeMo-cofactor having the Fe_6 prismatic cavity to accommodate N_2 [42]. Reactions of **13** in the presence of the proton donor (2,6-lutidinium triflate) and the reducing agent (Cp_2Co) afford NH_3 and N_2H_4, the combined yield of which indicates the transformation of the bridging N_2 to be almost quantitative. Interestingly, the Au product after the reaction has been identified as $[(LAu)_6]^{2+}$ for L=PPh_3. The lack in reactivity of $[(LAu)_6]^{2+}$ towards N_2, however, hampers the completion of the catalytic cycle. Nevertheless, it is to be noted that the N_2 ligand in **13** can be reduced and protonated to give NH_3 and N_2H_4 in one-pot reaction by the use of a combination of mild reducing agent/proton source [43]. More importantly, the oxidation state of the metals is preserved after the reaction if the N_2 ligand in **13** is regarded formally as neutral. This contrasts with the formation of NH_3 by the protonation of **2** briefly described in Sect. 2.1, in which the reaction proceeds in a stepwise manner only by using stronger inorganic acids as the proton source, concurrent with the electron flow from the metal center to the N_2 ligand. Hence, the highly oxidized Mo or W species are formed as the final metal product.

All of previous attempts to react N_2 complexes with H_2 have failed and resulted in the liberation of the N_2 ligand as a dinitrogen gas until the recent remarkable finding by Fryzuk et al. [44]. Thus, it is surprising that the dinuclear Zr complex with a side-on N_2 bridge $[\{(P_2N_2)Zr\}_2(\mu_2\text{-}\eta^2\text{:}\eta^2\text{-}N_2)]$ (**14**; P_2N_2=PhP$(CH_2SiMe_2NSiMe_2CH_2)_2$PPh) reacts with H_2 in toluene to give the spectroscopically characterized $[\{(P_2N_2)Zr\}_2(\mu_2\text{-}\eta^2\text{:}\eta^2\text{-NNH})(\mu_2\text{-H})]$ (**15**). A possible intermediate $[\{(P_2N_2)Zr\}_2(\mu_2\text{-}\eta^2\text{:}\eta^2\text{-}N_2)(\mu_2\text{-}\eta^2\text{:}\eta^2\text{-}H_2)]$ has also been isolated and fully characterized. The structure of **15** has been supported by the X-ray diffraction study of the related complex $[\{(P_2N_2)Zr\}_2(\mu_2\text{-}\eta^2\text{:}\eta^2\text{-NNSiH}_2\text{Bu}^n)(\mu_2\text{-H})]$ derived from the reaction of **14** with Bu^nSiH_3.

In the case of **1** and **2**, the N_2 ligand is not susceptible to any direct interaction with H_2. However, if the heterolytic cleavage of H_2 is realized in situ without evolution of N_2 from **1** or **2**, the N_2 ligand in these complexes may be transformed into nitrogen hydride species by interacting with the resulting H^+/H^- couple arising from H_2. A remarkable advance in the study performed in this context is our recent findings of a bimetallic reaction system. When **2** (M=W) is treated with an equilibrium mixture of $[RuCl(dppp)_2]X$ and $[RuCl(\eta^2\text{-}H_2)(dppp)_2]X$ (X=PF_6, BF_4, OSO_2CF_3; dppp=$Ph_2P(CH_2)_3PPh_2$) under H_2 (1 atm) at 55°C, NH_3 is formed in moderate yield [45]. The analogous reaction carried out in the presence of acetone produces acetone azine (Scheme 11). When treated similarly

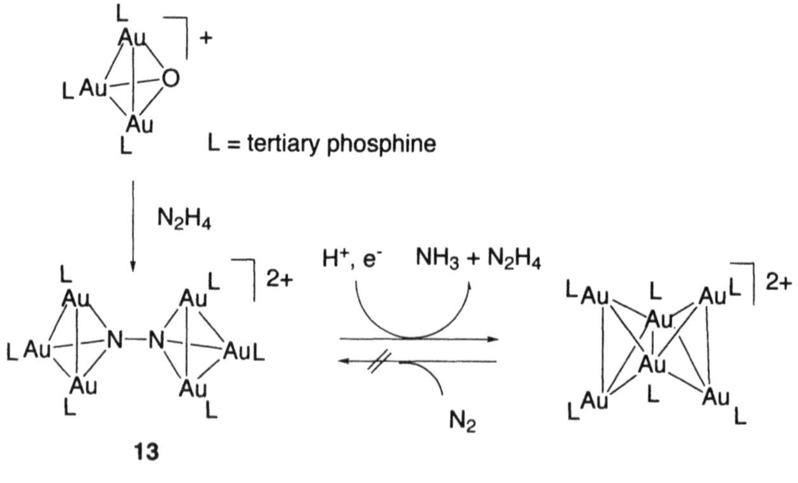

L = tertiary phosphine

Scheme 10.

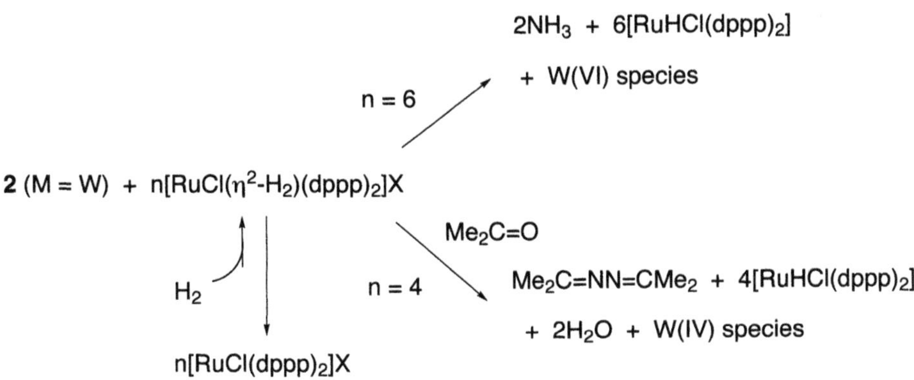

Scheme 11.

($X=PF_6$ or BF_4), **1** (M=W) gives the hydrazido(2-) complexes *trans*-[$WF(NNH_2)(dppp)_2$]X and the diazoalkane complexes *trans*-[$WF(NN{=}CMe_2)(dppp)_2$]X, respectively, indicating the presence of similar species as the intermediates in conversion of the N_2 ligand in **2** to NH_3 or the azine. These reactions closely relate to those of **2** with MeOH or a MeOH/acetone mixture (e.g., Eq. 2).

However, it is noteworthy that the reaction proceeds by the use of H_2 instead of MeOH as the proton source. After the reaction, **2** is transformed into an untractable high-valent W species, while the H^- resulting from the heterolytic splitting of H_2 is retained in [$RuHCl(dppp)_2$] as the hydride ligand. At present, the hydride ligand is not used for reduction of the high-valent W species leading to the formation of the starting compound **2**. It is to be noted that the heterolytic H_2 activation at a metal-sulfur site, viz., $M{-}S + H_2 \rightarrow M(H^-){-}SH^+$, has also been investigated by others with the aim of elucidating a catalytic N_2 reducing system [46].

References

1. Allen AD, Senoff CV (1965) J Chem Soc Chem Commun 621
2. (a) Hidai M, Mizobe Y (1995) Chem Rev 95:1115. (b) Richards RL (1996) Coord Chem Rev 154:83. (c) Bazhenova TA, Shilov AE (1995) Coord Chem Rev 144:69. (d) Gambarotta S (1995) J Organomet Chem 500:117
3. (a) Hidai M, Ishii Y (1996) Bull Chem Soc Jpn 69:819. (b) Mizobe Y, Ishii Y, Hidai M (1995) Coord Chem Rev 139:281. (c) Hidai M, Mizobe Y (1993) Molybdenum enzymes, cofactors, and model systems. American Chemical Society, Washington, DC, Chap 22. (d) Colquhoun (1984) Acc Chem Res 17:23
4. (a) Hidai M, Tominari K, Uchida Y, Misono A (1969) J Chem Soc Chem Commun 1392. (b) Hidai M, Tominari K, Uchida Y (1972) J Am Chem Soc 94:110. (c) Chatt J, Heath GA, Richards RL (1974) J Chem Soc Dalton Trans 2072
5. George TA, Seibold CD (1973) Inorg Chem 12:2544
6. (a) Leigh GJ (1992) Acc Chem Res 25:177. (b) Schrock RR (1997) Pure Appl Chem 69:2197

7. (a) Chatt J, Pearman AJ, Richards RL (1977) J Chem Soc Dalton Trans 1852. (b) Anderson SN, Fakley ME, Richards RL, Chatt J (1981) J Chem Soc Dalton Trans 1973. (c) Takahashi T, Mizobe Y, Sato M, Uchida Y, Hidai M (1980) J Am Chem Soc 102:7461
8. Chatt J, Diamantis AA, Heath GA, Hooper NE, Leigh GJ (1977) J Chem Soc Dalton Trans 688
9. (a) Hidai M, Komori K, Kodama T, Jin DM, Takahashi T, Sugiura S, Uchida Y, Mizobe Y (1984) J Organomet Chem 272:155. (b) Oshita H, Mizobe Y, Hidai M (1992) Organometallics 11:4116. (c) Oshita H, Mizobe Y, Hidai M (1993) J Organomet Chem 456:213
10. Street AC, Mizobe Y, Gotoh F, Mega I, Oshita H, Hidai M (1991) Chem Lett 383
11. Chatt J, Diamantis AA, Heath GA, Hooper NE, Leigh GJ (1977) J Chem Soc Dalton Trans 688
12. Ishii Y, Kawaguchi M, Ishino Y, Aoki T, Hidai M (1994) Organometallics 13:5062
13. Yoshida T, Adachi T, Ueda T, Kaminaka M, Sasaki N, Higuchi T, Aoshima T, Mega I, Mizobe Y, Hidai M (1989) Angew Chem Int Ed Engl 28:1040
14. (a) Chatt J, Head RA, Leigh GJ, Pickett CJ (1978) J Chem Soc Dalton Trans 1638. (b) Busby DC, George TA, Iske Jr. SD, Wagner SD (1981) Inorg Chem 20:22. (c) Bossard GE, Busby DC, Chang M, George TA, Iske Jr. SD (1980) J Am Chem Soc 102:1001
15. Chatt J, Hussain W, Leigh GJ, Terreros FP (1980) J Chem Soc Dalton Trans 1408
16. Bevan PC, Chatt J, Leigh GJ, Leelamani EG (1977) J Organomet Chem 139:C59
17. Pickett CJ, Leigh GJ (1981) J Chem Soc Chem Commun 1033
18. (a) Komori K, Oshita H, Mizobe Y, Hidai M (1989) J Am Chem Soc 111:1939. (b) Komori K, Sugiura S, Mizobe Y, Yamada M, Hidai M (1989) Bull Chem Soc Jpn 62:2953
19. Sellmann D, Weiss W (1978) J Organomet Chem 160:183
20. Evans WJ, Ulibarri TA, Ziller JW (1988) J Am Chem Soc 110:6877
21. Fryzuk MD, Haddad TS, Mylvaganam M, McConville DH, Rettig SJ (1993) J Am Chem Soc 115:2782
22. Henderson RA (1990) Transition Metal Chem 15:330
23. Mizobe Y, Yokobayashi Y, Oshita H, Takahashi T, Hidai M (1994) Organometallics 13:3764
24. Zanotti-Gerosa A, Solari E, Giannini L, Floriani C, Chiesi-Villa A, Rizzoli C (1998) J Am Chem Soc 120:437
25. (a) Schrock RR, Wesolek M, Liu AH, Wallace KC, Dewan JC (1988) Inorg Chem 27:2050. (b) Rocklage SM, Schrock RR (1982) J Am Chem Soc 104:3077
26. (a) Hidai M, Mizobe Y, Sato M, Kodama T, Uchida Y (1978) J Am Chem Soc 100:5740. (b) Bevan PC, Chatt J, Hidai M, Leigh GJ (1978) J Organomet Chem 160:165. (c) Mizobe Y, Uchida Y, Hidai M (1980) Bull Chem Soc Jpn 53:1781. (d) Mizobe Y, Ono R, Uchida Y, Hidai M, Tezuka M, Moue S, Tsuchiya A (1981) J Organomet Chem 204:377
27. For syntheses of related diazoalkane complexes, see also: (a) Hidai M, Aramaki S, Yoshida K, Kodama T, Takahashi T, Uchida Y, Mizobe Y (1986) J Am Chem Soc 108:1562. (b) Ishii Y, Miyagi H, Jitsukuni S, Seino H, Harkness BS, Hidai M (1992) J Am Chem Soc 114:9890. (c) Oshita H, Mizobe Y, Hidai M (1993) J Organomet Chem 461:43. (d) Aoshima T, Tamura T, Mizobe Y, Hidai M (1992) J Organomet Chem 435:85. (e) Aoshima T, Tanase T, Mizobe Y, Yamamoto Y, Hidai M (1992) J Chem Soc Chem Commun 586. (f) Harada Y, Mizobe Y, Aoshima T, Oshita H, Hidai M (1998) Bull Chem Soc Jpn 71:183
28. (a) Watakabe A, Takahashi T, Jin DM, Yokotake I, Uchida Y, Hidai M (1983) J Organomet Chem 254:75. (b) Hidai M, Kurano M, Mizobe Y (1985) Bull Chem Soc Jpn 58:2719
29. Seino H, Ishii Y, Sasagawa T, Hidai M (1995) J Am Chem Soc 117:12181
30. Ishii Y, Tokunaga S, Seino H, Hidai M (1996) Inorg Chem 35:5118
31. Seino H, Ishii Y, Hidai M (1997) Inorg Chem 36:161
32. Aoshima T, Mizobe Y, Hidai M, Tsuchiya J (1992) J Organomet Chem 423:39
33. Iwanami K, Mizobe Y, Takahashi T, Kodama T, Uchida Y, Hidai M (1981) Bull Chem Soc Jpn 54:1773

34. (a) Laplaza CE, Cummins CC (1995) Science 268:861. (b) Laplaza CE, Johnson MJA, Peters JC, Odom AL, Kim E, Cummins CC, George GN, Pickering IJ (1996) J Am Chem Soc 118:8623
35. (a) Kim J, Rees DC (1992) Science 257:1677. (b) Chen J, Christiansen J, Campobasso N, Bolin JT, Tittsworth RC, Hales BJ, Rehr JJ, Cramer SP (1993) Angew Chem Int Ed Engl 32:1592. (c) Howard JB, Rees DC (1996) Chem Rev 96:2965
36. Yamamoto A, Go S, Ookawa M, Takahashi M, Ikeda S, Keii T (1972) Bull Chem Soc Jpn 45:3110
37. Sobota P, Jezowska-Trzebiatowska B, Janas Z (1976) J Organomet Chem 118:253
38. (a) Mori M, Uozumi Y, Shibasaki M (1990) J Organomet Chem 395:255. (b) Mori M, Uozumi Y, Shibasaki M (1987) Tetrahedron Lett 28:6187
39. (a) Uozumi Y, Kawasaki N, Mori e, Mori M, Shibasaki M (1989) J Am Chem Soc 111:3725. (b) Uozumi Y, Mori E, Mori M, Shibasaki M (1990) J Organomet Chem 399:93
40. (a) Mori M, Kawaguchi M, Hori M, Hamaoka S (1994) Heterocycles 39:729. (b) Hori M, Mori M (1995) J Org Chem 60:1480. (c) Mori M, Hori K, Akashi M, Hori M, Sato Y, Nishida M (1998) Angew Chem Int Ed Engl 37:636
41. Shan H, Yang Y, James AJ, Sharp PR (1997) Science 275:1460
42. (a) Rees DC, Chan MK, Kim J (1994) Adv Inorg Chem 40:89. (b) Stravrev KK, Zerner MC (1996) Chem Eur J 2:83
43. The mixture of lutidinium cation/Cp_2Co has previously been used for the stoichiometric and catalytic reduction of hydrazine into ammonia by using the complexes such as $[(C_5Me_5)WMe_3(NH_2NH_2)]$ and $MoFe_3S_4$-polycarboxylate cluster anions: (a) Schrock RR, Glassman TE, Vale MG, Kol M (1993) J Am Chem Soc 115:1760. (b) Demadis KD, Malinak SM, Coucouvanis D (1996) Inorg Chem 35:4038
44. Fryzuk MD, Love JB, Rettig SJ, Young VG (1997) Science 275:1445
45. Nishibayashi Y, Iwai S, Hidai M (1998) Science 279:540
46. Sellmann D, Sutter AJ (1997) Acc Chem Res 30:460

Metal Reagents for Activation and Functionalization of Carbon–Fluorine Bonds

Thomas G. Richmond

Department of Chemistry, University of Utah, Salt Lake City, UT 84112-0850 USA
E-mail: Richmond@chemistry.utah.edu

The distinctive physical properties of fluorocarbons derived from the great strength of the C–F bond often result in chemical compounds with unique properties and technologically useful applications. However, this same factor conspires to make fluorocarbons unreactive under most conditions. In the past decade it has become clear that metal reagents provide a unique tool for activation and functionalization of the C–F bond under mild conditions. Fundamental approaches toward the activation of C–F bonds are discussed with an emphasis on transition metal reagents. Intramolecular systems provided the intellectual foundation for the development of intermolecular C–F bond activation chemistry. Mechanistic studies of model systems give an indication of the scope of this process. Exciting recent advances in the catalytic chemistry of the C–F bond in fluorinated aromatic and aliphatic systems will be described. Selective C–F bond activation is a key requirement for the application of these reagents in organic synthesis. Theoretical and gas phase studies which provide insight into the mechanisms of these reactions will be discussed.

Keywords: Carbon-fluorine, Bond activation, Transition metals, Catalysis, Oxidative addition, Fluorocarbons, CFCs, Electron transfer, Halocarbon coordination

1 **Introduction** 244

2 **Fundamental Approaches to the Activation of C–F Bonds** 245

2.1 Coordination Chemistry of the C–F Bond 245
2.2 Electrophilic Attack 246
2.3 Nucleophilic Attack 246
2.4 Oxidative Addition 247
2.5 Electron Transfer (Reduction) 248
2.6 Photochemistry 249

3 **Transition Metal Reagents for Aromatic C–F Bond Activation** . . . 250

3.1 Ligand Based Reactions Leading to C–F Bond Functionalization . . 250
3.2 Intermolecular C–F Bond Activation 252
3.3 Catalytic Chemistry of Aromatic C–F Bonds 255

4 **Reactions of Aliphatic C–F Bonds with Metal Reagents** 258

4.1 Stoichiometric Transformations 258

4.2 Catalytic Reactions of Saturated Perfluorocarbons 261
4.3 C–F Bond Activation in the Service of Organic Synthesis 264

5 Theoretical Approaches to C–F Bond Activation 265

6 Future Perspectives 266

References 266

1 Introduction

The most characteristic chemical property of fluorocarbons is their decided *lack* of reactivity [1]. This is a consequence of the great strength of the C–F bond that arises from the small size and high electronegativity of the fluorine atom. Many of the technological uses of fluorinated materials are based on the chemical and thermal stability of this class of compounds [2]. Replacement of hydrogen by fluorine in organic molecules can lead to useful pharmaceuticals or agricultural chemicals and the role of fluorine in bioorganic chemistry has been reviewed [3]. The long atmospheric lifetime of chlorofluorocarbons is attributed to the chemical inertness of the C–F bond and perfluorocarbons may be the most unreactive synthetic compounds known [4]. An understanding of the fundamental reaction chemistry of the C–F bond may be useful in designing new catalysts for the synthesis of chlorofluorocarbon replacements or destruction of existing stockpiles of these compounds [5]. The organic chemistry of the C–F bond has been the subject of extensive study [6].

An exciting recent development in this field is the application of the unusual solubility properties of perfluorocarbons as solvents in so-called "fluorous" biphase chemistry [7]. These novel synthetic procedures enable ready separation of suitably designed catalysts and reagents by simple phase separation of many hydrocarbon solvents (containing the product) from the fluorous phase containing the catalyst. Success in these procedures is of course contingent on the inert nature of the fluorous phase under the reaction conditions.

The interplay between the fields of transition metal organometallic and fluorocarbon chemistries was initially profitable because of the ability of fluorinated organic ligands to impart greater stability to numerous organometallic compounds relative to hydrocarbon analogous. Several reviews have summarized the synthesis, structure and bonding of these compounds [8]. Interest in a related class of materials has been kindled by the search for "least-coordinating" or "superweak" anions for fundamental studies of coordinatively unsaturated cationic metal complexes [9]. The chemical stability of the C–F bond protects highly fluorinated anions from many modes of chemical degradation and also serves to delocalize the negative charge in the anion. Highly electrophilic cationic complexes stabilized by these anions are also of practical importance as olefin polymerization catalysts and in certain C–H bond activation reactions [10].

Consistent with the title of this volume, this chapter will describe recent developments in the *reaction chemistry* of the normally inert C–F bond that have been discovered largely in the past decade. Much of this work relies on transition metals to assist in the activation of the C–F bond under mild conditions. An indication of the rapid progress in this field is the discovery of metal reagents which act as catalysts for chemistry at the C–F bond and may ultimately prove useful in the synthesis of highly fluorinated organic compounds. A comprehensive review of C–F bond activation by metals was published by Kiplinger and colleagues [11] in 1994 and further progress in the field was updated by Burdeniuc and coworkers [12] in 1997. The main focus of the present work will be on the fundamental pathways, many of which have their roots in the concepts of organic chemistry, that lead to C–F bond cleavage and the application of these reactions in synthetic and catalytic chemistry involving transition metals.

2 Fundamental Approaches to the Activation of C–F Bonds

There are several conceptual approaches that might lead to attack and successful activation of the strongest single bond to carbon [11, 12]. Examples of each strategy, which are chosen to represent the conceptual development of the field, will be considered below along with possible limitations of each technique. Although most work has focused on discovering new reactivity patterns, it is becoming increasingly apparent that the selectivity of these reactions will be of paramount importance in their applications. An insightful lecture-demonstration involving the highly exothermic reaction of polytetrafluoroethene with magnesium metal illustrates the point that simple reagents are indeed capable of activating C–F bonds, albeit with little control [13]! These methods are not mutually exclusive and some of the most effective systems for C–F activation and functionalization employ more than one of the concepts discussed below.

2.1 Coordination Chemistry of the C–F Bond

Interaction of a C–F bond with a metal ion might provide a way to weaken and ultimately cleave a C–F bond. As a consequence of the low σ-basicity of the fluorine lone pairs, neutral fluorocarbons are very poor ligands [14], even relative to the lower halocarbons [15], as well as traditional nitrogen and oxygen based ligands to most transition metals. The chemistry is dominated by simple fluorocarbon ligand dissociation if bound at all. Stronger interactions are observed between cationic metal centers and fluorinated anions where ion-pairing interactions become important [9, 16]. Normally these interactions do not lead to well-defined C–F bond cleavage under mild conditions. The elegant recent work by Plenio [17] utilizing carefully designed crown ethers and cryptands containing a CF binding site has demonstrated the efficacy of fluoroarene binding to alkali metals. In favorable cases, the fluoroarene ligand forms shorter bonds to the alkali metals cation than does the ether oxygen and clearly contributes to the thermodynamic stability of the complex [18]. Under chemical vapor deposition conditions, deposition of NaF and BaF_2 has been reported from precursors which show short Na···FC and Ba····FC contacts in the solid state [19].

2.2 Electrophilic Attack

Suitably chosen main group Lewis acids have proven to be much more effective than their transition metal counter parts in activating C–F bonds if the initial (weak) coordination can be coupled with a thermodynamically favored chemical reaction [11]. Classic examples in organic chemistry include the reaction of trifluoromethylnapthalene with aluminum chloride to afford trichloromethylnapthalene and aluminum fluoride and the acid catalyzed hydrolysis of trifluoromethylbenzene to afford benzoic acid [20]. These reactions are driven by the formation of Al–F and HF bonds, respectively. Similar exchange reactions with saturated chlorofluorocarbons occur but are usually not selective. Carbocations have been demonstrated to participate in intramolecular fluoride shifts between carbon atoms and in intermolecular fluoride abstractions [21]. Conceptually similar reactions of fluorinated ligands coordinated to transition metals discovered by research groups led by Shriver [20] and Roper [22] have provided facile entry into transition metal halocarbene complexes via C–F activation under exceedingly mild conditions. Compared to wholly organic systems, the ability of transition metals to stabilize carbene intermediates results in selective halide exchange or hydrolysis. Hydrolysis is selective α to the transition metal and hydrolysis of terminal CF_3 groups affords the CO ligand.

$$\mathrm{CpMo(CO)_3CCl_2CF_3} \xleftarrow[\text{- BClF}_2]{\text{BCl}_3} \mathrm{CpMo(CO)_3CF_2CF_3} \xrightarrow[\text{- 2 HF}]{\text{H}^+/\text{H}_2\text{O}} \mathrm{Cp(CO)_3Mo{-}C({=}O){-}CF_3} \quad (1)$$

As discussed below, electrophiles play a key role in several diverse C–F activation processes.

2.3 Nucleophilic Attack

The polarity of the C–F bond makes the carbon electrophilic and thus susceptible to attack by nucleophiles. This is especially important in aromatic chemistry and significantly enhanced by the presence of electron withdrawing groups as noted in the enhanced reactivity of Sanger's reagent, 2,4-dinitrofluorobenzene, as the classic reagent for reaction with N-terminal proteins. In contrast, fluorobenzene is essentially inert to nucleophilic aromatic substitution. In this context, hexafluorobenzene is also quite reactive with a host of nucleophilic reagents including several transition metal anions [11].

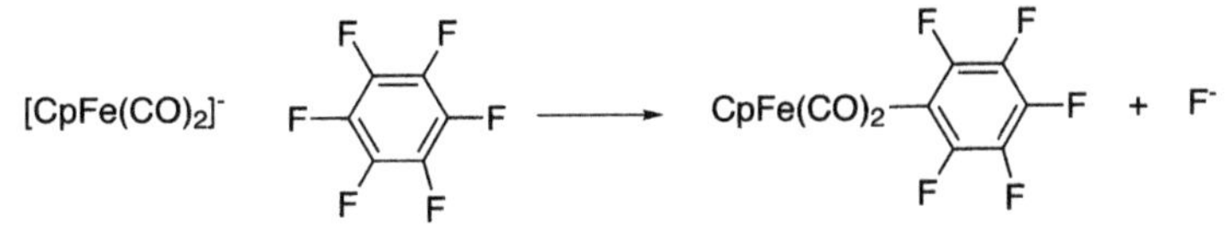

$[CpFe(CO)_2]^- + C_6F_6 \longrightarrow CpFe(CO)_2C_6F_5 + F^-$ (2)

The high reactivity of hexafluorobenzene makes it the most popular substrate for scouting new reagents for C–F activation.

2.4 Oxidative Addition

After a quiescent period from the late 1960s to the mid-1980s, the resurgence of interest in the organometallic chemistry of fluorocarbons [11, 12, 23] can be traced to the discovery that suitably designed ligands containing fluoroarenes can react with electron rich transition metals undergoing net insertion of the metal into the C–F bond [24]. This is illustrated for the room temperature oxidative addition of a dangling pentafluorophenyl group to W(0) with concomitant oxidation of the metal to W(II).

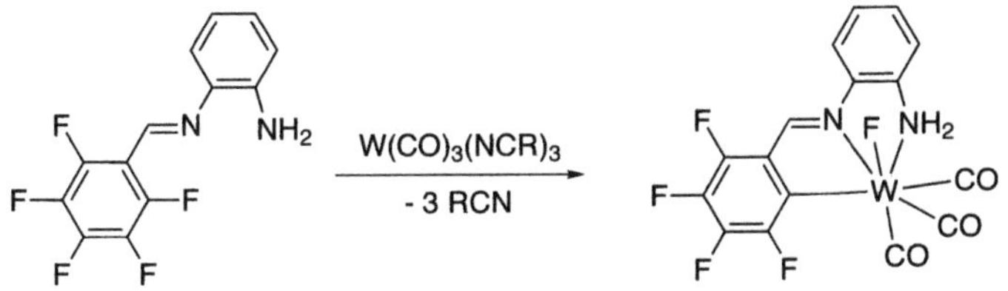

$W(CO)_3(NCR)_3$, $-3\ RCN$ (3)

Similar chemistry has been reported utilizing Mo(0)/Mo(II), Ni(0)/Ni(II) and Pt(II)/Pt(IV) redox couples [25, 26].

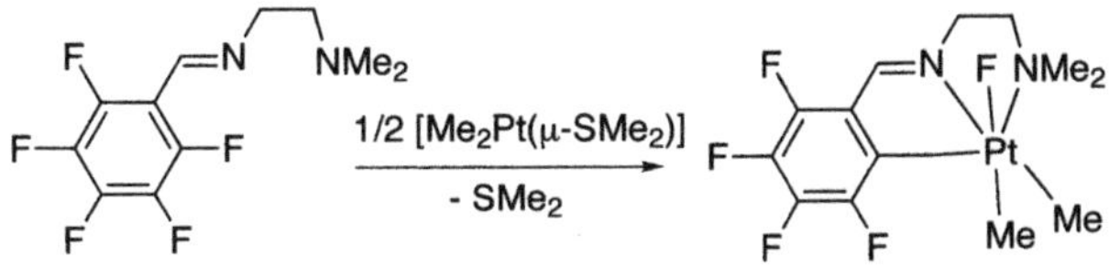

$1/2\ [Me_2Pt(\mu\text{-}SMe_2)]$, $-SMe_2$ (4)

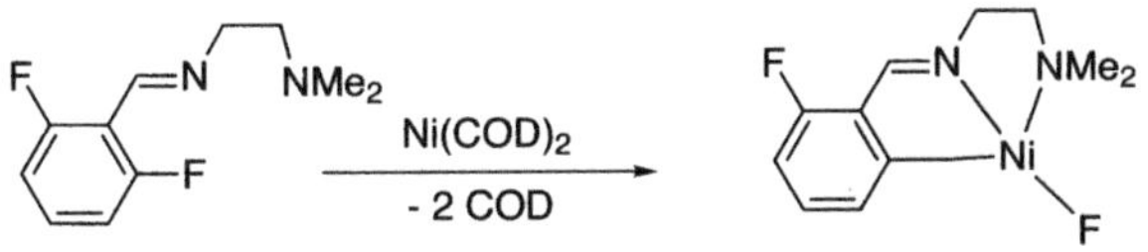

$Ni(COD)_2$, $-2\ COD$ (5)

Detailed investigations show that the structure of the ligand is critical in these reactions [27]. The rate of C–F activation decreases with decreasing fluorination of the ring as might be expected for the metal acting a nucleophile to attack the C–F bond and then trap the fluoride leaving group in the coordination sphere of the metal. Consecutive single electron transfer pathways cannot be ruled out. In the case of tungsten, C–F activation is favored over C–H activation probably because of the thermodynamic stability of the W–F bond formed [28].

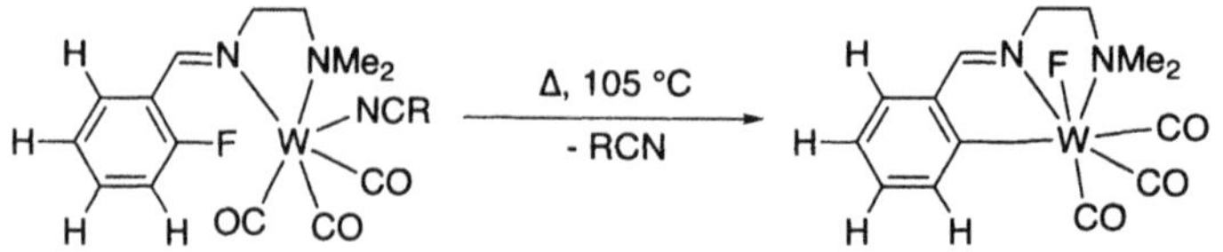

Δ, 105 °C; - RCN (6)

In contrast, C–H activation, followed by reductive elimination of methane, is favored over C–F activation at Pt(II) [29].

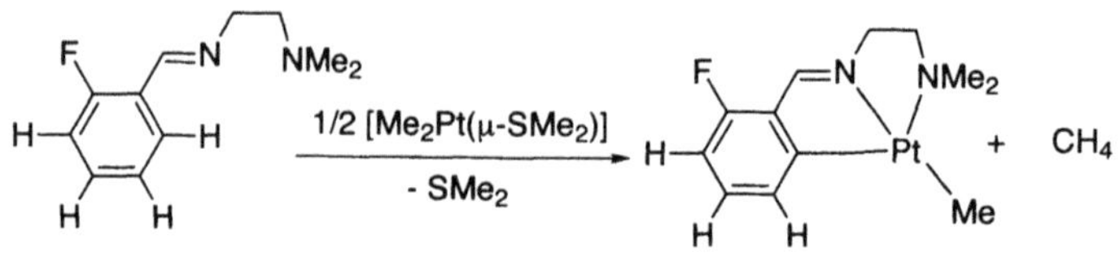

1/2 $[Me_2Pt(\mu\text{-}SMe_2)]$; - SMe_2; + CH_4 (7)

An important early example of bimolecular C–F activation of hexafluorobenzene was noted using a Pt(0) complex of a bulky chelating phosphine [30].

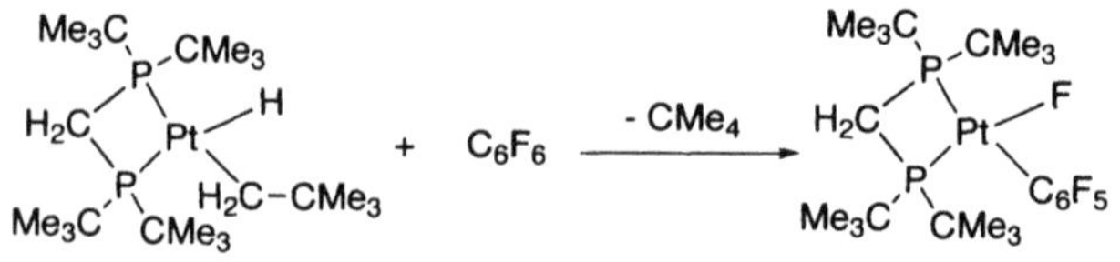

+ C_6F_6; - CMe_4 (8)

2.5
Electron Transfer (Reduction)

Interest in the use of reducing agents to activate C–F bonds was revived when systems that enabled selective removal of fluoride from perfluorocarbons were discovered in contrast to the complete (destructive) removal of halides as in sodium fusion reactions employed in methods for halogen analysis [31]. (Interest in this latter area has also had resurgence in the search for methods capable of destroying chlorofluorocarbons [32].) The key breakthrough was the report by MacNicol and Robertson that hexakis(thiophenoxy)napthalene can be prepared by treatment of perfluorodecalin with excess sodium phenylthiolate for 10 days at 70°C in DMF [33].

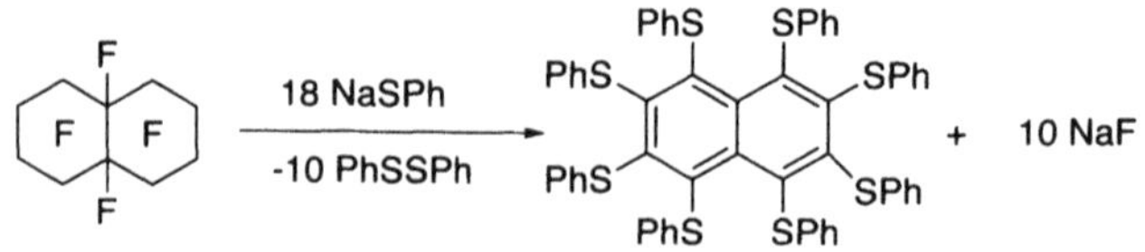

18 NaSPh; -10 PhSSPh; + 10 NaF (9)

The ability of the thiolate to function as a nucleophile and a reducing agent is required for this transformation. Later Pez and coworkers [34] showed that careful addition of sodium benzophenone radical anion to perfluorodecalin affords perfluoronapthalene. This led Harrison [35] to show that the organometallic anion $[CpFe(CO)_2]^-$ also defluorinates perfluorodecalin and perfluoromethylcyclohexane to give a mixture of perfluoroaromatics bound to $[CpFe(CO)_2]^-$.

$$\text{C}_6\text{F}_{11}\text{CF}_3 \xrightarrow[\text{-Fp}_2,\ \text{- NaF}]{\text{Na}^+\,\text{Fp}^-\text{(excess)}} \text{(CF}_3\text{)C}_6\text{F}_3\text{H(Fp)} + \text{(CF}_3\text{)C}_6\text{F}_4\text{(Fp)} + \text{(CF}_3\text{)C}_6\text{F}_2\text{H}_2\text{(Fp)} \qquad (10)$$

$Fp = CpFe(CO)_2^-$

Divalent lanthanide metallocenes also activate C–F bonds by reductive pathways as demonstrated by Burns and Andersen [36], who used formation of the mixed valent dimer $[(C_5Me_5)_2Yb]_2F$ as evidence for C–F abstraction from a variety of substrates including C_6F_6, C_6H_5F, C_2F_4, but not C_2F_6. Contemporaneously, Watson [37] showed that perfluorodienes could be prepared from perfluorolefins with the same reagent and visible light photolysis accelerated the rate of the process. Gas phase activation of CF_4 by the praseodymium cation is proposed to proceed by a homolytic pathway to generate CF_3 radicals [38] and related studies of gas phase ion-molecule reactions of this type promise to shed insight on the intimate nature of these transformations.

2.6
Photochemistry

Photochemical methods provide another route to provide the energy necessary to activate a C–F bond. ArF (193 nm) laser photolysis [39] of liquid hexafluorobenzene leads to a multitude of organic products including perfluorohexadiene, octafluoroindene, perfluorodiphenylmethane, decafluorobiphenyl, perfluoronapthalene, nonafluorobiphenyl, *m*- and *p*-perfluoroterphenyl, and several unidentified isomers of the formula C_6F_8 and $C_{12}F_{12}$. These products provide evidence for homolytic C–F cleavage and formation of pentafluorophenyl radicals as one mechanistic path for this complex transformation. The black solid obtained was characterized as graphitic and fluorine-containing carbon thought to arise from polymerization of transient fluorinated alkynes [39]. More selective photochemical transformations of perfluorocarbons have been discovered including the use of organic photosensitizers such as triphenylamine in the defluorination of perfluorodecalin to afford perfluorooctalin [40]. As described in more detail below, perfluorocarbons react with ammonia under catalytic Hg photosensitization conditions [41] to afford several interesting nitrogen substituted perfluorocarbon derivatives. These reactions are similar in nature to the analogous C–H activation chemistry in alkanes that was also developed in the laboratory of Crabtree [42]. In another metal system that has its roots in C–H bond activation chemistry, Jones, Perutz and coworkers have demonstrated that near UV photolysis (λ>285 nm) of $(C_5Me_5)Rh(PMe_3)(\eta^2\text{-}C_6F_6)$ affords the oxidative addition product $Cp^*Rh(PMe_3)(C_6F_5)F$. Interestingly, no oxidative addition is observed under thermal (110°C) conditions[43]. The less basic cyclopentadienyl analogue does not exhibit thermal or photochemical C–F bond activation chemistry [43]. In the case of partially fluorinated arenes, C–H bond activation is always observed in preference to C–F cleavage in this system.

3 Transition Metal Reagents for Aromatic C–F Bond Activation

3.1 Ligand Based Reactions Leading to C–F Bond Functionalization

Successful examples of chelate assisted activation of aromatic C–F bonds provided the foundation for further exploration of the organometallic chemistry of fluorocarbons. While early examples were limited to perfluorinated aromatic systems, the scope of this process is now fully defined using $W(CO)_3$ and $PtMe_2$ metal fragments [44, 29]. In addition to these amine, imine based ligand systems, several new examples of C–F activation have been achieved using a variety of later transition metals.

Shaw and coworkers [45] showed that cyclometallation of a C–F bond of a pentafluorobenzaldehyde azine phosphine ligand takes places at Ir(I). Coordinated fluorinated ligands continue to provide opportunities to discover unusual reaction chemistry. Saunders and coworkers [46] discovered several interesting reactions involving C–F activation in the chelating phosphine $(C_6F_5)_2PCH_2CH_2P(C_6F_5)_2$ upon treatment with $\{M(\eta^5\text{-}C_5Me_5)Cl(\mu\text{-}Cl)\}_2$ (M=Rh or Ir) in refluxing benzene (Scheme 1).

Activation of two C–H bonds of the pentamethylcyclopentadienyl ligand followed by formation of two C–C bonds affords the final product with net elimination of HF. In the absence of ethanol, thermolysis of $[M(\eta^5\text{-}C_5Me_5)Cl)(\eta^2\text{-}(C_6F_5)_2PCH_2CH_2P(C_6F_5)_2)]^+$ does not lead to C–F activation, suggesting an open coordination site on the metal is required for this transformation. Ethanol could also act as a reducing agent to generate coordinatively unsaturated Ir(I) or Rh(I) complexes. Interestingly, the less heavily fluorinated phosphine $(C_6H_3F_2\text{-}2,6)_2PCH_2CH_2P(C_6H_3F_2\text{-}2,6)_2$ exhibits similar C–F activation and functionalization chemistry as described above for M=Rh, but not for M=Ir where only the simple phosphine coordination complex is formed [47]. The reduced reactivity of the 2,6-difluoro substituted system is expected on electronic grounds but the reasons for the different behavior between Rh and Ir are unknown. Perhaps this difference can be ascribed to the enhanced kinetic lability of second row transition metals coupled with the increased thermodynamic stability of metal ligand bonds in the third row. These transformations are of interest since they also provide examples of net functionalization of a C–F bond to form a C–C bond.

As in the chemistry of the C–H bond, an important problem in organometallic fluorocarbon chemistry is to discover if further reaction chemistry is possible once a C–F bond is successfully attacked. Kiplinger and coworkers demonstrated [48] that the tungsten(II) metallacycles obtained from C–F activation can be

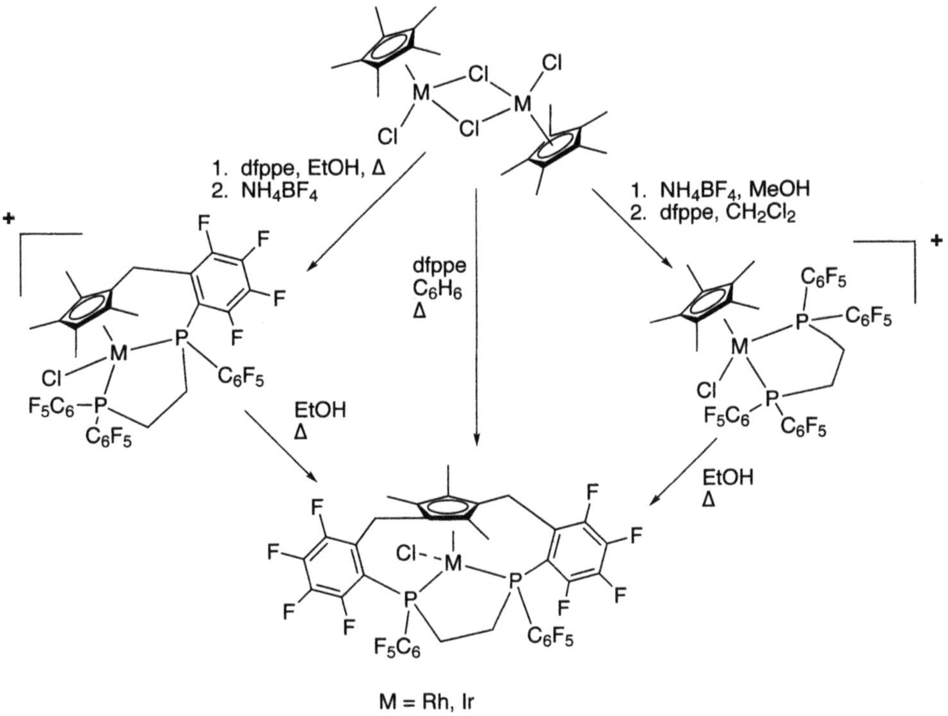

M = Rh, Ir

Scheme 1.

further functionalized by treatment with alkynes under mild conditions to afford η^2-vinyl or metallacyclopropene complexes in good yields.

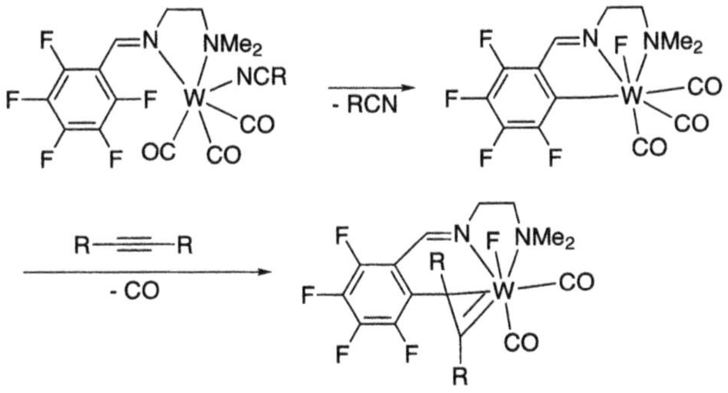

(11)

Electron poor alkynes are readily trapped by migratory insertion of the highly fluorinated phenyl ligand but competitive formation of 4-electron donor alkyne complexes is observed for electron rich alkynes [48]. If a CO ligand is removed by photolysis, migratory insertion is rapid at room temperature and detailed kinetic studies of the thermal reaction [49] have been reported. The above chemistry demonstrates that C–F bond functionalization reactions can be accomplished but unfortunately the products are even more tightly bound to the metal making liberation of the unusual organic products an unlikely prospect.

3.2
Intermolecular C–F Bond Activation

Reductive defluorination of polyfluorinated arenes has recently been achieved by reaction with zinc in aqueous ammonia [50]. Fahey and Mahan [51] provided probably the first report of C–F oxidative addition by reacting hexafluorobenzene with Ni(0) in the presence of PEt_3 but the product *trans*-$Ni(PEt_3)_2(C_6F_5)F$ was reported to be thermally unstable and isolated in low yield (7%). Perutz and coworkers [52] have now placed this chemistry on a firm footing including full spectroscopic and crystallographic characterization of the product which was isolated in 48% yield limited only by the high solubility of the product (Scheme 2).

The structure has the expected square planar arrangement about Ni with a Ni-F bond distance of 1.836(2) Å. This reaction is sluggish at room temperature

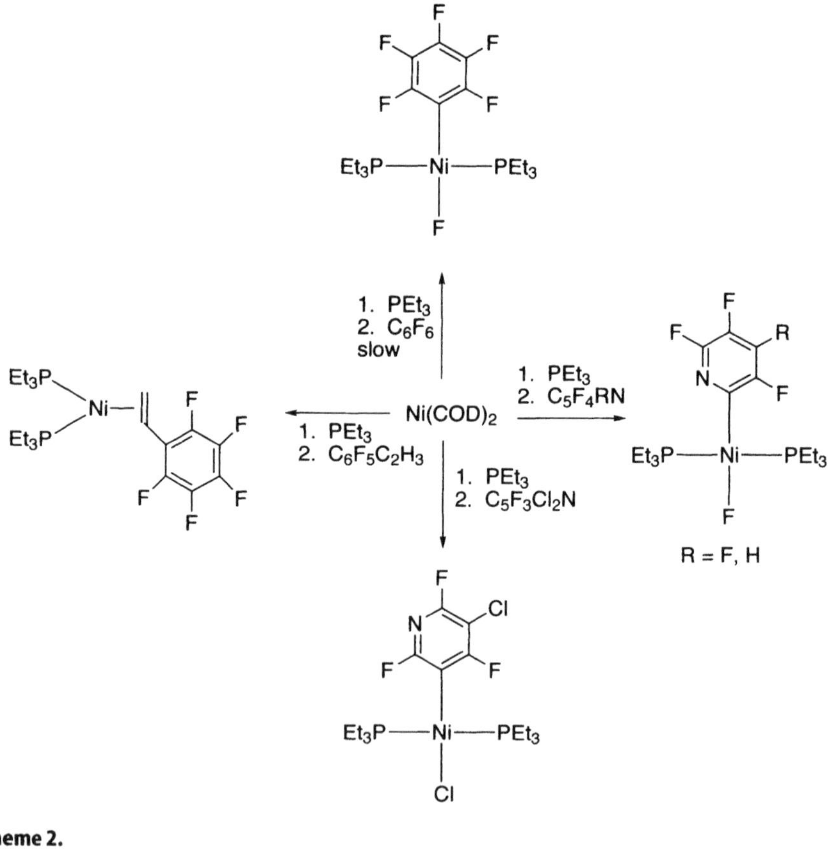

$Ni(COD)_2$

1. PEt_3 2. C_6F_6 slow

1. PEt_3 2. C_5F_4RN

1. PEt_3 2. $C_6F_5C_2H_3$

1. PEt_3 2. $C_5F_3Cl_2N$

R = F, H

Scheme 2.

and attempts to speed the reaction by variation of temperature and solvent were not fruitful. Excess hexafluorobenzene led to a side reaction with the phosphine to form some difluorophosphoranes. Importantly, the reaction of pentafluoropyridine with $Ni(PEt_3)_3$ was rapid to afford predominantly the 2-tetrafluoropyridyl isomer of the three possible C–F activation products. In the case of 2,3,5,6-tetrafluoropyridine, the 2-metallated product of C–F, not C–H, bond activation was the major product and its identity confirmed crystallographically with a Ni-F bond distance of 1.856(2) Å. Initial coordination of the poor σ-base pyridine nitrogen may be enhanced by the π-acceptor properties of pentafluoropyridine ring and serve to direct the C–F activation regioselectively to the *ortho* position. The propensity for ortho-C–F activation makes the observation of high regioselectivity noted for the 2,3,5,6-tetrafluoropyridine less useful in distinguishing between the propensity for this complex to attack C–F versus C–H bonds. However, only C–F activation was noted for the reaction of pentafluorobenzene, and quantities of 1,2,4,5-$C_6F_4H_2$ were also detected in the reaction mixture [52].

Recent reports of two other C–F activation reactions of hexafluorobenzene at Ni(0) have appeared. An η^2-C_6F_6 has been implicated as an intermediate in the preparation of L_2Ni $(C_6F_5)F$ where L_2 is a bulky bidentante phosphine ligand [53].

(12)

Low temperature experiments [54] provide evidence for precoordination of C_6F_6 to $Ni(bipy)Et_2$ prior to formation of $Ni(bipy)(C_6F_5)_2$.

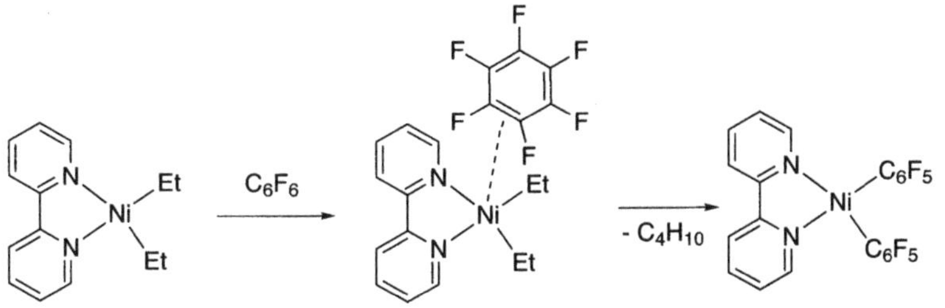

(13)

As noted above (Sect. 2.2), Lewis acids are useful reagents for electrophilic activation of C–F bonds and have a long and continuing fascinating history in organometallic chemistry. Continuing this trend, Green and coworkers [55] have discovered an unusual C–F bond activation reaction of tris(pentafluorophenyl)boron with transition metal alkyls as illustrated below.

$$CpFe(CO)_2Me + B(C_6F_5)_3 \xrightarrow[-\,B(C_6F_5)_2F]{} \qquad (14)$$

In the products, which were completely characterized by spectroscopic methods and X-ray crystallography, the tetrafluorophenyl group is bound to the metal as part of a five-membered acetylaryl metallacycle. A mechanism involving Lewis acid assisted alkyl migration to generate a coordinatively unsaturated metal center which participated in the process of C–F bond cleavage was proposed. The cleaved fluorine is bound to boron in $(C_6F_5)_2BF$. Photolysis in the presence of PMe_3 promotes unexpected net hydrogenolysis of the C–F bond *trans* to the acyl group in moderate yield [55]. Deuterium labeling experiments would be useful to determine the source of the hydrogen atom and whether the carbon has radical or anionic character after C–F activation.

Several examples of electron rich metal hydrides that react with highly fluorinated aromatics have been noted. Reaction of pentafluorobenzonitrile with *trans*-$Pt(PCy_3)_2H_2$ was reported to proceed by an electron transfer pathway to afford tetrafluorobenzonitrile and a Pt aryl complex in moderate yield [56]. Even more remarkable is the reaction of *cis*-$[Ru(dmpe)_2H_2]$ with hexafluorobenzene at –78°C which occurred to form *trans*-$[Ru(dmpe)_2(C_6F_5)H]$ as determined by X-ray crystallography and interpretation of NMR spectroscopic data [57].

$$\xrightarrow[-\,HF]{C_6F_6,\ -78\ ^\circ C} \qquad (15)$$

Importantly, less heavily fluorinated aromatics react to give only products of C–F activation. A radical mechanism is proposed and the thermodynamic driving force is formation of HF. Further investigation [58] of this system reveals that this chemistry can also lead to the formation of a coordinated bifluoride complex that can be suppressed by addition of Et_3N.

In an effort to understand the high reactivity of electron rich transition metal hydrides, Edelbach and Jones [59] have studied the mechanism of the reaction of $(C_5Me_5)Rh(PMe_3)H_2$ with perfluoroaromatics in pyridine/benzene solution.

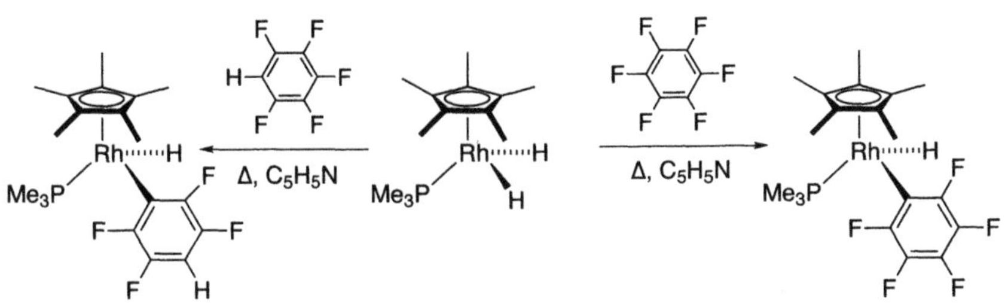

(16)

The presence of pyridine was necessary for high yields (and to sequester the HF by-product) and also increase the rate of reaction. Reaction with hexafluorobenzene has a half-life of approximately 12 h at 85°C but proceeds cleanly in high yield. Pentafluorobenzene is activated exclusively at the *para*-position and perfluoronaphthalene at the β-position. Qualitatively, the rate of reaction varies by a factor of about 25 for different arenes and also is dependent on the arene concentration. Kinetic plots do not follow simple first or second order behavior but appear to be autocatalytic in nature. The key observation is that the rate of reaction is greatly increased in the presence of added fluoride ion with the half-life at room temperature reduced by nearly an order of magnitude in the presence of 0.21 M fluoride. A mechanism involving deprotonation of the metal hydride to afford $[(C_5Me_5)Rh(PMe_3)H]^-$ and rate determining nucleophilic aromatic substitution on the perfluoroarene as the key C–F cleavage step was proposed.

$$(C_5Me_5)Rh(PMe_3)H_2 + \text{base} \xrightleftharpoons{\text{fast}} [(C_5Me_5)Rh(PMe_3)H]^- + \text{baseH}^+$$

$$[(C_5Me_5)Rh(PMe_3)H]^- + C_6F_6 \xrightarrow{\text{slow}} [(C_5Me_5)Rh(PMe_3)(C_6F_5)H] + F^-$$

$$F^- + \text{baseH}^+ \xrightleftharpoons{\text{fast}} \text{base}\cdot\text{HF}$$

(17)

Independent generation of the anion showed it to be kinetically competent for this scheme and the rate was unaffected by free radical traps such as 9,10-dihydroanthracene. Although catalytic chemistry was not achieved in this system, the authors suggest [59] this mechanism may be operative in related metal hydride systems that catalytically hydrogenate C–F bonds.

3.3 Catalytic Chemistry of Aromatic C–F Bonds

A major driving force for the the study of the organometallic chemistry of fluorocarbons is the potential for development of catalysis for C–F bond functionalization. The stability imparted by perfluorinated ligands in organometallic com-

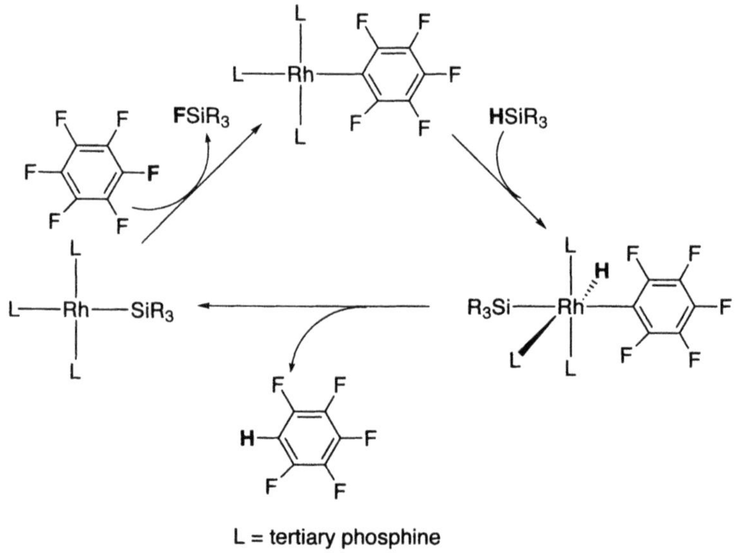

L = tertiary phosphine

Scheme 3.

plexes is often at odds with this endeavor. Aizenberg and Milstein [60] reported the first homogeneous catalyst for selective hydrogenolysis of C–F bonds in hexafluorobenzene and pentafluorobenzene under relatively mild conditions in comparison to heterogeneous C–F hydrogenation reactions (Scheme 3).

In the presence of a trialkylsilane as the terminal fluoride acceptor and hydride source, $(PMe_3)_3RhSiR_3$ promotes sequential C–F bond activation, silyl-hydride addition, and C–H reductive elimination of C_6F_6 to afford C_6F_5H and then more slowly 2,3,5,6-$C_6F_4H_2$ at 95°C. The regioselectivity is typical for C_6F_5H and each step of the proposed catalytic cycle was independently demonstrated. A related system based on $(PMe_3)_3RhC_6F_5$ as the catalyst utilizes molecular hydrogen (85 psi) and base to remove HF to effect the same transformations [61].

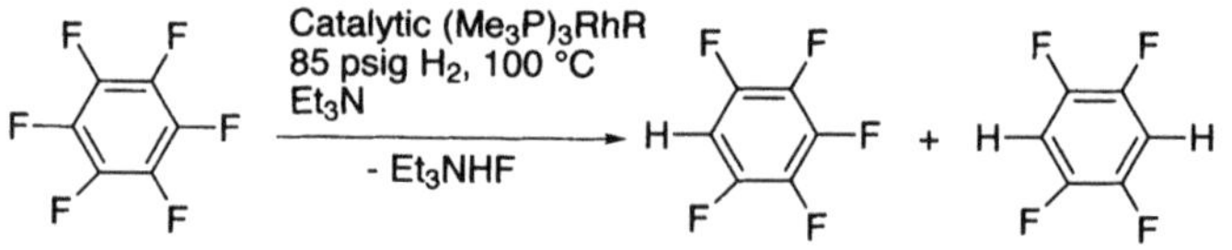

Catalytic $(Me_3P)_3RhR$, 85 psig H_2, 100 °C, Et_3N; $-\ Et_3NHF$ (18)

Reactions are run in neat hexafluorobenzene as solvent with turnover numbers as high as 114 reported for a 36-h reaction at 32% conversion. The catalyst precursor could be isolated in 52% yield from a typical reaction but some PMe_3 is consumed by reaction with fluoride to afford F_2PMe_3.

Another rhodium based system has been reported by Murai and coworkers [62] to be effective for catalyzing Si-F exchange between aromatic C–F bonds adjacent to acyl or oxazoline functionality.

$$C_6F_5C(O)Me \xrightarrow[10\%\ Rh(COD)_2BF_4]{Me_3SiSiMe_3,\ 130\ ^\circ C} C_6F_4(SiMe_3)C(O)Me + FSiMe_3 \qquad (19)$$

Reaction of pentafluoroacetophenone with hexamethyldisilane for 20 h in toluene in a 130°C oil bath in the presence of 10 mol% $Rh(cod)_2BF_4$ gave a 79–88% yield of 2,3,4,5-tetrafluoro-6-trimethylsilylacetophenone. Not surprisingly, 2,6-difluoroacetophenone affords the mono-Me_3Si-F exchange product in somewhat lower yields (33–48%). In the case of a related oxazoline derivative some disilylation accompanied the mono-substituted product. The authors propose a chelate assisted mechanism for the initial C–F activation step [62]. It is interesting to note that these catalytic reactions all involve the later transition metal rhodium with a relatively labile Rh-F bond removed as R_3Si-F or HF. However, related catalytic reactions of aromatic C–F bonds have also been discovered for early transition metals and even in the very electropositive lanthanide series.

Taking advantage of similar structural features in the Murai system [62], Deacon and coworkers earlier found that $Cp_2Yb(dme)$ acts as a catalyst to cleave ortho-C–F bonds of fluorinated benzoic acids utilizing magnesium as the terminal reductant [63]. Kiplinger and Richmond [64] discovered that catalytic reaction chemistry of fluorocarbons was not precluded by the great strength of the metal-fluoride bond. Room temperature hydrogenolysis of C_6F_6 was achieved by in situ generation of low valent zirconocene by $Mg/HgCl_2$ reduction of Cp_2ZrCl_2 to afford C_6F_5H and $C_6F_4H_2$ in a stepwise fashion [65]. Deuterium labeling studies showed that THF solvent was the hydrogen atom donor suggesting a radical process for H for F exchange. Control experiments demonstrated that the metallocene was necessary for these reactions. Two C–F bonds in octafluoronapthalene were also replaced by C–H bonds in a stepwise fashion.

Together these examples demonstrate that catalytic chemistry is possible for aromatic C–F bonds at least in the case of these relatively simple transformations, which might be better termed "defunctionalization" reactions since a fluorine is replaced by hydrogen. In addition all of these systems rely on highly fluorinated substrates which make them particularly susceptible to attack by either electron transfer or classical nucelophilic aromatic substition mechanisms. Promoting reactions in saturated perfluorocarbons presents an even greater challenge to the chemist.

4
Reactions of Aliphatic C–F Bonds with Metal Reagents

4.1
Stoichiometric Transformations

The trifluoromethyl group provides a convenient model for the behavior of larger perfluorocarbons. Thus it is appropriate to mention a system [66] where a ligand designed to promote C–F cleavage instead resulted in carbon-carbon bond activation at Rh(I). No evidence for the anticipated C–F activation product, which is known for the hydrocarbon analogue [67], was detected. In contrast to the known reactions of the CF_3 group with electrophiles (Sect. 2.2), the electron rich metal center is unable to effect C–F activation in this instance [66].

Two fascinating reports from Hughes and coworkers provide new perspectives on electrophilic attack on saturated fluorocarbons attached to cyclopentadienyl rhodium complexes. The first report [68] describes a perplexing observation in which Tl(I) salts selectively abstract fluoride from a tertiary C–F bond on a cyclopentadienyl ligand whereas Ag(I) abstracts iodide from the rhodium metal center.

F_3C CF_3 F Rh—H Me_3P PMe_3 (+) $\xleftarrow{Ag^+}$ F_3C CF_3 C–F H Rh—I Me_3P PMe_3 $\xrightarrow{Tl^+}$ F_3C CF_3 H Rh—I Me_3P PMe_3 (+) (20)

Fluoride abstraction requires a tertiary C–F bond which has been termed the "Achilles heel" of saturated fluorocarbons [68]. The reasons for this puzzling difference in reactivity are not yet clear but this observation is important to consider given the widespread use of these reagents to abstract halides in organic and organometallic synthesis.

Electrophilic activation and hydrolysis of α-CF bonds in transition metal complexes is well precedented with strong Lewis or protic acids (Sect. 2.2). Abstraction of halide using $AgBF_4$ in moist dichloromethane from the perfluorobenzyl and perfluoropropyl complexes **1** and **2** affords the cationic aqua complexes **3**$^+$ and **4**$^+$ which exist as hydrogen bonded dimers in the solid state with tetrafluoroborate anions bridging the coordinated aqua ligands [69] (Scheme 4).

Upon dissolution in a noncoordinating solvent, the benzyl complex hydrolyzes readily with the carbon of the CF_2 group ultimately transformed to coordinated carbon monoxide. Coordination of the aqua ligand enhances its acidity to participate in this transformation. In the case of the less reactive perfluoropropyl system, replacement of $[BF_4]^-$ by the nonhydrogen bonding $[B(C_6H_3(CF_3)_2\text{-}3,5)_4]^-$ anion enhances the activity of the coordinated aqua ligand to induce hydrolysis of the α-CF_2 group of the perfluoroalkyl chain as well [69]. Thus coor-

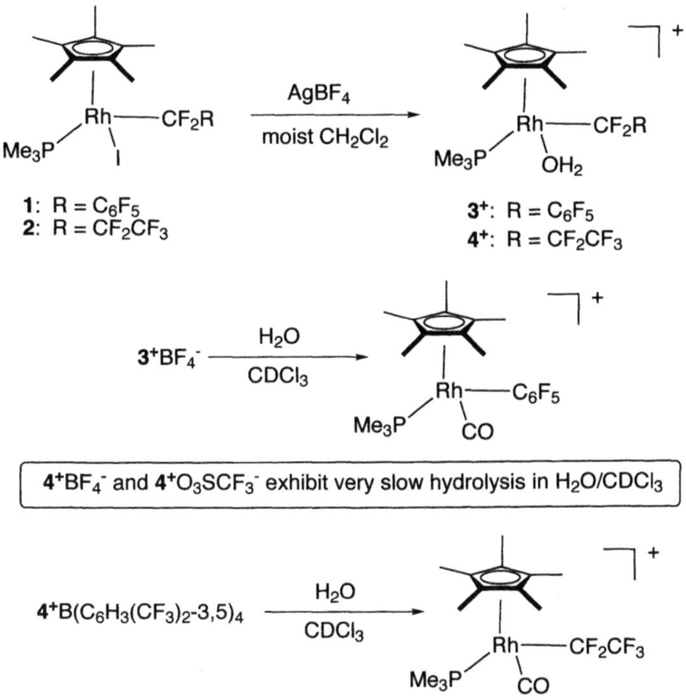

4^{+}BF$_4^-$ and 4^{+}O$_3$SCF$_3^-$ exhibit very slow hydrolysis in H$_2$O/CDCl$_3$

Scheme 4.

dination and hydrogen bonding are useful parameters in tuning the reactivity of water to control and enhance C–F bond hydrolysis under exceptionally mild conditions.

Despite the mature nature of this field [22], Huang and Caulton [70] have also discovered some unusual chemistry of the trifluoromethyl group bound to a coordinatively unsaturated Ru(II) phosphine complex. As shown below, treatment of RuHF(CO)L$_2$ (L=P^{t}Bu$_2$Me) with the trifluormethyl anion source Me$_3$SiCF$_3$/CsF affords a difluorocarbene complex by α-fluoro migration.

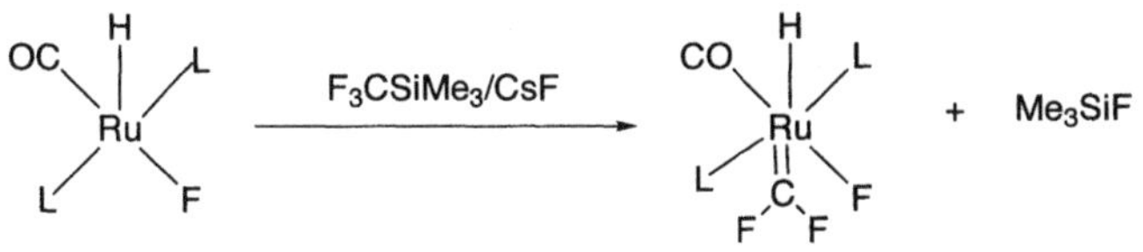

$$\text{RuHF(CO)L}_2 \xrightarrow{\text{F}_3\text{CSiMe}_3/\text{CsF}} \text{RuHF(CO)(=CF}_2\text{)L}_2 + \text{Me}_3\text{SiF} \quad (21)$$

NMR spin saturation transfer experiments indicate that this process is facile at 75°C and a η^2-CF$_3$ group is postulated as the transition state structure.

CO, H, L, Ru, L, CF_3 ⇌ [CO, H, L, Ru----F, L, C, F, F]‡ ⇌ CO, H, L, Ru, L, C, F, F, F (22)

The absence of π-orbitals in saturated perfluorocarbons eliminates the most common modes of attack noted for fluorinated alkenes and arenes. However, the high electron affinity of these compounds makes them susceptible to attack by reducing agents. Electron transfer to a C–F σ^* orbital can lead to facile C–F bond cleavage and loss of fluoride. Heterogeneous aromatization of cyclic perfluorocarbons takes place using Ni or Fe metals at 400–600°C in a flow tube to remove the products and prevent over reduction [71]. Under ultra high vacuum conditions [72], clean iron surfaces defluorinate trifluoromethylaryl ethers at –120°C. Homogenous systems have been discovered which enable selective defluorination under mild conditions. Bennett and coworkers [73] have shown that the relatively mild reducing agent cobaltocene reacts with perfluorodecalin at room temperature to liberate fluoride ions forming $[Cp_2Co]F$, which serves as a source of "naked" fluoride. In the presence of LiO_3SCF_3 as a fluoride ion acceptor, perfluoronapthalene is formed in high yield [73].

$$\text{perfluorodecalin} \xrightarrow[-10\ Cp_2CoO_3SCF_3]{10\ Cp_2Co,\ 10\ LiO_3SCF_3} C_{10}F_8 + 10\ LiF \quad (23)$$

Perfluoronapthalene appears to form a charge transfer complex with cobaltocene but no further C–F activation is detected [74]. The crystal structure of a ruby-colored ferrocene-perfluorophenanthrene molecular complex has recently been reported [75]. The presence of only one tertiary C–F bond is required for reactivity in this system as illustrated by the synthesis of perfluorotoluene from perfluoromethylcyclohexane under similar conditions.

$$\text{perfluoromethylcyclohexane} \xrightarrow[-6\ (C_5Me_5)_2CoO_3SCF_3]{6\ (C_5Me_5)_2Co,\ 6\ LiO_3SCF_3} C_6F_5CF_3 + 6\ LiF \quad (24)$$

The more strongly reducing $(Me_5C_5)_2Co$ reacts more rapidly than the parent cobaltocene, but the acidity of the methyl groups in the cation conspired against detection of the fluoride ion [73,74]. Although electron transfer is clearly important in the initial C–F activation step in these reactions, additional mechanistic study is required to assess the scope of these processes. In particular, the role of the lithium cation and possible intermediates such as perfluorinated anions need to be addressed.

A particularly active reductive defluorination system [64] prepared from $Cp_2ZrCl_2/Mg/HgCl_2$ is able to attack perfluorocyclohexane, a substrate without tertiary C–F bonds, with 2,3,5,6-tetrafluorobenzene obtained in 35% yield as the final organic product.

$$\text{C}_6\text{F}_{12} \xrightarrow[\text{THF-d}^8]{Cp_2ZrCl_2/Mg/HgCl_2} 1,4\text{-D}_2\text{-}2,3,5,6\text{-C}_6\text{F}_4 \quad (25)$$

Reaction in THF-d^8 reveals that solvent is the source of the two hydrogens in the final product. This reaction is especially significant since it shows that perfluorocarbons without tertiary C–F bonds are also subject to defluorination chemistry [64]. Remarkably, related transformations of perfluorocarbons can be carried out using solid sodium oxalate as the (rather weak) reducing agent [76] in a flow system at 425°C.

$$\text{perfluorodecalin} \xrightarrow[\text{- NaF}]{Na_2C_2O_4,\ 425\ ^\circ C} C_{10}F_8 + \text{perfluorotetralin} \quad (26)$$

Fluoride is trapped as NaF and under certain conditions the yield of perflorotetralin can be maximized. Lewis acidic sites in crystal surface defects are thought to assist in the C–F activation process and at 470°C the system is active for demineralization of chlorofluorocarbons to afford NaF, NaCl and carbon [76]. Detailed electrochemical studies of the reduction of C–F bonds in aryltrifluoromethanes and fluoroalkoxyarenes have been reported [77]. Free radicals cleave C–F bonds of fluorinated self-assembled monolayer surfaces [78].

4.2 Catalytic Reactions of Saturated Perfluorocarbons

Perhaps the most exciting recent advances in the organometallic chemistry of fluorocarbons have been the contemporaneous and complementary discoveries of examples of catalytic activation and functionalization of perfluorocarbons in laboratories led by Crabtree and Richmond [23]. Kiplinger and Richmond [64] showed that Group 4 metallocenes function as catalysts in the synthesis of perfluoronapthalene from perfluorodecalin using activated Mg or Al as the terminal reductant. Low valent "zirconocene" or "titanocene" species were postulated as intermediates in the catalytic cycle and control experiments showed the central role played by the metallocene in mediating electron transfer in these systems. Turnover numbers up to 12 (net removal of 120 fluorines/metallocene) were noted [64].

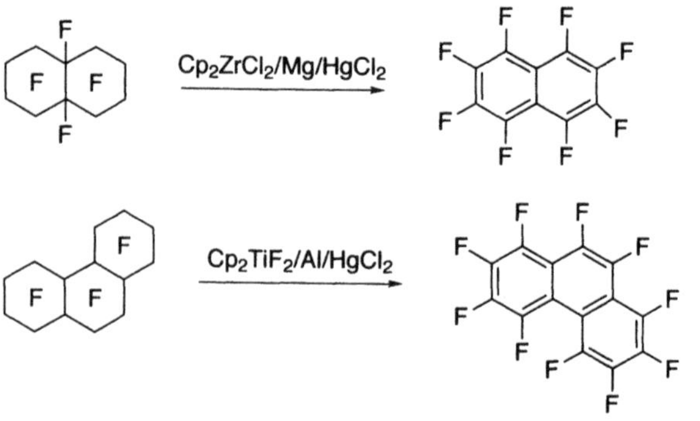

$Cp_2ZrCl_2/Mg/HgCl_2$

$Cp_2TiF_2/Al/HgCl_2$

(27)

Independently generated low valent "zirconocene" and "titanocene" were demonstrated to be competent reagents for defluorination in the absence of terminal reductant. The critical role played by the metallocenes is best illustrated by noting that, in the absence of metallocene, $Mg/HgCl_2$ causes exothermic reduction of perfluorodecalin to carbon. The milder aluminum system only functions in the presence of metallocene. Additional insight into the nature of this multistep process was obtained by showing that plausible flurocarbon intermediates react to afford perfluorodecalin [79].

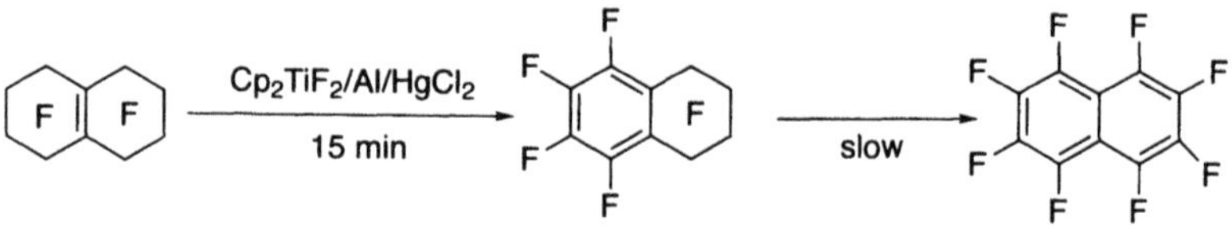

$Cp_2TiF_2/Al/HgCl_2$ 15 min

slow

(28)

Some catalyst could be recovered after completion of the reaction but the nature of catalyst deactivation requires further study [79]. The cobaltocene chemistry reported above (Sect. 4.1) can also be made catalytic with respect to the metallocene by using Al or Hg as the terminal reductant [74].

Building on related work in C–H bond activation chemistry [42], Crabtree and Burdeniuc [41] have discovered two metal based photochemical systems which are useful for the functionalization of perfluorocarbons. Under Hg photosensitization, saturated perfluorocarbons react with ammonia to afford amine, imine and cyano-derivatives [41] (Scheme 5).

Although a tertiary C–F bond is required, this chemistry is not limited to cyclic systems. Electron transfer is thought to occur from an exciplex such as $[Hg^*(NH_3)_2]^+$ which has a calculated ionization potential comparable to cesium metal.

Ultraviolet irradiation of perfluorocarbons in the presence of $(C_5Me_5)_2Fe$ and LiO_3SCF_3 as a fluoride acceptor provides an exceptionally mild system [80] to prepare perfluoroalkenes from perfluoroalkanes and the chemistry can be made catalytic with respect to iron by using zinc to reduce $[(C_5Me_5)_2Fe]^+$(Scheme 6).

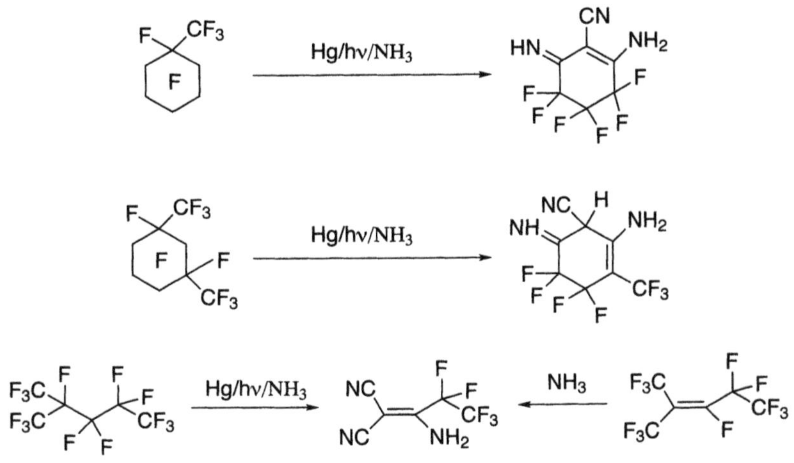

Scheme 5.

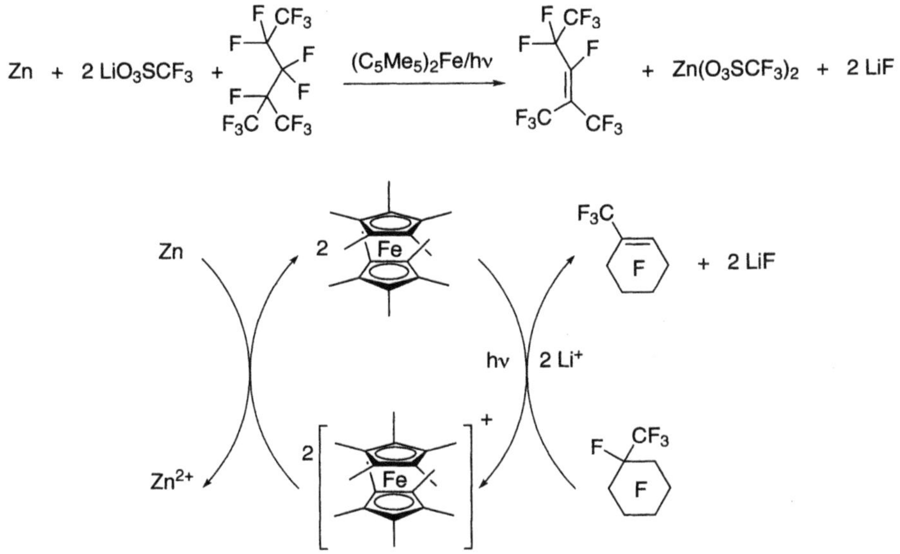

Scheme 6.

Although up to 110 turnovers were reported at 12% completion, the reaction cannot be forced to completion because of over-reduction problems and this makes product separation difficult. Since the perfluoroalkene products are much more reactive than the perfluoroalkane starting materials, this difficulty can be partially overcome by performing further transformations of the mixture in situ [80].

4.3 C–F Bond Activation in the Service of Organic Synthesis

Compared to their widespread use in modern organic synthesis, transition metal catalysts have only recently been applied to the chemistry of fluorocarbons. The research outlined above suggests that new methodologies will be discovered in the next millennium. The selectivity inherent in effective homogeneous catalytic systems will be particularly important in preparation of unsaturated or functionalized compounds which are often more reactive than perfluorocarbon starting materials. Given the high value of fluorinated materials, even stoichiometric transformations using metal reagents may be attractive. The increasing availablity of saturated perfluorocarbons from direct fluorination [81] of hydrocarbon precursors will provide numerous new substrates to test the limits of the defluorination technology discussed above.

Lagow [82] has utilized sodium benzophenone anion radical to defluorinate and intramolecularly couple perfluorodicyclohexyl ether to afford perfluorodibenzofuran in 60% yield.

$$\xrightarrow[-78 \rightarrow 20\ ^{\circ}C]{Na^+ Ph_2CO^-} \quad + 14\ NaF \tag{29}$$

This is the first high yield example of using reductive defluorination to induce C–C bond coupling. Perfluorodiphenyl ether is not an intermediate in this reaction and treatment of perfluorodicyclohexyl affords only perfluorodiphenyl rather than a fused ring system.

$$\xrightarrow[-78 \rightarrow 20\ ^{\circ}C]{Na^+ Ph_2CO^-} \quad + 12\ NaF \tag{30}$$

Utilizing magnesium anthracene, Crabtree and coworkers [83] have shown that aromatic fluorinated Grignard reagents can be prepared from either perfluorocyclohexane or perfluoro(methycyclohexane) albeit in poor yields (10–20%). Better success (34%) was noted for the synthesis of perfluorobenzoic acid from hexafluorobenzene using the Grignard generated by this method. In its present state this work represents a "proof of concept" rather than a useful synthetic method. Finally, Bennett [74] and Kiplinger [79] have noted a remarkable selectivity in the reaction of bis(benzene)chromium with perfluorodecalin to yield exclusively perfluorotetralin at room temperature.

$$\xrightarrow[-6\ (\eta^6\text{-}C_6H_6)_2Cr^+F^-]{excess\ (\eta^6\text{-}C_6H_6)_2Cr} \tag{31}$$

Further defluorination to octafluoronapthalene, which is often the product of similar reactions, does not occur. Coupled with terminal reductants such as aqueous dithionite this reaction can be performed under catalytic conditions with respect to $(C_6H_6)_2Cr$ [74]. Together these reactions provide a glimpse of possible future directions for research into the application of C–F bond activation chemistry to synthetic problems.

5 Theoretical Approaches to C–F Bond Activation

Many of the C–F activation reactions discussed in this account are complicated by multi-step reaction sequences and secondary reactions with solvent or fluoride ion generated in the course of the transformation. In catalytic systems, the additional reagents needed to achieve catalytic turnover may cause undesired side reactions to take place. Gas-phase ion molecule reactions provide one way to study model systems in the absence of these complicating factors to obtain fundamental information on the C–F activation process [84]. In addition, theoretical treatments of C–F bond activation are beginning to provide insight into these transformations.

Schwartz and colleagues have extensively studied the reaction of fluorocarbons with various metal ions in the gas phase. Generation of FeF^+ (from C_6H_5F) in an FT-ICR spectrometer and study of its reactions allowed the Fe-F bond energy to be bracketed between 86 and 101 kcal mol^{-1}, with theoretical calculations supporting the higher value [85]. A study of six representative lanthanide cations with several organic fluorides showed that metals with the lowest second ionization energy were most reactive [86]. For the Ca^+ cation, the rate of C–F activation increased in the series $CH_3F<CH_3CHF_2<C_6H_5F<C_6F_6$. No reaction was observed for CF_3H and CF_4 even though all of the reactions to form CaF^+ (BDE= 140.8 kcal mol^{-1}) and organic radical are thermodynamically favorable [87]. Ab initio MO and VB calculations suggest the importance of both electron transfer and metal-fluorocarbon bonding in the reductive defluorination process [87].

Chen and Freiser [88] have used similar techniques to study the chemistry of $FeCF_3^+$ and $CoCF_3^+$ with a series of alkanes and alkenes. Collision induced dissociation indicates the structures of these ions prepared from CF_3I are in fact best described as $[FM^+\cdots F_2C]$ ion dipole complexes, but a minimum is also noted in the potential energy surface for the fluoro-difluorocarbene structure $[F\text{-}M^+{=}CF_2]$. The calculated transition state for C–F activation involves an η^2-interaction of a CF bond of the trifluoromethyl group [88]. Interestingly, this transition state is similar to that proposed by Caulton (Eq. 21) [70] for α-fluoride migration in his Ru-CF_3 system. Note that these are *not* considered to be agostic interactions as found in binding a C–H bond to a metal since the fluorine lone pairs can participate in bonding, albeit weakly, in the transition state.

A similar transition state was computed by density functional theory for the hypothetical reaction of the 14 electron complex *trans*-$MX(PH_3)_2$, where M=Rh or Ir and X=H, Cl or CH_3 with fluoromethane [89]. An advantage of theory is that competitive C–H activation reactions could be ignored. All of the cases studied were thermodynamically favored by 23 kcal mol^{-1} (M=Rh, X=CH_3) to 51 kcal mol^{-1} (M=Ir, X=Cl). Surprisingly, Ir complexes have lower activation barriers than analogous Rh compounds. π-Donor ligands (X=Cl) were found to decrease the activation energy for C–F activation. The authors favor a concerted process, rather than radical intermediates for oxidative addition of the C–F bond.

6 Future Perspectives

In the past decade, C–F bond activation chemistry has progressed from a laboratory curiosity to the verge of becoming a useful synthetic technique in organic chemistry. The most important issue that needs to be addressed is that of selectivity. In addition, extension of work described above to less heavily fluorinated aromatics and to aliphatic fluorocarbons in the absence of tertiary C–F bonds would be desirable. A healthy debate concerning the mechanisms of this reactions has begun and further study may guide the discovery of new catalysts for fluorocarbon functionalization. Fluorocarbons may also serve as sources of carbon in the synthesis of new materials [39, 90]. Under appropriate conditions, fluorocarbons are indeed reactive molecules [91].

References

1. Richmond TG (1995) J Chem Educ 72:731 and references therein.
2. (a) Banks RE, Sharp DWA, Tatlow JC (eds) (1986) Fluorine – the first hundred years (1886–1986). Elsevier, New York. (b) Banks RE, Smart BE, Tatlow JC (eds) (1994) Organofluorine compounds: principles and commercial applications. Plenum, New York. (c) Olah GA, Chambers RD, Prakash GKS (eds) (1992) Synthetic fluorine chemistry. Wiley, New York
3. (a) Filler R (ed) (1976) Biochemistry involving the carbon-fluorine bond. American Chemical Society, Washington, DC. (b) Welch JT (ed) (1991) Selective fluorination in organic and bioorganic chemistry. American Chemical Society, Washington, DC
4. Ravishankara AR, Solomon S, Turnipseed AA, Warren RF (1993) Science 259:194
5. Zachariah MR, Dufaux DP (1997) Environ Sci Technol 31:2223
6. Chambers RD (1973) Fluorine in organic chemistry. Wiley, New York. For recent reviews in fluorine chemistry see the special issue, (1996) Chem Rev 96:1555
7. (a) Horváth IT, Rabái J (1994) Science 266:72. (b) Gladysz JA (1994) Science 266:55. (c) Horvath IT, Kiss G, Cook RA, Bond, JE, Stevens PA, Rabai J, Mozeleski EJ (1988) J Amer Chem Soc 120:3133. (d) Guillevic MA, Rocaboy C, Arif AM, Horvath IT, Gladysz JA (1998) Organometallics 17:707. (e) Studer A, Hadida S, Ferritto R, Kim SY, Jeger P, Wipf P, Curran DP (1997) Science 275:823
8. (a) Murphy EF, Murugavel R, Roesky HW (1997) Chem Rev 97:3425. (b) Hughes RP (1990) Adv Organomet Chem 31:183. (c) Doherty NM, Hoffman NW (1991) Chem Rev 91:553
9. (a) Reed CA (1998) Acc Chem Res 31:133. (b) Strauss SH (1993) Chem Rev 93:927
10. (a) See for example Karl J, Erker G, Fröhlich R (1997) J Amer Chem Soc 119:11165 and references therein. (b) Crabtree RH (1985) Chem Rev 85:245
11. Kiplinger JL, Richmond TG, Osterberg CE (1994) Chem Rev 94:373

12. Burdeniuc J, Jedlicka B, Crabtree RH (1997) Chem Ber/Recueil 130:145
13. Wright SW, Marx D (1994) J Chem Educ 71:251
14. Kulawiec RJ, Crabtree RH (1990) Coord Chem Rev 99:89
15. Harrison RG, Arif AM, Wulfsberg G, Lang R, Ju T, Kiss G, Hoff CD, Richmond TG (1992) J Chem Soc Chem Commun 1374
16. (a) Horton AD, Orpen AG (1991) Organometallics 10:3910. (b) Yang X, Stern CL, Marks TJ (1991) Organometallics 10:840
17. Plenio H (1997) Chem Rev 97:3363
18. (a) Plenio H, Hermann J, Diodone R (1997) Inorg Chem 36:5722. (b) Plenio H, Diodone R, Badura D (1977) Angew Chem Int Ed Engl 36:156
19. (a) Purdy AP, George CF, Callahan JH (1991) Inorg Chem 30:2812. (b) Samuels JA, Lobkovsky EB, Streib WE, Folting K, Huffman JC, Zwanziger JW, Caulton KG (1993) J Amer Chem Soc 115:5093
20. Richmond TG, Shriver DF (1984) Organometallics 3:305 and references therein.
21. Ferraris D, Cox C, Anand R, Lectka T (1997) J Amer Chem Soc 119:4319
22. (a) Brothers PJ, Roper WR (1988) Chem Rev 88:1293. (b) Gallop MA, Roper WR (1986) Adv Organomet Chem 25:121. (c) Roper WR (1986) J Organomet Chem 300:167
23. Saunders GC (1996) Angew Chem Int Ed Engl 35:2615
24. Richmond TG, Osterberg CE, Arif AM (1987) J Amer Chem Soc 109:8091
25. (a) Ceder RM, Granell J, Muller G, FontBardia M, Solans X (1996) Organometallics 15:4618. (b) Ceder RM, Granell J, Muller G, FontBardia M, Solans X (1995) Organometallics 14:5544
26. Anderson CM, Crespo M, Ferguson G, Lough AJ, Puddephatt RJ (1992) Organometallics 11:1177
27. (a) Crespo M, Martinez M, Sales J (1993) Organometallics 12:4297. (b) Poss MJ, Arif AM, Richmond TG (1988) Organometallics 7:1669
28. Lucht B, Poss MJ, King MA, Richmond TG (1991) J Chem Soc Chem Commun 400
29. López O, Crespo M, Font-Bardía M, Solans X (1997) Organometallics 16:1233
30. Hofmann P, Unfried G (1992) Chem Ber 125:659
31. (a) Simons JH, Block LP (1937) J Amer Chem Soc 59:1407. (b) Miller JF, Hunt M, McBee ET (1947) Anal Chem 19:148
32. Intergovernmental Panel on Climate Change (1990) Climate Change – The IPCC scientific assessment. Cambridge University Press, Cambridge, UK
33. (a) MacNicol DD, Robertson CD (1988) Nature 332:59. (b) MacNicol DD, McGregor WM, Mallinson PR, Robertson CD (1991) J Chem Soc Perkin Trans I:3380
34. Marsella JA, Gilicinski AG, Coughlin AM, Pez GP (1992) J Org Chem 57:2856
35. Harrison RG, Richmond TG (1993) J Amer Chem Soc 115:5303
36. Burns CJ, Andersen RA (1989) J Chem Soc Chem Commun 136
37. Watson PL, Tulip TH, Williams I (1990) Organometallics 9:1999
38. Heinemann C, Goldberg H, Tornieporth-Oetting IC, Klapötke TM, Schwarz H (1995) Angew Chem Int Ed Engl 34:213
39. Pola J, Urbanová M, Bastl Z, Plzák Z, Subrt J, Gregora I, Vorlícek V (1988) J Mater Chem 8:187
40. Kaprinidis NA, Turro NJ (1996) Tetrahedron Lett 37:2372
41. Burdeniuc J, Crabtree RH (1995) J Amer Chem Soc 117:10119
42. Muedas CA, Ferguson RR, Brown SH, Crabtree RH (1991) J Amer Chem Soc 113:2233 and references therein.
43. Chin RM, Dong L, Duckett SB, Partridge MG, Jones WD, Perutz RN (1993) J Amer Chem Soc 115:7685
44. (a) Richmond TG (1990) Coord Chem Rev 105:221. (b) Richmond TG, Osterberg CE (1994) ACS Symp Ser 555:392. (c) For C–F activation in a silica surface tethered ligand system see Looman CD, Richmond TG (1995) Inorg Chim Acta 240:479
45. Perera SD, Shaw BL, Thorntonpett M (1995) Inorg Chim Acta 233:103

46. (a) Atherton MJ, Fawcett J, Holloway JH, Hope, EG, Karaçar A, Russell DR, Saunders GC (1996) J Chem Soc Dalton Trans 3215. (b) Atherton MJ, Fawcett J, Holloway JH, Hope EG, Karaçar A, Martin SM, Saunders GC (1998) J Organomet Chem 555:67
47. Fawcett J, Friedrichs S, Holloway JH, Hope EG, McKee V, Nieuwenhuyzen M, Russell DR, Saunders GC (1998) J Chem Soc Dalton Trans 1477
48. (a) Kiplinger JL, King MA, Fechtenkötter A, Arif AM, Richmond TG (1996) Organometallics 15:5292. (b) Nitriles form η^2 (4e) donor complexes: Kiplinger JL, Arif AM, Richmond TG (1997) Organometallics 16:246. (c) Phosphaalkynes participate in a complex coupling reaction: Benvenutti MHA, Hitchcock PB, Kiplinger JL, Nixon, JF, Richmond TG (1997) Chem Commun 1539
49. Kiplinger JL, Richmond TG, Arif AM, Dücker-Benfer C, van Eldik, R (1996) Organometallics 15:1545
50. Laev SS, Shteingarts VD (1997) Tetrahedron Lett 38:3765
51. Fahey DR, Mahan JE (1977) J Amer Chem Soc 99:2501
52. Cronin L, Higgitt CL, Karch R, Perutz RN (1997) Organometallics 16:4920
53. Bach I, Porschke KR, Goddard R, Kopiske C, Kruger C, Rufinska A, Seevogel K (1996) Organometallics 15:4959
54. Yamamoto T, Abla M (1997) J. Organomet Chem 535:209
55. Chernega AN, Graham AM, Green MLH, Haggitt J, Lloyd J, Mehnert, CP, Metzler N, Souter J (1997) J Chem Soc Dalton Trans 2293
56. Hintermann S, Pregosin PS, Rüegger H, Clark HC (1992) J Organomet Chem 435:225
57. Whittlesey MK, Perutz RN, Moore MH (1996) Chem Commun 787
58. Whittlesey MK, Perutz RN, Greener B, Moore MH (1997) Chem Commun 187
59. Edelbach BL, Jones WD (1997) J Amer Chem Soc 119:7734
60. Aizenberg M, Milstein D (1994) Science 25:359
61. Aizenberg M, Milstein D (1995) J Amer Chem Soc 117:8674. For a very recent report of aromatic C–F bond hydrogenolysis catalyzed by homogeneous and heterogeneous rhodium systems see: Young Jr RJ, Grushin VV (1999) Organometallics 18:294
62. Ishii Y, Chatani N, Yorimitsu S, Murai S (1998) Chem Lett 157
63. (a) Deacon GB, Forsyth CM, Sun J (1994) Tetrahedron Lett 35:1095. (b) Deacon GB, Mackinnon PT, Tuong TD (1983) Aust J Chem 36:43
64. Kiplinger JL, Richmond TG (1996) J Amer Chem Soc 118:1805
65. Kipliinger JL, Richmond TG (1966) Chem Commun 1115
66. van der Boom ME, Ben-David Y, Milstein D (1988) Chem Commun 917
67. Rybtchinski B, Vigalok A, Ben-David Y, Milstein D (1996) J Amer Chem Soc 118:12046 and references therein
68. Hughes RP, Husebo TL, Maddock SM, Rheingold AL, Guzei IA (1997) J Amer Chem Soc 119:10231
69. Hughes RP, Linder DC, Rheingold, AL Liable-Sands LM (1997) J Amer Chem Soc 119:11544
70. Huang D, Caulton KG (1997) J Amer Chem Soc 119:3185
71. (a) Gething B, Patrick CR, Stacey M, Tatlow JC (1957) Nature 183:588. (b) Bailey J, Plevey RG, Tatlow JC (1987) J Fluorine Chem 37:1
72. (a) Napier ME, Stair PC (1991) J Vac Sci Technol A 9:649. (b) Napier ME, Stair PC (1992) J Vac Sci Technol A 10:2704
73. Bennett BK, Harrison RG, Richmond TG (1994) J Amer Chem Soc 116:11165
74. Bennett BK (1997) PhD thesis, University of Utah
75. Beck CM, Burdeniuc J, Crabtree RH, Rheingold AL, Yap GAP (1998) Inorg Chim Acta 270:559
76. Burdeniuc J, Crabtree RH (1996) Science 271:340
77. (a) Andrieux CP, Combellas C, Kanoufi F, Savéant JM, Thiébault A (1997) J Amer Chem Soc 119:9527. (b) Combellas C, Kanoufi F, Thiébault A (1997) J Electroanal Chem 432:181

78. Shen J, Grill V, Cook, RG (1998) J Amer Chem Soc 120:4254
79. Kiplinger JL (1996) PhD thesis, University of Utah
80. (a) Burdeniuc J, Crabtree RH (1996) J Amer Chem Soc 118:2525. (b) Burdeniuc J, Crabtree RH (1998) Organometallics 17:1582
81. Lagow RJ (1995) Encyclopedia of chemical technology 11:482
82. Sung K, Lagow RJ (1998) J Chem Soc Perkin Trans 637
83. Beck CM, Park YJ, Crabtree RH (1988) Chem Commun 693
84. Pradeep T, Riederer DE, Hoke SH, Ast T, Cooks RG, Linford MR (1994) J Amer Chem Soc 116:8658 and references therein
85. Schröder D, Hrusák J, Schwarz, H (1992) Helv Chim Acta 75:2215
86. Cornehl HH, Hornung G, Schwarz H (1996) J Amer Chem Soc 118:9960
87. Harvey JN, Schröder D, Koch W, Danovich D, Shaik S, Schwarz, H (1997) Chem Phys Lett 278:391
88. Chen Q, Freiser BS (1998) J Phys Chem A 102:3343
89. Su MD, Chu SY (1997) J Amer Chem Soc 119:10178
90. Li Y, Quia Y, Lio H, Ding Y, Yang L, Xu C, Li F, Zhou, G (1998) Science 281:246
91. I am grateful for the creative intellectual contributions and experimental efforts of several talented former graduate students (B.K. Bennett, R.G. Harrison, J.L. Kiplinger and C.E. Osterberg) who have carried out the bulk of the research involving C–F bond activation and functionalization at Utah. Financial support from the National Science Foundation and the Donors of the Petroleum Research Fund Administered by the American Chemical Society is gratefully acknowledged.

Author Index Volumes 1–3

Alper H see Grushin VV (1999) *3*:193–225
Anwander R (1999) Principles in Organolanthanide Chemistry. *2*: 1–62

Dowdy EC see Molander G (1999) *2*: 119–154

Fürstner A (1998) Ruthenium-Catalyzed Metathesis Reactions in Organic Synthesis. *1*: 37–72

Gibson SE (née Thomas), Keen SP (1998) Cross-Metathesis. *1*: 155–181
Gossage RA, van Koten G (1999) A General Survey and Recent Advances in the Activation of Unreactive Bonds by Metal Complexes. *3*:1–8
Gröger H see Shibasaki M (1999) *2*: 199–232
Grushin VV, Alper H (1999) Activation of Otherwise Unreactive C–Cl Bonds. *3*:193–225

He Y see Nicolaou KC, King NP (1998) *1*: 73–104
Hidai M, Mizobe Y (1999) Activation of the N–N Triple Bond in Molecular Nitrogen: Toward its Chemical Transformation into Organo-Nitrogen Compounds. *3*:227–241
Hou Z, Wakatsuki Y (1999) Reactions of Ketones with Low-Valent Lanthanides: Isolation and Reactivity of Lanthanide Ketyl and Ketone Dianion Complexes. *2*: 233–253
Hoveyda AH (1998) Catalytic Ring-Closing Metathesis and the Development of Enantioselective Processes. *1*: 105–132

Ito Y see Murakami M (1999) *3*:97–130
Ito Y see Suginome M (1999) *3*:131–159

Jones WD (1999) Activation of C–H Bonds: Stoichiometric Reactions. *3*:9–46

Kagan H, Namy JL (1999) Influence of Solvents or Additives on the Organic Chemistry Mediated by Diiodosamarium. *2*: 155–198
Kakiuchi F, Murai S (1999) Activation of C–H Bonds: Catalytic Reactions. *3*:47–79
Keen SP see Gibson SE (née Thomas) (1998) *1*: 155–181
Kiessling LL, Strong LE (1998) Bioactive Polymers. *1*: 199–231
King NP see Nicolaou KC, He Y (1998) *1*: 73–104
Kobayashi S (1999) Lanthanide Triflate-Catalyzed Carbon-Carbon Bond-Forming Reactions in Organic Synthesis. *2*: 63–118
Kobayashi S (1999) Polymer-Supported Rare Earth Catalysts Used in Organic Synthesis. *2*: 285–305
Koten G van see Gossage RA (1999) *3*:1–8

Lin Y-S, Yamamoto A (1999) Activation of C–O Bonds: Stoichiometric and Catalytic Reactions. *3*:161–192

Mizobe Y see Hidai M (1999) *3*:227–241
Molander G, Dowdy EC (1999) Lanthanide- and Group 3 Metallocene Catalysis in Small Molecule Synthesis. *2*: 119–154
Mori M (1998) Enyne Metathesis. *1*: 133–154
Murai S see Kakiuchi F (1999) *3*:47–79
Murakami M, Ito Y (1999) Cleavage of Carbon–Carbon Single Bonds by Transition Metals. *3*:97–130

Namy JL see Kagan H (1999) *2*: 155–198
Nicolaou KC, King NP, He Y (1998) Ring-Closing Metathesis in the Synthesis of Epothilones and Polyether Natural Products. *1*: 73–104

Pawlow JH see Tindall D, Wagener KB (1998) *1*: 183–198

Richmond TG (1999) Metal Reagents for Activation and Functionalization of Carbon–Fluorine Bonds. *3*:243–269

Schrock RR (1998) Olefin Metathesis by Well-Defined Complexes of Molybdenum and Tungsten. *1*: 1–36
Sen A (1999) Catalytic Activation of Methane and Ethane by Metal Compounds. *3*:81–95
Shibasaki M, Gröger H (1999) Chiral Heterobimetallic Lanthanoid Complexes: Highly Efficient Multifunctional Catalysts for the Asymmetric Formation of C-C, C-O and C-P Bonds. *2*: 199–232
Strong LE see Kiessling LL (1998) *1*: 199–231
Suginome M, Ito Y (1999) Activation of Si–Si Bonds by Transition-Metal Complexes. *3*:131–159

Tindall D, Pawlow JH, Wagener KB (1998) Recent Advances in ADMET Chemistry. *1*: 183–198

Wagener KB see Tindall D, Pawlow JH (1998) *1*: 183–198
Wakatsuki Y see Hou Z (1999) *2*: 233–253
Yamamoto A see Lin Y-S (1999) *3*:161–192
Yasuda H (1999) Organo Rare Earth Metal Catalysis for the Living Polymerizations of Polar and Nonpolar Monomers. *2*: 255–283

Springer and the environment

At Springer we firmly believe that an international science publisher has a special obligation to the environment, and our corporate policies consistently reflect this conviction.

We also expect our business partners – paper mills, printers, packaging manufacturers, etc. – to commit themselves to using materials and production processes that do not harm the environment. The paper in this book is made from low- or no-chlorine pulp and is acid free, in conformance with international standards for paper permanency.

GPSR Compliance
The European Union's (EU) General Product Safety Regulation (GPSR) is a set of rules that requires consumer products to be safe and our obligations to ensure this.

If you have any concerns about our products, you can contact us on

ProductSafety@springernature.com

In case Publisher is established outside the EU, the EU authorized representative is:

Springer Nature Customer Service Center GmbH
Europaplatz 3
69115 Heidelberg, Germany

www.ingramcontent.com/pod-product-compliance
Ingram Content Group UK Ltd.
Pitfield, Milton Keynes, MK11 3LW, UK
UKHW021832190726
13853UKWH00003B/1281

* 9 7 8 3 6 6 2 1 4 2 5 1 6 *